International Conference "Welding in Russia 2019: State-of-the-Art and Perspectives" (WR 2019)

IOP Conference Series: Materials Science and Engineering
Volume 681

Tomsk, Russia
3 - 7 September 2019

Editor:

Yuriy Saraev

ISBN: 978-1-7138-0933-3
ISSN: 1757-8981

Printed from e-media with permission by:

Curran Associates, Inc.
57 Morehouse Lane
Red Hook, NY 12571

Some format issues inherent in the e-media version may also appear in this print version.

Institute of Physics
Dirac House, Temple Back
Bristol BS1 6BE UK

Phone: 44 1 17 929 7481
Fax: 44 1 17 920 0979

techtracking@iop.org

Additional copies of this publication are available from:

Curran Associates, Inc.
57 Morehouse Lane
Red Hook, NY 12571 USA
Phone: 845-758-0400
Fax: 845-758-2633
Email: curran@proceedings.com
Web: www.proceedings.com

TABLE OF CONTENTS

IOP Conf. Series: Materials Science and Engineering **681** (2019) 011001 doi:10.1088/1757-899X/681/1/011001

These proceedings of the International Conference 'Welding in Russia 2019: State-of-the-Art and Perspectives , which was held from 3 to 7 September 2019 in Tomsk (Russia), are generalization of the presented and discussed achievements as well as problems in this area of science. It was the first large-scale meeting in the post-Soviet region covering all parts of welding production including international cooperation aimed at strengthening ties between leading scientists and specialists. Representatives of large corporations and industrial associations, scientific and educational institutions took part in the Conference. Many important issues were discussed: new materials and equipment for welding; repair of critical structures operated under extreme conditions, their reliability, reducing risks of man-made accidents as well as negative impact on the environment.

Welding, as the main technological process for production of metal structures for various industrial purposes, is a multi-disciplinary science covering a wide range of subjects such as materials science, physics, mathematics, chemistry, electrical and computer engineering. Its development requires the results of numerous investigations because the main drawback affecting the performance of the welded structure is application of materials that do not meet operation conditions, ineffective technologies for their joining or incorrect parameters of welding modes. As a result, various micro- and macro-discontinuities, stresses and deformations are formed in welds. This contributes to different problems such as low strength and corrosion resistance, brittle fracture under load (especially at low temperatures), etc. Therefore, special attention should be paid to heat input control during welding. This may significantly improve quality of welded joints. In order to develop knowledge about the listed welding features, the following sections are presented in the proceedings:

- ✓ New materials for production and repair of critical structures and special-purpose equipment operated under extreme loads and low climatic temperatures of the Far North and the Arctic.
- ✓ Directions and trends in application of conventional heat sources (electric arc, plasma, etc.) as well as high-performance ones (laser and electron beams) significantly expanding the boundaries of physical capabilities to control processing of metals and alloys, stresses and deformations formed in complex-shaped products.
- ✓ Welding and technological properties of materials and equipment: registration methods, assessment of reliability, expert assessment.
- ✓ Reliability of critical structures, reducing risks from industrial accidents, as well as negative environmental impacts.

The presented papers may be interesting to a wide range of specialists in the fields of materials science, hydrodynamics, chemistry and mechanics of solids, reliability and technogenic safety, welding control in the manufacture of structures for critical purposes.

The Editor of the Proceedings,
Professor, Doctor of Technical Science, Yuriy Saraev
Institute of Strength Physics and Materials Science, Siberian Branch, Russian Academy of Science
Tomsk, Russia
litsin@ispms.tsc.ru

IOP Conf. Series: Materials Science and Engineering **681** (2019) 011002 doi:10.1088/1757-899X/681/1/011002

Peer review statement

All papers published in this volume of *IOP Conference Series: Materials Science and Engineering* have been peer reviewed through processes administered by the proceedings Editors. Reviews were conducted by expert referees to the professional and scientific standards expected of a proceedings journal published by IOP Publishing.

IOP Conf. Series: Materials Science and Engineering **681** (2019) 012001 doi:10.1088/1757-899X/681/1/012001

Comparison of the resistance to cavitation and abrasive wear of deposited layers prepared from steels 06X19H9T and Fe-Cr-Ti-Al

H L Alwan[1,3], Yu S Korobov[1,2], N O Bessonov[1], M S Smolentsev[1], N V Lezhnin[2], N M Razikov[1], M N Razikov[1]

[1]Ural Federal University named after the first President of Russia B.N. Yeltsin, Yekaterinburg, Russian Federation;
[2]M.N. Miheev Institute of Metal Physics, Ural Branch of the Russian Academy of Sciences, Yekaterinburg, Russian Federation;
[3]University of Technology, Baghdad, Iraq;

E-mail: lefta.hussam@gmail.com, yukorobov@gmail.com

Abstract. The resistance of deposited layers to cavitation erosion and abrasive wear was evaluated for the steels AISI 321 (known as 06X19H9T manufactured according to the Russian standard GOST 18143-72) and Fe-Cr-Ti-Al (two materials namely Ural AS-3 and PPM-6). The Gas Tungsten Arc Welding (GTAW or TIG) process was utilized to deposit these welding material wires onto a medium carbon steel substrate AISI 1040. Cavitation test was conducted by using ultrasonic vibratory method to induce the erosion. In addition, a three-body abrasion test was used to evaluate the resistance of the studied materials to abrasive wear. The material loss criterion and wear rate of each material as a function to testing time were evaluated and discussed. The cavitation and abrasive wear tests have shown similar results regarding the PPM-6 steel. Among the tested material, PPM-6 exhibited the better resistance to cavitation and abrasion. The 06X19H9T stainless steel exhibited a higher resistance to cavitation comparing with the Ural AS-3. With respect to the abrasive wear, Ural AS-3 was to some degree better than the 06X19H9T in resistance to abrasion.

1. Introduction

Wear, in general, is common to occur in machinery components that serve many industrial applications in the fields of oil, mining and mineral processing. Abrasive wear is one of the most important types of wear in which the damage involves a progressive loss of material due to the presence of hard particles rolling between two sliding surfaces. Abrasion wear resistance is very important, therefore; the development of materials and technologies to protect the surfaces of components and increase the wear resistance are necessary for the performance and lifetime of machinery components [1-6]. Hardness affects the abrasive wear resistance of metals such that higher hardness imparts a higher wear resistance, and it can be used as a useful indicator to measure the wear resistance [3, 4]. Many laboratory testing methods can be considered to evaluate the wear resistance. The three-body abrasion test is the most common. The rotating rubber wheel, test specimen, and the solid hard particles are the basic elements in the test which is considered as a low-stress method [7].

Beside the abrasive wear, a recurrent problem which is the wear caused by cavitation occurs frequently in vital components such as pump impellers, turbine blades, valves and marine propellers

[8-10]. Cavitation wear affects certainly the performance and service life of these component. Avoiding or reducing the cavitation effect is necessary from the economic point of view, since cavitation phenomenon would increase maintenance operations and the repairing costs of components [11]. It is defined as formation and collapsing bubbles produced as a result of relative motion between the solid component and a liquid. Collapsing of bubbles near the solid surface is usually accompanied by generating a pulse stress and high velocity micro-jets of liquid, causing fatigue failure and material loss eventually [12]. In this study, the cavitation test condition applied is a combination of mechanical and electrochemical action, since a certain voltage is used along with water as a test liquid. Applying the voltage is to form this interaction of mechanical and electrochemical effect as well as accelerate forming the cavities [13]. The purpose of the current study is to conduct a rubber wheel abrasion test and an ultrasonic vibratory cavitation test for evaluating the resistance of 06X19H9T and Fe-Cr-Ti-Al steels.

2. Experimental procedure

2.1 Deposited Materials. Two types of material systems Fe-Cr-Ti-Al (from now on Ural AS-3 and PPM-6) as well as the austenitic stainless steel 06X19H9T were deposited onto a medium carbon steel substrate AISI 1040. The Gas Tungsten Arc Welding process (GTAW or TIG) was utilized to deposit all the welding wire materials. The 06X19H9T is as a solid wire, while both Ural AS-3 and PPM-6 are in form of cored wire. The chemical compositions of the deposited-welded materials are given in Table 1. The final thickness of the deposited layer was approximately 4-5 mm.

Table 1. The chemical compositions of deposited welding wire materials [wt. %].

Material	C	Cr	Ni	Si	Mn	Al	P	S	Ti	Fe
06X19H9T	Max. 0.08	18.0-20.0	8.0-10.0	0.40-1.00	1.00-2.00	-	Max. 0.03	Max. 0.015	Min. 5×C	Bal.
Ural AS-3	0.83	25.0	0.12	0.5	0.4	1.15	-	-	2.4	Bal.
PPM-6	0.6	8.0	-	-	-	1.5	-	-	1.0	Bal.

2.2 Abrasive wear test. The experiments of the three-body abrasive wear test were carried out on a rubber wheel abrasion test shown in Fig.1. The abrasive particles flow from the container to roll between the test specimen and the rotating rubber wheel. The rubber wheel rotates in the same direction of abrasive particles flow. Abrasive particles used was aluminium oxide of size 125-180 μm. A specific load is placed on the lever arm to form a compressive force towards the rubber wheel and to keep a face contact between the specimen and rubber wheel. The size of test specimen is 70 × 18 × 8 mm^3. The conditions, under which the abrasive wear tests were performed, can be summarized in Table 2.

Table 2. Parameters of the three-body abrasion test

Parameter	value
Load [Kg]	5.0
Rubber wheel rotating speed [rev/min]	75
Rubber wheel diameter [cm]	20
Flow rate of grits [g/min]	110
Total testing time [min]	10
Testing time interval [min]	1.0
Weight distance [cm]	30

The measurements of material weight loss were taken every minute using an analytical weight device with an error 1.0 mg.

2.3 Cavitation test. The test specimens were prepared according to the requirements of ASTM standard G32–10 [14]. The cavitation test conditions applied in this study to conduct the experiments of ultrasonic vibratory tests are shown in Table 3. Water was used as a test solution and a certain voltage (12 V) has been applied with water to accelerate the cavitation and give a combined mechanical-electrochemical effect. The description of working principle of the cavitation test developed by authors [15] is used in this study, as shown in Fig. 2. The total exposure time of cavitation test was 330 minutes. The weight device with an error of 0.5 mg was used to evaluate the weight loss of test specimens.

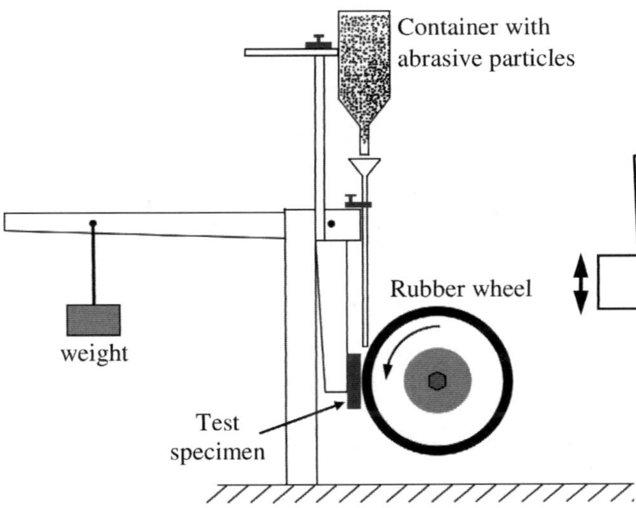

Figure 1. Schematic diagram of Brinell-Hovart rubber wheel abrasion wear test.

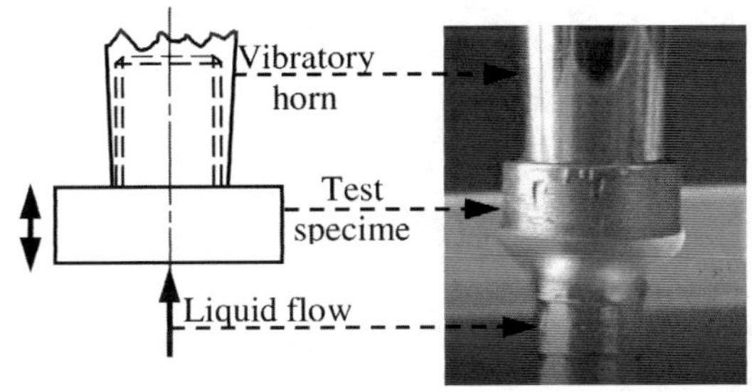

Figure 2. The work principle of the cavitation test.

Table 3. Cavitation test conditions.

Frequency of vibration	20±0.1 [kHz]
Peak-to-peak of amplitude displacement	55±3 [μm]
Power of the ultrasonic generator	500 [W]
Cavitation medium	water
Applied voltage	12 [V]
Total testing time	330 [min]

3. Results and Discussion

3.1 Abrasive wear results. The results of the abrasion test represented by mass loss and wear rate as a function of testing time for all deposited materials and substrate are shown in Figs. 3 and 4, respectively. The difference in mass loss between the deposited materials and substrate can be easily compared. There is a slight difference in mass loss between Ural AS-3 and the 06X19H9T so that Ural AS-3 is better than the 06X19H9T by 10% approximately. The best material among all the studied materials was PPM-6. It exhibited a higher resistance to the abrasive wear comparing with the Ural AS-3 and 06X19H9T by approximately 41% and 56%, respectively. With respect to the speed of mass loss, Fig. 4 represents the abrasive wear rate as a function of testing time for all tested materials. A little difference can be recognized between Ural AS-3 and 06X19H9T particularly at the beginning of test, while a distinct difference can be seen comparing with the PPM-6. To compare the wear rate at each interval of testing time, Fig. 5 shows the abrasive wear rates of all tested materials during the test.

The maximum abrasive wear rate had been recorded for the substrate material AISI 1040, which was of 90 mg/min, while the minimum abrasive wear rate was of 31 mg/min for PPM-6 material. The abrasive wear rate is calculated using Eq. 1.

$$\text{Wear rate} = \frac{\Delta w}{\Delta t} \tag{1}$$

where: Δw is the difference in mass of test specimen between two sequential periods in mg, and Δt is the difference in testing time between these two periods in minutes.

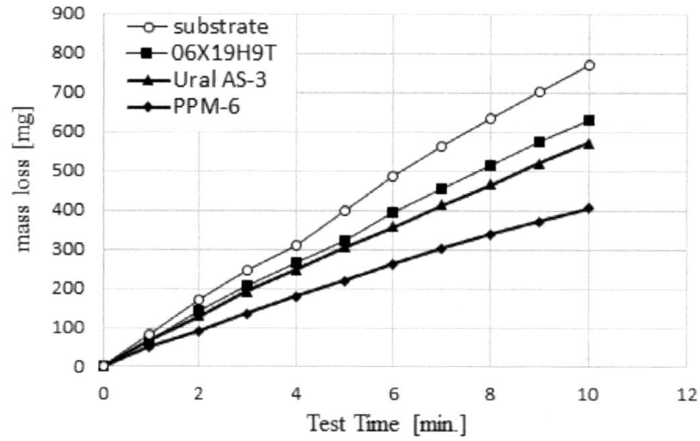

Figure 3. Results of three-body abrasion test of all studied materials (mass loss).

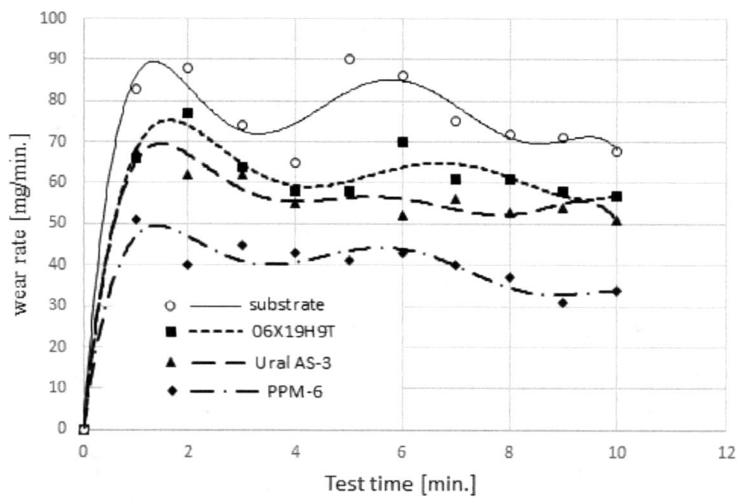

Figure 4. Results of three-body abrasion test of all studied materials (wear rate).

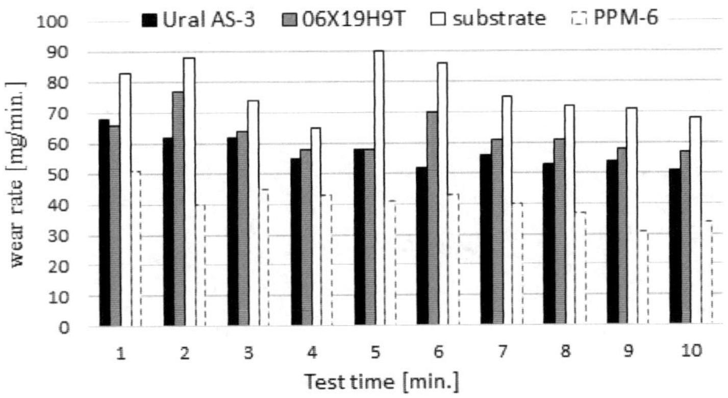

Figure 5. Wear rates during the three-body abrasion test for the tested materials.

The HV_{10} was measured for the deposited materials before and after the abrasive test. The results are labeled in Table 3. It is expected that the hardness after the abrasive test is somewhat higher than of that before testing due to work hardening in the affected area. However, this is not observed for the Ural AS-3 maybe due to presence of defects on the tested area. Nevertheless, it exhibited greater resistance to abrasive wear comparing with the 06X19H9T.

Table 4. The HV_{10} of the deposited materials.

Tested material	HV_{10} before testing	HV_{10} after testing
Ural AS-3	256 ± 2	244 ± 8
06X19H9T	242 ± 12	331 ± 26
PPM-6	685 ± 57	764 ± 5

3.2 Cavitation curves. The ultrasonic vibratory tests were conducted for all the studied materials by applying a combination of mechanical-electrochemical interaction, and the results are shown in Fig. 6. The cumulative mass loss-time curve was attained to evaluate the resistance of tested materials to cavitation erosion-corrosion. It can be easily distinguished the difference in mass loss between the substrate AISI 1040 steel and the deposited materials. Further, no significant difference can be noticed between the PPM-6 and the 06X19H9T steels. However, the best cavitation results are shown by the PPM-6 which exhibited higher resistance to cavitation followed by 06X19H9T, Ural AS-3 and substrate AISI 1040, respectively. In addition, the behavior of the studied materials during the cavitation test can be illustrated by the erosion rates calculated depending on the material loss at each interval of time, as shown in Fig. 7. Moreover, the maximum and minimum erosion rates attained during the cavitation tests for each tested material can be represented by bars graph shown in Fig. 8. It can be seen that the PPM-6 has the minimum erosion rate, in contrast, the substrate has the maximum erosion rate among all the tested materials. The minimum erosion rate achieved for the PPM-6 steel is 0.012 mg/min, while the maximum one is 0.434 mg/min for the AISI 1040 substrate steel. Moreover, the maximum erosion rate for both 06X19H9T and PPM-6 steels are very close to each other as illustrated in Fig.7, and they occurred at the last of the cavitation test.

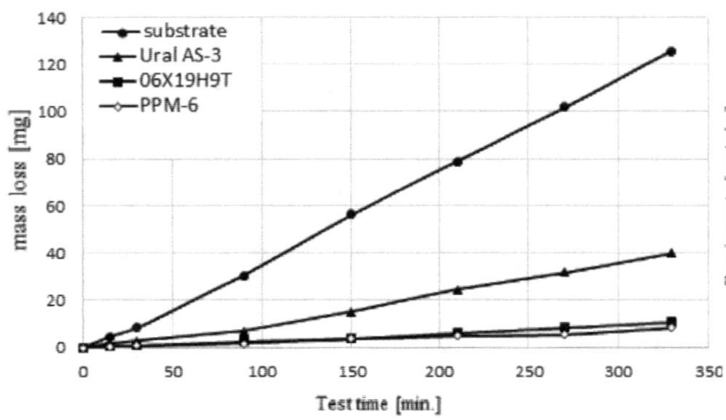

Figure 6. Cavitation results (mass loss) of all the studied materials.

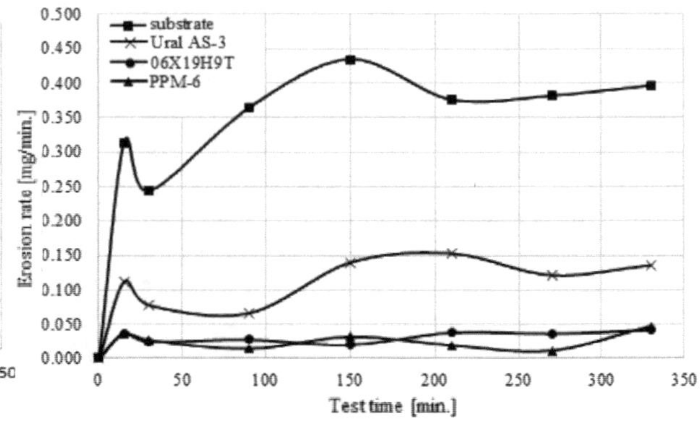

Figure 7. Behavior of the studied materials during the test (erosion rate).

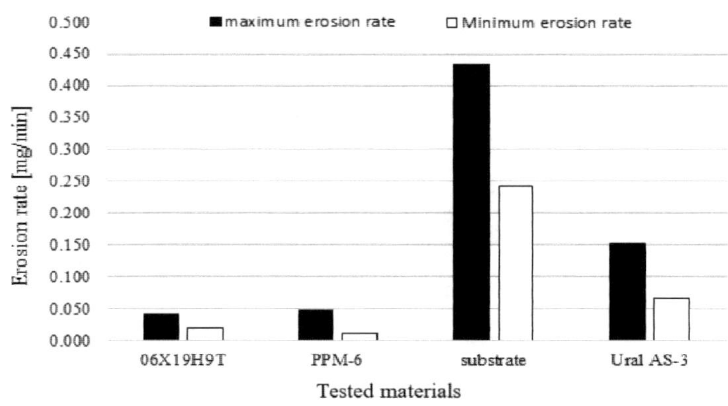

Figure 8. Maximum and minimum erosion rates of the studied materials during the cavitation test.

4. Conclusions

A three-body abrasion and an ultrasonic vibratory tests were performed to evaluate the resistance to abrasive wear and cavitation erosion-corrosion of deposited layers prepared from steels 06X19H9T and Fe-Cr-Ti-Al. The results showed the following:

- All the deposited materials exhibited a higher resistance to abrasive wear and cavitation than the substrate material AISI 1040 steel.
- Among all the tested materials, the PPM-6 showed the better resistance to the abrasive wear and cavitation erosion-corrosion.
- The cavitation results showed that both PPM-6 and 06X19H9T are almost identical.
- The 06X19H9T was exhibited a better resistance to cavitation comparing with the Ural AS-3. In contrast Ural AS-3 showed to some degree a better resistance than the 06X19H9T regarding the abrasive wear.
- For abrasive wear test, the PPM-6 has a better resistance than the substrate AISI 1040 steel, 06X19H9T, and Ural AS-3 by 90%, 56% and 41%, respectively. As for cavitation erosion-corrosion, the PPM-6 was exhibited a better resistance than the 06X19H9T by 30%, 5 times higher than Ural AS-3, and 15 times as high comparing with the AISI 1040 substrate steel.

Acknowledgement
This work was done within the state order of IMP UB RAS on the subject no. AAAA-A18-118020190116-6, within the state order of IMP UB RAS on the subject “Laser”, and IES UB RAS on the subject no. AAAA-A18-118020790147-4. The present study was supported by FASIE, program Development-NTI 2017, project No. 0035960.

References
[1] Ian Hutchings and Philip Shipway, Tribology, Friction and Wear of Engineering Materials, second edition, Published by Elsevier Ltd, 2017.
[2] A.P. Harsha and U.S. Tewari, Two-body and three-body abrasive wear behaviour of polyaryletherketone composites, Polymer Testing 22 (2003) p 403-418.
[3] A. Vencl, A. Rac, B. Ivkovic, Investigation of Abrasive Wear Resistance of Ferrous-Based Coatings with Scratch Tester, Tribology in industry 29 (2007) 13-16.
[4] A. Sundström, J. Rendón, M. Olsson, Wear behaviour of some low alloyed steels under combined impact/abrasion contact conditions, Wear 250 (2001) 744–754
[5] L. Fu, L. Li, D.Y. Li, Further look at correlation between ASTM G65 rubber wheel abrasion and pin-on-disc wear tests for data conversion, Tribology 7 (2013) 109–113.
[6] A. Karimi and J. L. Martin, Cavitation erosion of materials, Int. Met. Rev. 31 (1986) 1–26.
[7] K.G. Budinski and S.T. Budinski, On replacing three-body abrasion testing with two-body abrasion testing, Wear 376-377 (2017) 1859–1865.
[8] P. Kumar and R.P. Saini, Study of cavitation in hydro turbines—A review, Renewable and Sustainable Energy Reviews 14 (2010) 374–383.
[9] A. Karabenciov, A.D. Jurchela, I. Bordeaşu, M. Popoviciu, N. Birău, A. Lustyan, Considerations upon the cavitation erosion resistance of stainless steel with variable Chromium and Nickel content, IOP Conference Series: Earth and Environmental Science 12 (2010).
[10] L Tôn-Thât, Experimental comparison of cavitation erosion rates of different steels used in hydraulic turbines, IOP Conference Series: Earth and Environmental Science 12 (2010).
[11] L.A. Espitia and A. Toro, Cavitation resistance, microstructure and surface topography of materials used for hydraulic components, Tribology International 43 (2010) 2037–2045.
[12] C.T. Kwok, F.T. Cheng, H.C. Man, Synergistic effect of cavitation erosion and corrosion of various engineering alloys in 3.5% NaCl solution, Materials Science and Engineering A290 (2000) 145–154.
[13] Y. Zheng, S. Luo, W. Ke, Effect of passivity on electrochemical corrosion behavior of alloys during cavitation in aqueous solutions, Wear 262 (2007) 1308–1314.
[14] ASTM, Standard Test Method for Cavitation Erosion Using Vibratory Apparatus, G 32 - 10, (2011) 1–19.
[15] Yu.S Korobov, H.L. Alwan, M. Barbosa, N.V. Lezhnin, N.N. Soboleva, A.V. Makarov, M.S. Deviatiarov, and A.Yu. Davydov, "Cavitation erosion-corrosion resistance of WC–CoCr and WC–NiCr HVAF coatings, Bulletin PNRPU. Mechanical engineering, materials science 21(2019) 20–27.

IOP Conf. Series: Materials Science and Engineering **681** (2019) 012002 doi:10.1088/1757-899X/681/1/012002

Behavior of a welded-deposited stainless steel tested at different cavitation test conditions

H L Alwan[1,2,a], Yu S Korobov[2,3,b], N N Soboleva[2,4,c], N V Lezhnin[3,d], A V Makarov[2,3,4,e], M S Deviatiarov [5,f], D M Elkind[2]

[1]University of Technology, Baghdad, Iraq;
[2]Ural Federal University named after the first President of Russia B.N. Yeltsin, Yekaterinburg, Russian Federation;
[3]M.N. Miheev Institute of Metal Physics, Ural Branch of the Russian Academy of Sciences, Yekaterinburg, Russian Federation;
[4]Institute of Engineering Science, Ural Branch of the Russian Academy of Sciences, Yekaterinburg, Russian Federation
[5]Ural Welding Institute-Metallurgy, Yekaterinburg, Russian Federation

E-mail: [a]lefta.hussam@gmail.com, [b]yukorobov@gmail.com, [c]natashasoboleva@list.ru [d]nlezhnin@bk.ru, [e]av-mak@yandex.ru, [f]dms@mashprom.ru.

Abstract. Two different ultrasonic vibratory-cavitation test conditions have been applied to a welded-deposited austenitic stainless steel AISI 321 to evaluate the resistance of deposited layer to cavitation erosion-corrosion. The cavitation test was conducted utilizing two test fluids; water and 3.5% NaCl solution. In addition, a certain voltage difference has been applied between the test specimen and water to form a combination effect. The welding wire of the AISI 321 stainless steel was deposited onto AISI 1040 steel substrate by using tungsten inert gas welding process. To evaluate and compare the behavior of the deposited material, the cumulative mass loss curves were attained and discussed. Moreover, the surface topography and scanning electron microscope (SEM) micrographs were utilized to characterize the worn surface after the cavitation tests. The results showed that the surface subjected to cavitation was more affected when applying water-voltage condition comparing with the 3.5% NaCl solution condition. The results of material loss, surface roughness and scanning electron microscope are fairly consistent with each other. This study highlights the effect of electrochemical-mechanical combinations on resistance to cavitation erosion-corrosion.

1. Introduction

Cavitation is a type of wear that deteriorates material surface for the components subjected to underwater erosion such as hydro turbines, valves, pumps impeller. When a solid component moves in high speed through a liquid, high and low pressure regions form due to the relative motion between the component and liquid. This leads to frequent formation and collapsing bubbles near the solid surface, and so, high pressure shock waves and micro-jets can produce and damage the surface. Maximum pressure was found to be at the end stage of bubble collapse. It can reach to1750 MPa and the velocity of liquid jet released ranges from 120 m/s to several of 100 m/s. Consequently, loss of material takes place and small pits or cavities appear over the surface, affecting the performance of components and reduction their lifetime [1-7].

It is known that the damaged surfaces of components due to cavitation can be restored by using the basic methods such as Metal Arc Welding (MAW) and Gas Tungsten Arc Welding (GTAW). Depositing the protective materials onto the origins to repair the defects or to combat cavitation erosion attack, can be achieved by those methods [8,9]. The behavior of material in the cavitation test differs by the test modes applied [10]. Various test liquids are usually used by the research works to evaluate the resistance to cavitation erosion such fresh water [11] and tap water [12]. The 3.5% NaCl solution is commonly used as a test liquid [11-14]. In addition, applying a voltage to the test forms a combined effect of mechanical and electrochemical interaction as well as accelerate forming the cavities [15]. In the present study, the water-voltage and 3.5%NaCl solution conditions were applied to evaluate the cavitation erosion–corrosion resistance of deposited AISI 321 stainless steel wire. Research works have been focused on wear resistance of this type of steel more than, in particular, on resistance to cavitation [16, 17]. Thus, in the present study, the deposited AISI 321 material was studied to evaluate its performance in resistance to cavitation erosion–corrosion.

2. Experimental Procedure

2.1 Tested materials and depositing technique. The solid welding wire of stainless steel AISI 321 (equivalent to steel 06KH19N9T or "06X19H9T" manufactured according to the Russian standard GOST 18143-72) was deposited onto AISI 1040 steel substrate. The gas tungsten arc welding or tungsten inert gas welding (GTAW or TIG) process was utilized to deposit the 1.6 mm diameter welding wire with current of 120 A and voltage of 10-12 V. Three layers of stainless steel wire have been deposited onto substrate, and the final deposition thickness about 4 mm. The chemical compositions of the deposited material and substrate are shown in Table 1.

Table 1. Chemical compositions of deposited material and substrate [wt. %]

Material	C	Cr	Ni	Si	Mn	Cu	P	S	Ti	Fe
AISI 321	Max. 0.08	18.0-20.0	8.0-10.0	0.40-1.00	1.00-2.00	-	Max. 0.03	Max. 0.015	Min. 5×C	Bal.
AISI 1040	0.404	-	-	0.283	0.6	0.142	0.0053	<0.001	-	Bal.

2.2 Specimens and cavitation test. The cavitation test specimens were prepared according to the requirements of performing the cavitation erosion test, which is the ASTM standard G32–10[18]. Using cylindrical substrate blocks with dimensions $\phi 24\times 20$ mm, the stainless steel wire was firstly deposited on them, and then, the substrate blocks were machined to the final required dimensions. All the machined specimens were ground together by silicon carbide grinding wheel to obtain the same surface state. The evaluation of surface resistance of the deposited material was conducted by obtaining the curves of cumulative weight loss for the two test conditions used in this study. Cavitation test was conducted under vibration frequency of 20 kHz, peak-to-peak amplitude of 50 μm, power of 500 W and applying voltage of 12 V. Fig. 1 shows the working principle of the cavitation test developed by authors [19], in which, the test specimen is attached to the vibratory horn. The specimens were exposed to 270 minutes of test. The procedure of conducting cavitation experiments included cleaning the test specimen by acetone, drying by hot air for 30-40 seconds, and weighing the specimen using a weight device with an error of 0.5 mg.

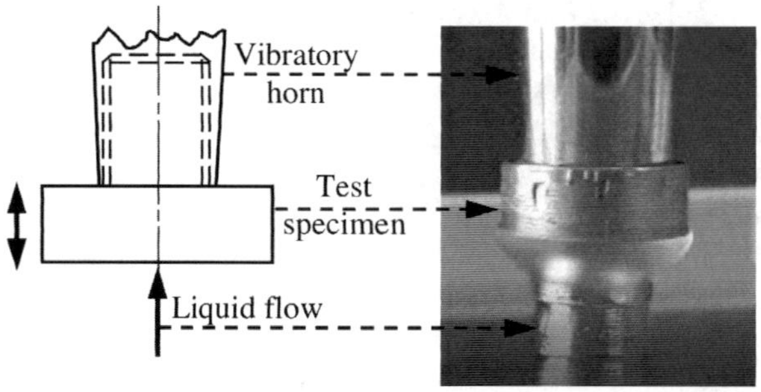

Figure 1. Work principle of the cavitation test facility.

2.3 Equipment. The necessary equipments that have been used for evaluating the eroded surface, in addition to the cavitation vibratory device, included the optical profilometer Wyko NT 1100 and scanning electron microscope (SEM) TescanVEGA II XMU. Surface roughness was measured several times over the intended surface to determine the value of the arithmetic mean deviation of the roughness profile Ra. In addition, Shimadzu HMV-G21DT microhardness tester was used for measuring the microhardness of the surface exposed to cavitation under a load of 0.49 N as an arithmetic mean of 10 measurements.

3. Results and Discussion

3.1 Cavitation curves. The ultrasonic vibratory method was used to induce forming cavities over the tested surface of the deposited material. The relationship between the exposure time and the mass loss was plotted for a number of periods. This relationship is one of indications to surface resistance to cavitation erosion. Fig. 2 shows the results of mass loss for the two test conditions. Although the material loss produced by the water-voltage condition is more than that of 3.5% NaCl solution, the difference in material loss is not significant during the cavitation test. The same conditions (water-voltage and 3.5% NaCl solution) were applied on AISI 1040 in a previous study [10], and there was a significant difference in material loss, which was estimated approximately 4 times at the last of the test. However, the material loss of AISI 321 was about 8.3 mg when applying the water-voltage condition, while it was about 6.2 mg using 3.5% NaCl solution condition. It was compared with the substrate AISI 1040 which has been tested at similar cavitation test parameters in the previous research work [20], where the cumulative material loss of AISI 1040 was about 125 mg. Fig. 3 shows the cumulative mass loss curves for the AISI 1040 and the AISI 321 steels.

3.2 Surface roughness and microhardness. Surface roughness is an important factor in resistance to cavitation, and it can be considered as a useful indicator in evaluating the eroded surface. Table 2 represents the arithmetic mean deviation of the roughness profile Ra for five measures taken on the surface of the AISI 321 stainless steel at the same area of measurement. Regarding the microhardness, the $HV_{0.05}$ had been measured for the surface to be exposed to cavitation effect and was found to be 420 ± 40. The good mechanical properties of stainless steel have certainly positive effects in resistance to cavitation erosion. Many research works have shown that the mechanical properties, e.g. hardness, affect the resistance to cavitation in stainless steels. It has been reported that the hardness is inversely proportional to cavitation erosion [21, 22].

Table 2. The arithmetic mean deviation of the roughness profile Ra for the deposited AISI 321 stainless steel before and after applying the two cavitation test conditions.

Condition	Value of Ra [μm]
Before cavitation test	0.71
Water-Voltage condition	1.06
3.5% NaCl solution condition	0.91

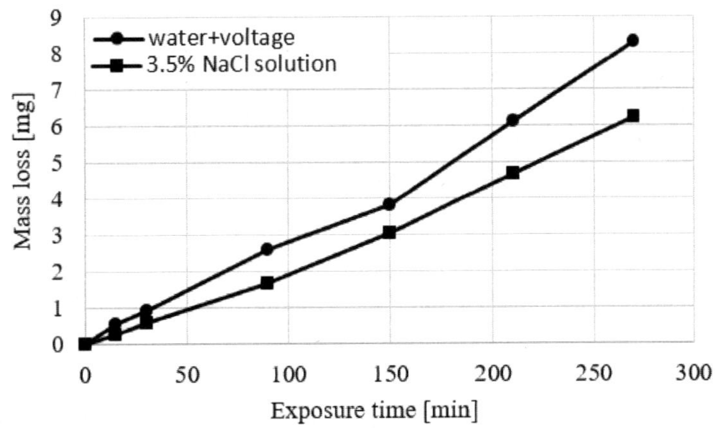

Figure 2. Results of cavitation vibratory test ("mass loss – time" curve) for the AISI 321.

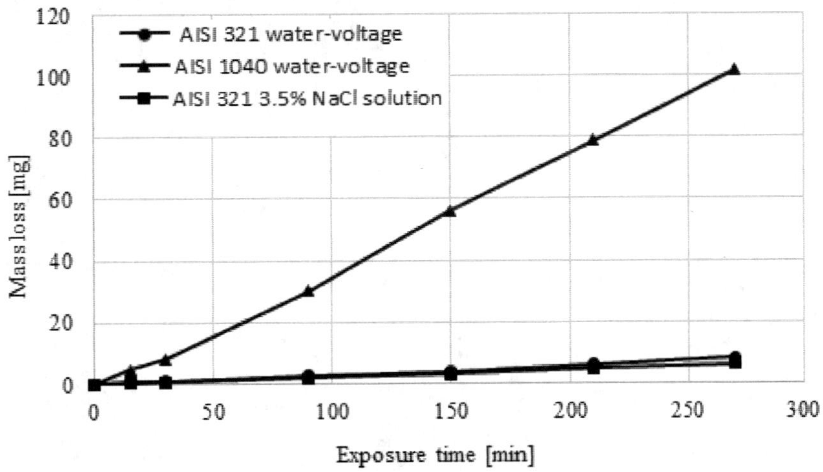

Figure 3. Cumulative mass loss-time curve for the deposited AISI 321 stainless steel and AISI 1040 substrate.

3.3 Topographical and SEM images. Surface topographical images and scanning electron microscope (SEM) micrographs have been utilized to evaluate the eroded surfaces of the welded-deposit stainless steel AISI 321. Fig. 4 shows the SEM micrographs of the eroded surfaces at different magnification for the two cavitation test conditions. It can be noticed that the surface exposed to water-voltage condition is more affected by cavitation, as shown in Figs. a, b, and c, than of that subjected to the 3.5% NaCl solution as shown in Figs. d, e, and f. Comparing the size and appearance of cavities in the eroded surface, they appear to some degree to be big in size, deep, and dense as shown in Fig. 4c. This is due to the effect of voltage applied which forms a combination action, although the second cavitation test condition (3.5% NaCl solution) is not considered as a pure cavitation erosion.

Regarding the surface roughness and texture of the deposited material, Fig. 5 shows three-dimensional topographical images of the original surface before testing and the eroded surfaces when applying the two cavitation test conditions. The same indication can be considered when evaluating the topographical images of the eroded surfaces and the values of surface roughness Ra. The average values of the arithmetic mean deviation of the roughness profile Ra resulted from five measures taken for the same area of measure, were found to be as represented in Table 2. The maximum value of Ra

IOP Conf. Series: Materials Science and Engineering **681** (2019) 012002 doi:10.1088/1757-899X/681/1/012002

was as a result of applying water-voltage condition, which was of 1.06 μm, while it was of 0.91 for the 3.5% NaCl solution condition. This gives an indication that the damage caused by water-voltage condition is more than of that caused by the 3.5% NaCl solution. As for the texture of the original and eroded surfaces shown in Fig. 5, it can be clearly seen the surface topography for each one. The grinding lines appear on the original surface to be tested as shown in Fig. 5a, and this surface is the initial state for comparison. The eroded surface shown in Fig. 5a represents the surface exposed to water-voltage cavitation test condition, and it appears to be more affected comparing with the others. Fig. 5c is the eroded surface when applying the 3.5% NaCl solution condition, and it seems to be less damaged than of that shown in Fig. 5b. It can be noticed the grinding line along the center of the surface shown in Fig. 5c. It can be evidence to ensure that the surface is less affected by the cavitation. These results are consistent with the results of cavitation tests, surface roughness values Ra, and the SEM micrographs of the eroded surfaces.

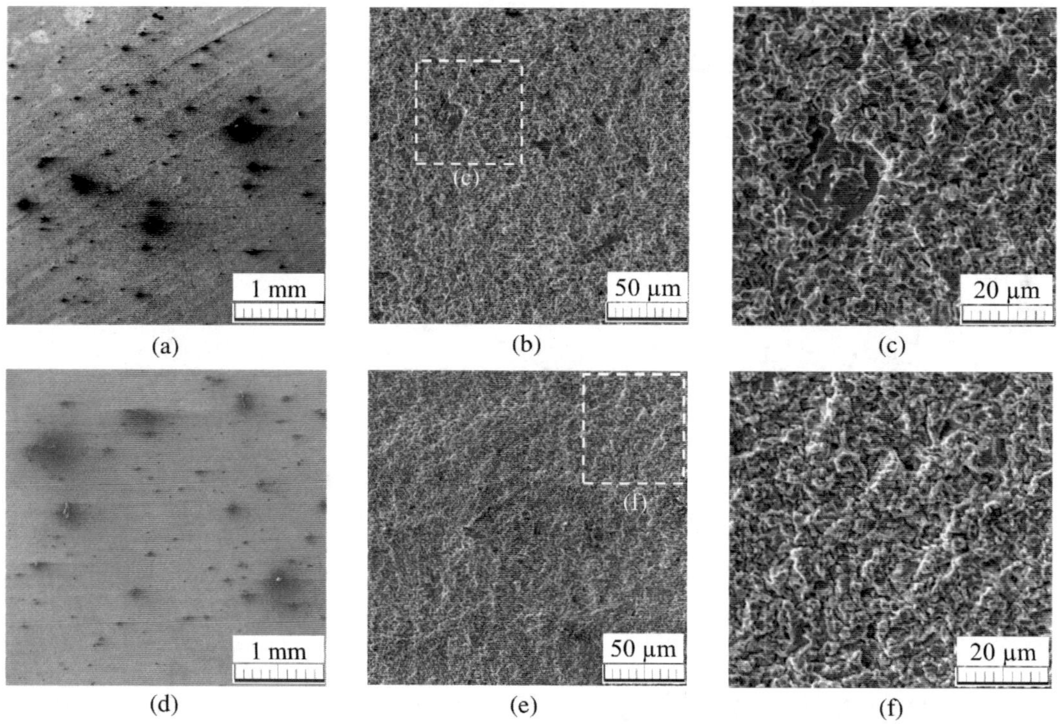

Figure 4. SEM micrographs of the eroded surfaces of AISI 321 for the two cavitation test conditions: a, b, c – water-voltage condition; d, e, f – 3.5% NaCl solution condition.

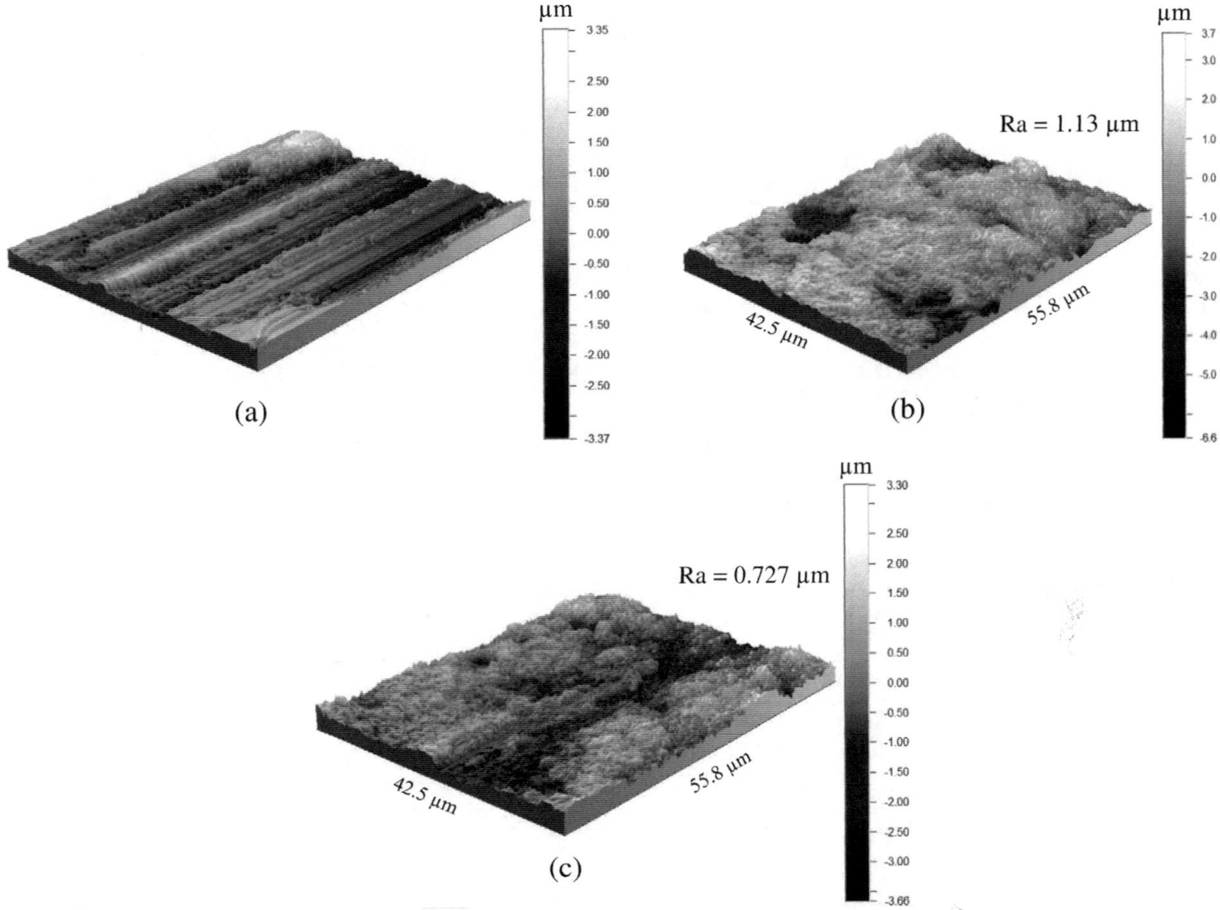

Figure 5. Surface topography and Ra values of the AISI 321 before and after testing: a – original surface before cavitation test, b – water-voltage condition, c – 3.5% NaCl solution condition.

4. Conclusion

The welding wire of AISI 321 stainless steel was deposited by using gas tungsten arc welding process onto the AISI 1040 substrate. The ultrasonic vibratory tests, surface topography and scanning electron microscope - SEM micrographs were used to evaluate the cavitation erosion-corrosion resistance of the deposited layer. The main results are as follows:

- The deposited layer was more damaged when applying the water-voltage condition comparing with applying the 3.5% NaCl solution condition. This is due to the significant effect of voltage which induces an electrochemical action.
- Ultrasonics vibratory cavitation tests showed that the mass loss for the water-voltage condition was higher comparing with using the 3.5% NaCl solution.
- The results of surface topography and scanning electron microscope micrographs are consistent with cavitation tests in terms of the eroded surfaces and the values of Ra.
- Cavitation erosion-corrosion resistance of deposited AISI 321stainless steel layer is 12 times as high comparing with the 1040 steel substrate.

Acknowledgement

This work was done within the state order of IMP UB RAS on the subject no. AAAA-A18-118020190116-6, within the state order of IMP UB RAS on the subject “Laser”, and IES UB RAS on the subject no. AAAA-A18-118020790147-4. The present study was supported by FASIE, program Development-NTI 2017, project No. 0035960. Microhardness, surface roughness and SEM were done on the equipment installed at the Plastometriya collective use center of IES UB RAS.

References

[1] G.W. Stachowiak and A.W. Batchelor, Engineering Tribology, Butterworth-Heinemann, 2013.
[2] J-P Franc, M. Riondet, A. Karimi, G. L. Chahine, Material and velocity effects on cavitation erosion pitting, Wear 274– 275 (2012) 248– 259.
[3] L.A. Espitia and A. Toro, Cavitation resistance, microstructure and surface topography materials used for hydraulic components, Tribology International 43 (2010) 2037–2045.
[4] J.F. Santa, J.A. Blanco, J.E. Giraldo, A. Toro, Cavitation erosion of martensitic and austenitic stainless steel welded coatings, Wear 271 (2011) 1445–1453.
[5] E.A. Brujan, T. Ikedab, Y. Matsumoto, Shock wave emission from a cloud of bubbles, Soft Matter 8 (2012) 5777–5783.
[6] W. Lauterborn and H. Bolle, Experimental investigation of cavitation bubble collapse in the neighborhood of a solid boundary, J. Fluid Mech. 72 (1975) 391–399.
[7] M. Dular, B. Bachert, B. Stoffel, B. Širok, Relationship between cavitation structures and cavitation damage, Wear 257 (2004) 1176–1184.
[8] K. Nakata, M. Oishi, M. Koshiishi, T. Hashimoto, H. Anzai, Y. Saito, W. Kono, Re-weldability of neutron-irradiated stainless steels studied by multi-pass TIG welding, J. Nucl. Mater. 307 (2002) 1578–1583.
[9] A. Shah, M. Rayjada, S. Gupta, Investigating mechanical properties of SAW and SMAW welded joints & the effect of SMAW-repair of saw welded specimen, IJMPE 3 (2015) 66–70.
[10] H.L. Alwan, N. Lezhnin, Yu. Korobov, Analysis of the external conditions affecting on the cavitation resistance of a steel, Trans Tech Publications, Switzerland 946 (2019) 31–36.
[11] S. Han, J.H. Lin, J.J. Kuo, J.L. He, H.C. Shih, The cavitation-erosion phenomenon of chromium nitride coatings deposited using cathodic arc plasma deposition on steel, Surf. Coatings Technol., 161 (2002) 20–25.
[12] S.A. Karrab, M.A. Doheim, M.S. Aboraia, S.M. Ahmed, Examination of Cavitation Erosion Particles Morphology in Corrosive Waters, J. Eng. Sci. Assiut Univ. 40 (2012) 1793–1814.
[13] C.T. Kwok, H.C. Man, F.T. Cheng, Cavitation erosion of duplex and super duplex stainless steels, Scr. Mater. 39 (1998) 1229–1236.
[14] Y.X. Qiao, X. Cai, C. Ouyang, Y.G. Zheng, Effect of Hydrogen on Cavitation Erosion Behaviour of High Strength Steel, Int. J. Electrochem. Sci. 11 (2016) 10329–10346.
[15] Y. Zheng, S. Luo, W. Ke, Effect of passivity on electrochemical corrosion behavior of alloys during cavitation in aqueous solutions, Wear 262 (2007) 1308–1314.
[16] A.N. Grigorchik, V.A. Kukareko, M.A. Belotserkovskii, A.V. Belyib, Structure and wear of sprayed 06X19H9T steel coatingsafter ion-beam nitriding, Steel in Translation 44 (2014) 126–29.
[17] A.V. Byeli, M.A. Belotserkovskii, V.A. Kukareko, Microstructure and wear resistance of thermal sprayed steel coatings ion beam implanted with nitrogen, Wear 267 (2009) 1757–1761.
[18] ASTM, Standard Test Method for Cavitation Erosion Using Vibratory Apparatus, G 32 - 10, (2011) 1–19.
[19] Yu.S Korobov, H.L. Alwan, M. Barbosa, N.V. Lezhnin, N.N. Soboleva, A.V. Makarov, M.S. Deviatiarov, A.Yu. Davydov, Cavitation erosion-corrosion resistance of WC–CoCr and WC–NiCr HVAF coatings, Bulletin PNRPU. Mechanical engineering, materials science 21 (2019) 20–27.
[20] H.L. Alwan, Yu.S. Korobov, N.N. Soboleva, N.V. Lezhnin, A.V. Makarov, E.P. Nikolaeva, M.S. Deviatiarov, Cavitation erosion-corrosion resistance of deposited austenitic stainless

steel/E308L-17 electrode, International conference on Industrial Engineering, Sochi, Russian Federation 25-29 March 2019 Journal Solid State Phenomena, in press.

[21] A. Selokar, U. Prakash, D.B. Goel, B.V.M. Kumar, Cavitation erosion and solid particle erosion behaviour of a nitrogen alloyed austenitic stainless steel, ISIJ International 55 (2015) 1123–1130.

[22] A.W. Selokar, D.B. Goel, U. Prakash, A comparative study of cavitation erosive behaviour of 23/8N nitronic steel and 13/4 martensitic stainless steel, Advanced Materials Research 585 (2012) 554–558.

IOP Conf. Series: Materials Science and Engineering **681** (2019) 012003 doi:10.1088/1757-899X/681/1/012003

Effect of a gas-vortex stabilization method on the kinematic properties of the jet in metal-cutting plasmatrons

S V Anakhov[1,a], A V Matushkin[2,b], Yu A Pyckin[3,c]

[1]Russian State Vocational Professional University, Department of mathematical and natural sciences, 620012 Russia, Yekaterinburg, ul. Mashinostroiteley, 11
[2]Ural Federal University, Department of welding technology, 620078 Russia, Yekaterinburg, ul. Mira,19
[3]Ural State Forest Engineering University, Department of physical and chemical technologies for biosphere protection, 620038 Yekaterinburg, Sibirskiy trakt, 37

E-mail: [a]sergej.anahov@rsvpu.ru, [b]227433@rambler.ru, [c]yappoligon@mail.ru

Abstract. Various methods of gas-vortex stabilization in plasma cutting torches are investigated. The results of the study demonstrate the dependence of the angle of gas injection into the nozzle on the kinematic and kinetic characteristics of plasma jet and, consequently, on the cutting parameters. Increased radial velocity component at the swirler output provides better uniformity of the velocity distribution and kinetic properties of the jet in the affected zone of a cut metal. As a conclusion, it is recommended to use the suggested narrow-jet-plasma technology for cutting thin sheet metals.

1. Introduction

The main criteria of efficiency in the design of plasma torches for air-plasma cutting of metals are productivity, cutting quality and reliability of plasma equipment. High values of these indicators depend, first of all, on the operation of the gas-dynamic stabilization system. The efficiency of this system determines the high degree of heat input and kinetic properties of the plasma arc (jet), the durability of the cathode and nozzle of the plasma torch [1, 2]. In modern plasma torches for metal cutting, as a rule, a gas-vortex stabilization method is used. This method implies a tangential supply of plasma-forming gas (PFG) to the nozzle and additional compression of the plasma arc (jet) at the nozzle outlet by a vortex flow of stabilizing gas (in narrow-jet-plasma technology) [3, 4]. Tangentially twisted gas flow is formed in the vortex chambers (swirlers). The number, design and location of swirlers in the plasma torch appreciably affects the efficiency of metal cutting.

The authors` investigations [5] showed that the method of gas-vortex stabilization with the use of single-flow scheme of PFG supplying to the nozzle via a single swirler does not provide uniform distribution of gas flow over the cross section of the air-gas path (AGP) and in the nozzle channel of the plasma torch. This problem arises due to the asymmetrical position of the gas supply relatively to the expansion chamber located in front of the swirler and due to the limited size of the chamber. The uniformity coefficient of gas flow distribution can be raised by increasing the chamber size [5] or using a system of gas dynamic filters as additional obstacles in the expansion chamber [6]. However, this increases the complexity of manufacturing and the cost of the plasmatron. In this regard, it is necessary to investigate other possibilities of constructive improvement of the gas-vortex stabilization system of metal-cutting plasmatrons.

IOP Conf. Series: Materials Science and Engineering **681** (2019) 012003 doi:10.1088/1757-899X/681/1/012003

2. Research Methods

Two plasma torches for metal cutting, developed by the authors' group and manufactured by the Polygon OJSC (Yekaterinburg, Russia) were taken for the study:

- PMVR-5.3 – from the series of single-flow plasma torches PMVR-5 with the built-in stepwise gas-dynamic filter and increased productivity for operation at currents up to 200 A (Fig. 1);
- PMVR-9.1 – from the series of double-flow plasma torches PMVR-9 with the built-in gas-dynamic filter and a system of the plasma arc double compression (narrow-jet-plasma technology) (Fig. 2).

Figure 1. Plasma torch PMVR-5.3

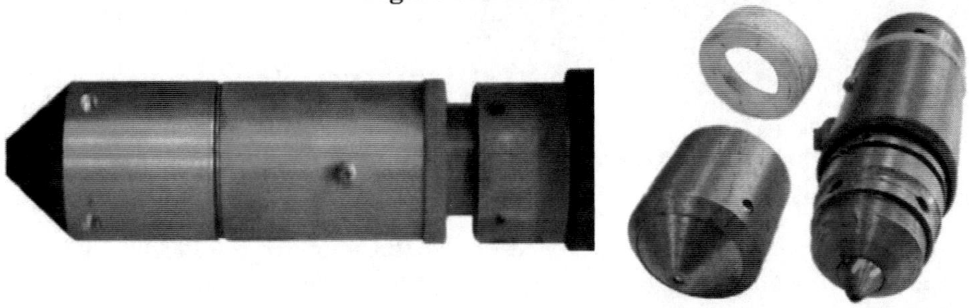

Figure 2. Plasma torch PMVR-9.1

The design of the AGP in plasmatron PMVR-5.3 is based on the principle of stepwise alignment of the PFG velocity across the channel section (i.e., the gas comes from 1 channel to 2, then from 2 to 4, etc.). The final element in this chain is the main swirler. The design of the plasma torch used commercially available cathode A-141 and a series of nozzles for A-141 with the outlet diameter from 1.4 to 1.8 mm. Initially, a serial swirler installed on the cathode A-141 (Fig. 3a) was used for testing the plasma torch. The peculiarity of this swirler is that the flow of PFG passing through the swirler is supplied tangentially to the cathode surface in a plane perpendicular to the cathode. Thus, the flow of PFG at the output of this swirler has only a radial component of the velocity (Fig. 3b).

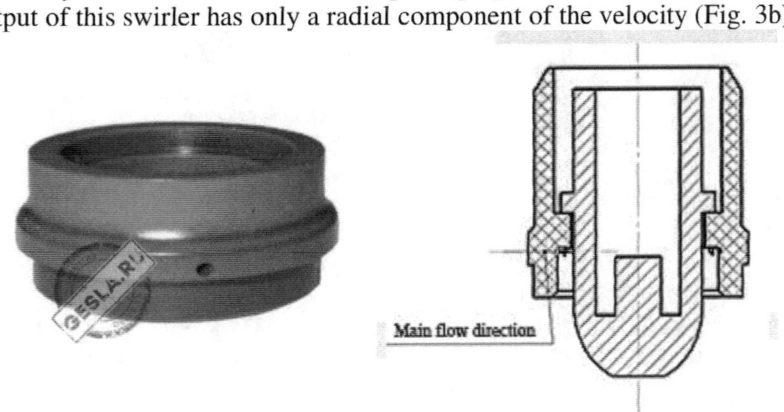

a b

Figure 3. Serial swirler for the plasma torch PMVR-5.3:
a) visual appearance, b) application scheme

IOP Conf. Series: Materials Science and Engineering **681** (2019) 012003 doi:10.1088/1757-899X/681/1/012003

At the second stage an upgraded swirler instead of the serial swirler was used (Fig. 4a) for the testing of plasma torch. Distinctive features of the new swirler are the change in the number of holes (in accordance with the principle described above) and additional inclination of the holes to the axis of the plasma torch. The flow of PFG at the output of such swirler has not only a radial, but also an axial component of the flow velocity (Figs. 4b and 4c).

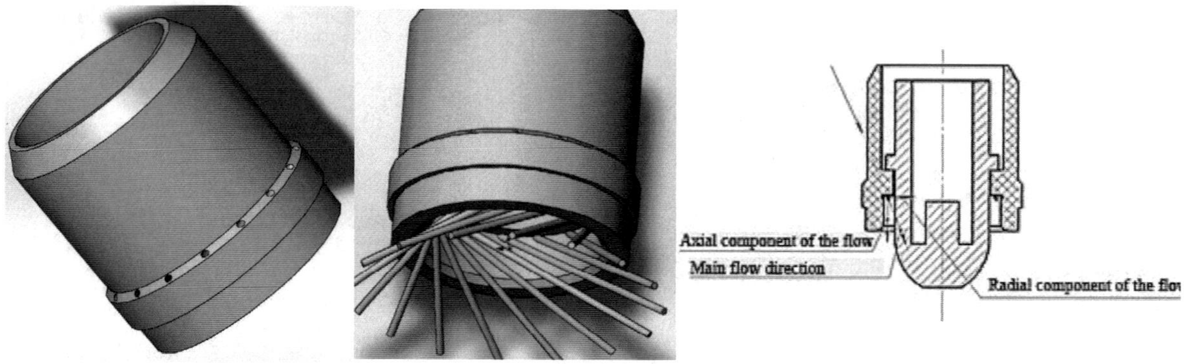

a b c

Figure 4. Upgraded swirler for the plasma torch PMVR-5.3:
a) visual appearance, b) direction of the PFG flow in the swirler, c) scheme of application.

The design of the plasma torch PMVR-9.1 is based on the separation of the PFG flow (plasma-forming and stabilizing flows) with separate gas supply to the nozzle. The plasma arc, formed by the main plasma-forming vortex flow, in this case, is further compressed (stabilized) due to an additional twisted gas at the outlet of the main nozzle of the plasma torch (Fig. 5a). The system of individual gas-vortex stabilization of the PFG flow use the standard swirler of Hypertherm company. The swirler of secondary (stabilizing) flow has its own design (Fig. 5b).

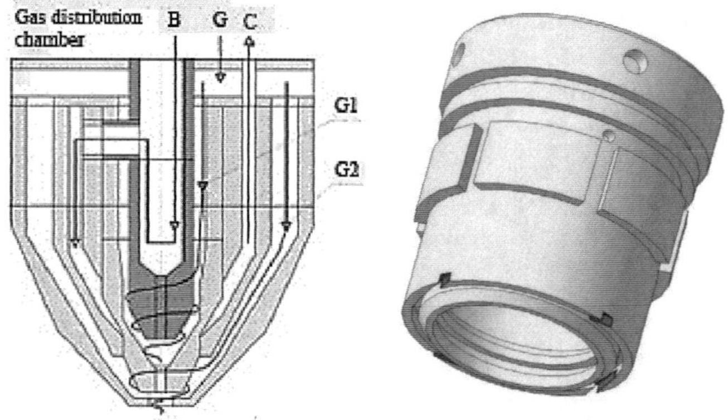

a b

Figure 5. System of gas-vortex stabilization in plasma torch PMVR-9.1: a) scheme of AGP, B, C – inlet and outlet of the cooling system, G – gas supply into the gas distribution chamber, G1, G2 – plasma-forming and stabilizing flows; b) swirler of secondary (stabilizing) flow

The kinematic properties (local velocity) of the plasma jet were calculated to assess the effect of various methods of gas-vortex stabilization on the efficiency of the plasma torches. The studies were performed in the FlowWorks application of the SolidWorks software. They were based on a simulation of the gas flow through the channels of the plasma torch and in the area of its expiration. Technical passport data of such plasma torches for the mass flow of gas (common value is 0.011 kg/s) were used as initial data. Environmental parameters with normal atmospheric pressure were used at the

IOP Conf. Series: Materials Science and Engineering **681** (2019) 012003 doi:10.1088/1757-899X/681/1/012003

outlet of the plasma torch nozzle. In order to abstract from the thermodynamic effects, the calculation was carried out for a cold gas jet in disregard of the heating of the gas and the walls of the AGP channels by the plasma arc. The velocity was calculated at different distances from the axis of the nozzle channel and the cathode section of the plasma torch. The calculations were made along the lines parallel to the axis within the cylindrical region limited by the diameter of the nozzle channel (Fig. 6).

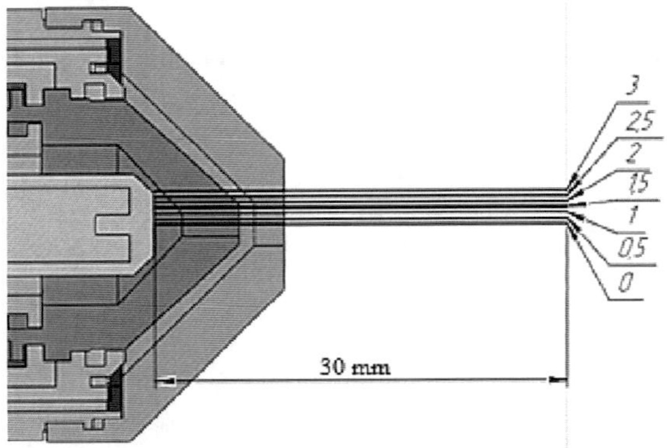

Figure 6. Calculation scheme for determination of kinematic characteristics of jet for the plasma torch PMVR-5.3

3 statistical parameters were used to analyze the flow rates V distribution:

1. Scope of variation $R=V_{max}-V_{min}$,
2. Ratio of variation $L= V_{max}/V_{min}$,
3. Coefficient of variation $F=S/(\overline{V})$, where S is dispersion of V, $\overline{V}$ is average flow rate.

The combination of these parameters allows comparing the characteristics of flows both from the point of view on the kinetic properties and with regard to the uniformity of the velocity distribution in the cross section of the jet.

3. Results and Discussion

The results of the velocities calculation for different methods of gas-vortex stabilization in single-and double-flow plasma torches are demonstrated in Fig. 7. The analysis of the presented images allows drawing a conclusion about the differences in the velocity distribution, which determines the shape, length, kinematic and energy parameters of the plasma jet.

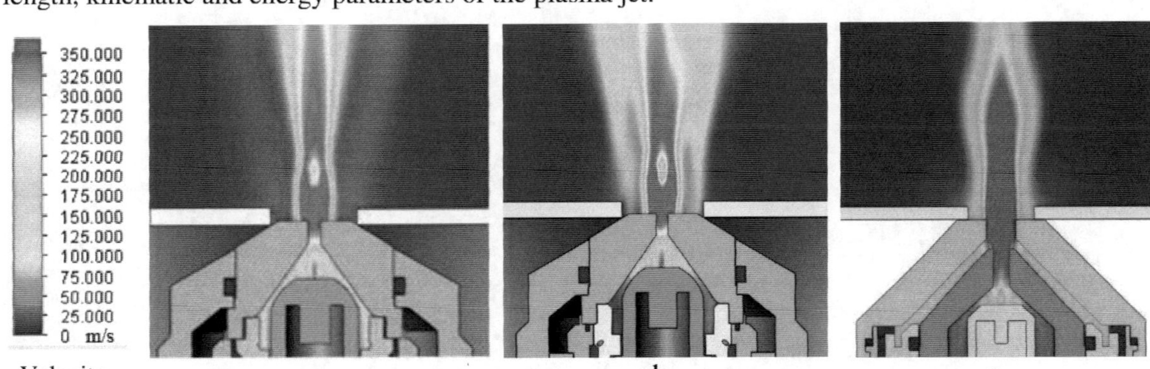

Figure 7. Velocity distribution in the jet of plasma torches: a) PMVR-5.3 with a serial swirler, b) PMVR-5.3 with the new swirler, c) PMVR-9.1 (double nozzle)

IOP Conf. Series: Materials Science and Engineering **681** (2019) 012003 doi:10.1088/1757-899X/681/1/012003

Figures 8, 9 and 10 demonstrate the results of the velocity calculation, executed in accordance with the scheme shown in Fig. 6 (for the upper axisymmetric part of the computational domain). Attention should be paid to the geometry of the nozzle, which determines the distance from the cathode plane to the nozzle section about 7 mm.

The analysis of the velocity distributions presented in Figs. 8 and 9 allows concluding about the effect of the new swirler applied in the PMVR-5.3 plasma torch on the kinematic properties of the jet. The nonuniformity of the velocity distribution along the channel section becomes significantly less expressed as it moves away from the jet expiration plane. In the jet of the plasma torch PMVR-5.3 with the new swirler, a smaller dispersion of velocities is also observed at a maximum estimated distance compared to the jet of plasma torch PMVR-5.3 with a serial swirler. The maximum velocities are approximately equal, and the average cross-section velocities have higher values in the jet of the PMVR-5.3 plasma torch with a new swirler. In fact, this indicates the preservation of higher energy properties over a larger range of plasma jet lengths, which, in turn, should have a positive effect on the current-voltage characteristics of the arc and its ability to cut metal of greater thickness at an increased speed.

Data calculations for the double-flow plasmatron PMVR-9.1 (Fig. 9) indicate a significantly more uniform distribution of velocities in the jet, both in the plane of expiration and at a distance from this plane. However, kinematic and energy properties of the jet, determined by the maximum and average velocity values, are maintained at a shorter (less than 25 mm) distance than in the jets generated by the single-flow plasmatron PMVR-5.3. This is also evidenced by the comparative graphs (Fig. 11). The last factor determines not only the technological conditions for cutting by this plasma torch, but also affects the acoustic properties of the jet. The size of the acoustic core of the jet with maximum velocities, as known [7], determines the level of noise emission, so the reduction of the core length should have a positive effect on the acoustic safety of this technology.

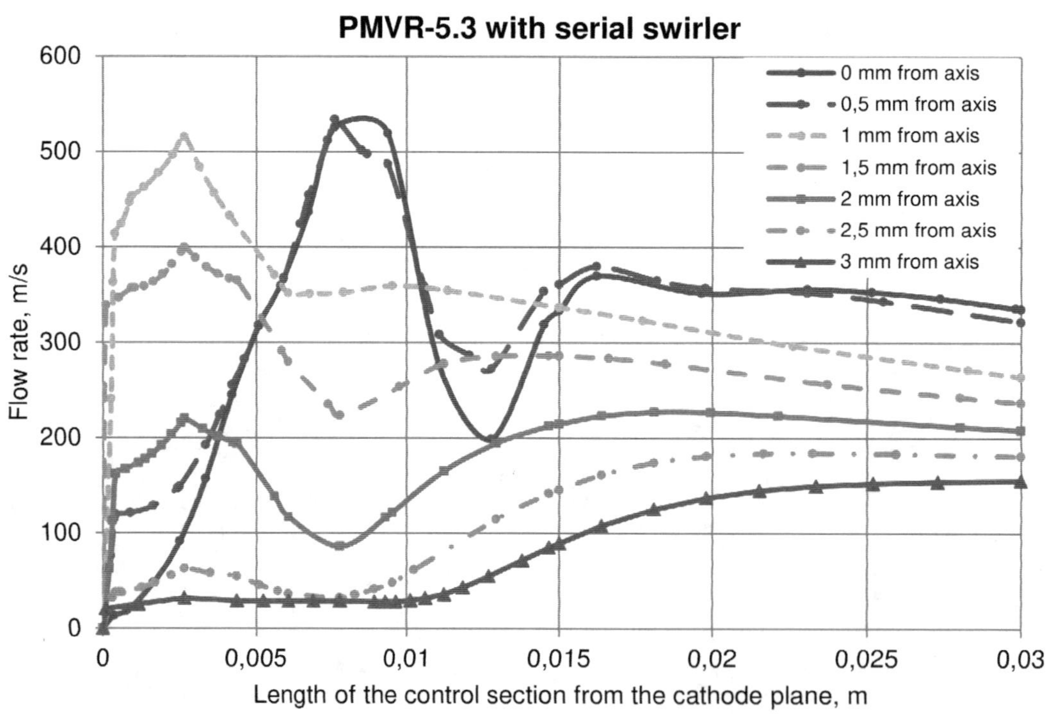

Figure 8. Velocity distribution in the jet of plasma torch PMVR-5.3 with serial swirler

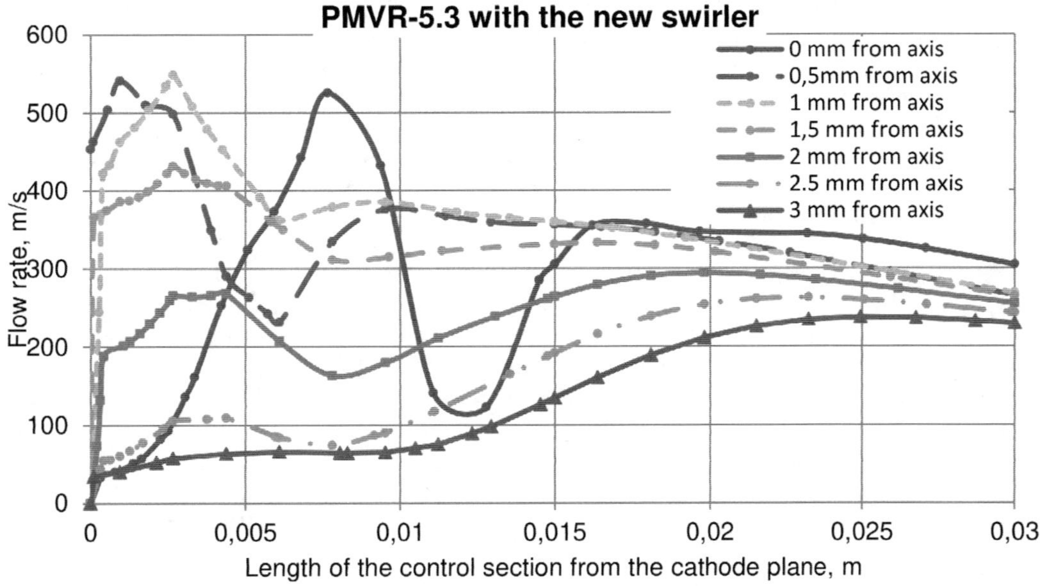

Figure 9. Velocity distribution in the jet of plasma torch PMVR-5.3 with the new swirler

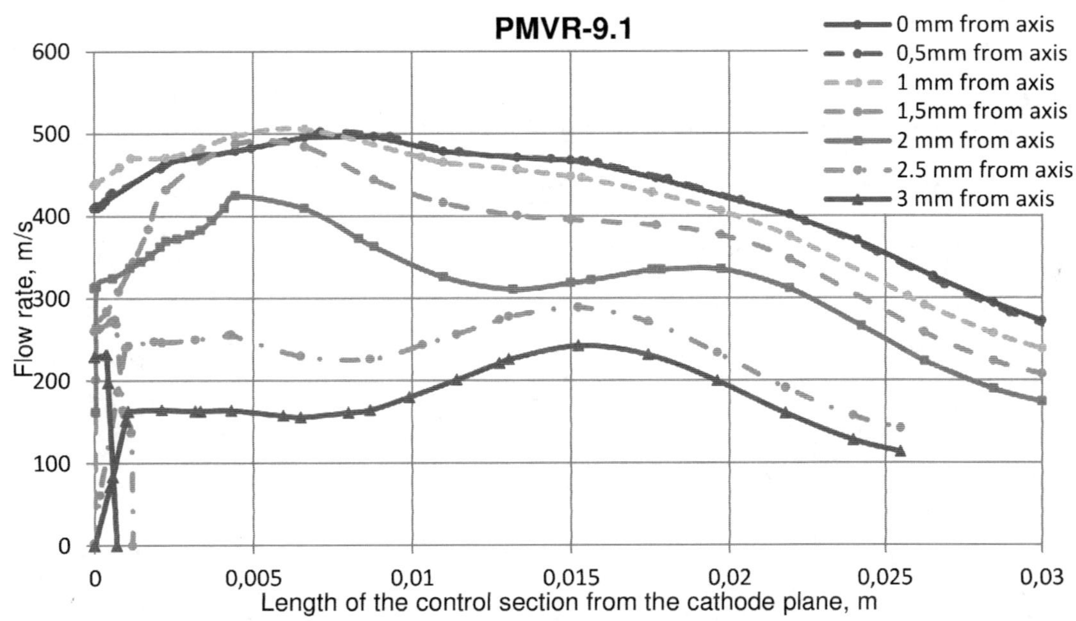

Figure 10. Velocity distribution in the jet of plasma torch PMVR-9.1

Fig. 12 shows the velocity distribution around the calculated cross section of the jet, which, in general, confirms earlier conclusions. Additional compression of the main PFG in the plasma torch PMVR-9.1 can significantly reduce nonuniformity of the velocity distribution in the plasma torch PMVR-5.3, arising on the nozzle section due to the disruption of the sharply expanding flow. In addition, the jet of the plasma torch PMVR-9.1 has higher kinetic properties (higher top speed) than the jet of the plasma torch PMVR-5.3. However, such advantages are observed only within a distance of 15-20 mm from the cathode section. At large distances, the PMVR-5.3 plasma torch with a new swirler has the higher indices of uniformity and kinetic properties, which is confirmed by the graphs of statistical distributions of kinematic characteristics (Fig. 13). The growth of these indicators, in turn, should affect the increase in productivity and quality of cutting.

The last conclusion is confirmed by the data obtained in the process of plasma torch PMVR-5.3 cutting. The maximum cutting speed for 10-mm thick 09G2S steel plates at a current of 90 A was not more than 1.1-1.2 m/min. The use of the upgraded swirler increased the maximum cutting speed up to 1.5-1.6 m/min.

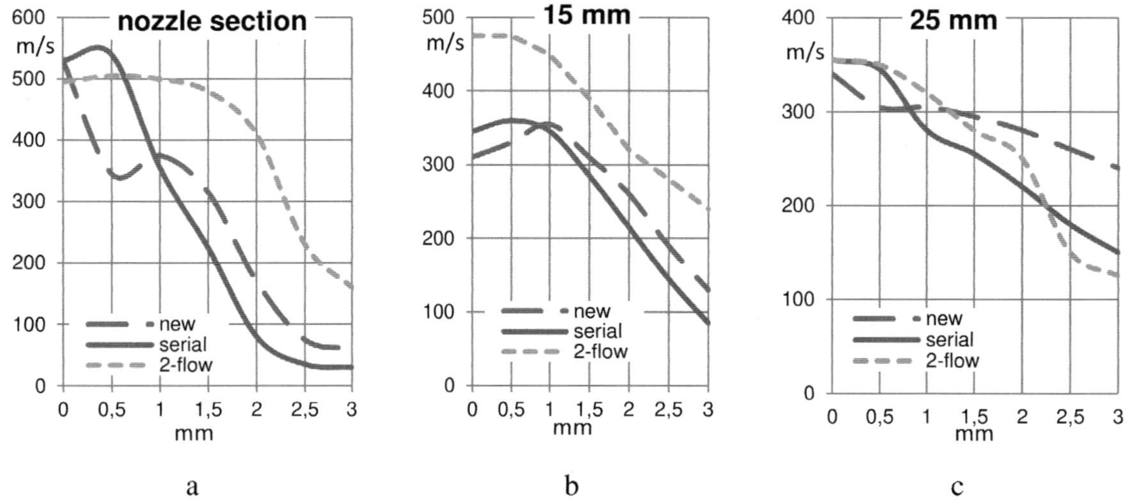

Figure 11. The velocity distribution over the cross section of the jet depending on the distance from the nozzle output of the plasma torch (PMVR-5.3 with serial and new swirlers and 2-flow PMVR-9.1)

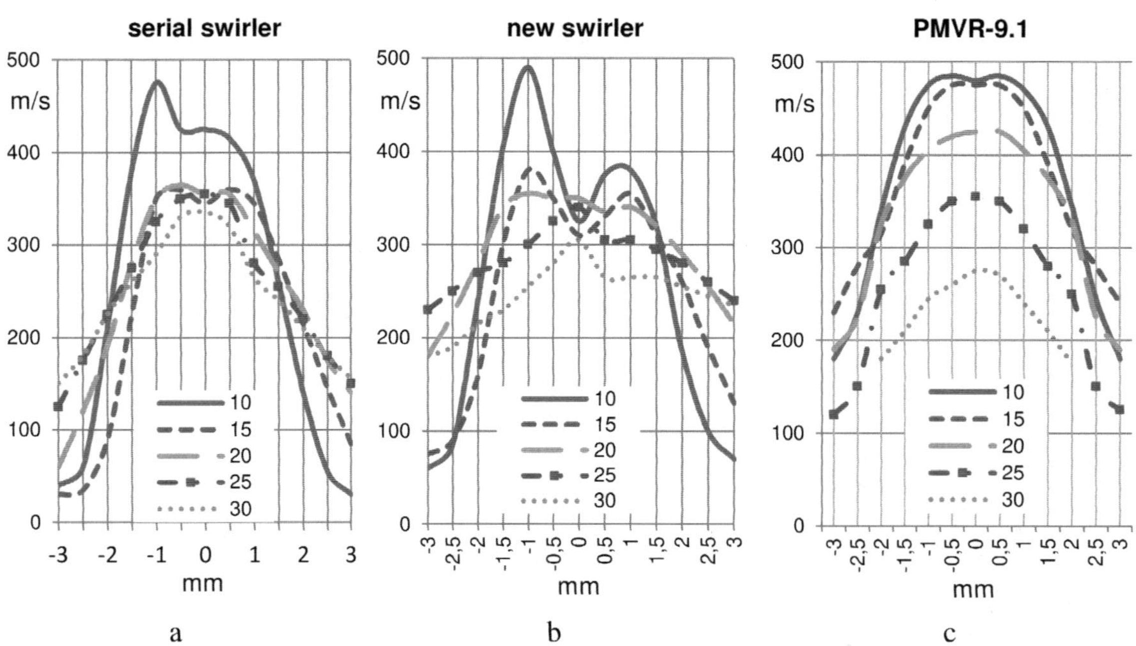

Figure 12. The velocity distribution over the cross section of the jet depending on the distance from the nozzle output of the plasma torch (PMVR-5.3 with serial and new swirlers and 2-flow PMVR-9.1)

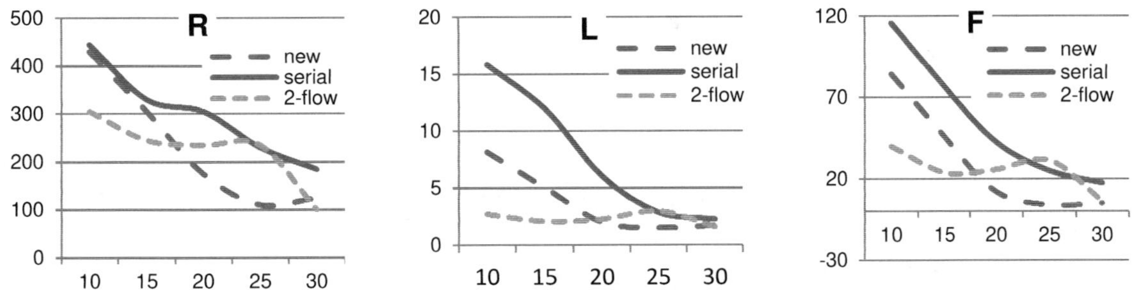

a b c

Figure 13. Statistical distributions of kinematic characteristics depending on the distance from the plasma torch nozzle output (PMVR-5.3 with serial and new swirlers and 2-flow PMVR-9.1): a) scope of variation R, b) ratio of variation L, c) coefficient of variation F

4. Conclusions

The results of the study demonstrate the dependence of the angle of PFG injection into the nozzle on the kinematic and kinetic characteristics of plasma jet and, consequently, the cutting parameters. This conclusion provides the basis for further research to determine the optimal angle of PFG injection into the nozzle and its impact on the productivity, cutting quality and durability of the plasma torch. As a conclusion, it is recommended to use the suggested narrow-jet-plasma technology for cutting thin sheet metals.

Acknowledgements

The work was executed under the support of state assignment of the Ministry of Science and Higher Education of the Russian Federation No. 13.10317.2018/11.12.

References

[1] Cherednichenko V.S., Anshakov A.S., Kuzmin M.G. Plasma electro-technological installations, Novosibirsk, NSTU, 2011.

[2] Kaydalov A.A. Modern technologies of thermal and remote cutting of structural materials, Kiev, Ecotechnology, 2007.

[3] Anakhov S.V. Principles and methods of plasma torch designing, RSVPU, Ekaternburg, 2018.

[4] Anakhov S.V, Pykin Yu.A., Matushkin A.V. Narrow Jet Plasma as the Energy Efficient and Safe Technology for Metal Cutting // Solid State Phenomena (Material Science Forum). 2016, Volume 870, p. 523-527

[5] Anakhov S.V., Pykin Yu.A., Matushkin A.V. Investigation of systems for gas vortex stabilization of plasma torches // Welding International. 2016, Volume 30, Issue 4, p. 310-314

[6] Anakhov S.V., Pykin Yu.A., Matushkin A.V. Gas vortex stabilization in plasma torches: new solutions // Welding International. 2016, Volume 30, Issue 5, p. 408-412

[7] Kuznetsov V.M. Fundamentals of the theory of turbulent jets noise, Moscow, FIZMATLIT, 2008.

Substantiation of the narrow-jet plasma technology for finishing cutting of steel sheets

S V Anakhov[1], N B Pugacheva[2], T M Bykova[2], B N Guzanov[1], A V Matushkin[3] and Y A Pykin[3]

[1] State Vocational Professional University, 11 Mashinostroiteley St., Ekaterinburg, 11620012, Russia
[2] Institute of Engineering Science, Ural Branch of the Russian Academy of Sciences, 34 Komsomolskaya St., 620049, Ekaterinburg, Russia,
[3] B.N. Yeltsin Ural Federal University, 19 Mira St., Ekaterinburg, 620002, Russia

E-mail: nat@imach.uran.ru

Abstract. The paper presents a study of the effect of the PMVR-5.2 plasma torch inclination angle on the cut surface profile, the structure and hardness of the material of the edge on 10 mm thick 09G2S steel plates. The plasma torch implements the narrow-jet technology. The heat-affected zones prove to have a gradient structure. When cutting is performed at an angle of 30°, there appear ~500 μm thick edges with a 15 to 30 μm thick surface zone of structureless martensite having a microhardness of 900 HV 0.025 and a zone of fine-lamellar pearlite with a hardness of 350 HV 1. This requires mechanical removal of a 40 μm thick layer from the cut surface before subsequent welding. When the plasma torch is inclined at an angle of 90°, the quality of the edge material requires no mechanical removal of the heat-affected zone for subsequent welding.

1. Introduction

Plasma cutting of metal is currently one of the most popular technologies in metallurgy and mechanical engineering. Today, there is a wide choice of metal-cutting plasma torches of various functional purposes and designs, of both domestic and foreign production. A line of research in plasma technologies is their safe application. In most cases, the use of plasma torches requires monitoring and ensuring a safe level of noise emission, gas-aerosol pollution, heat, light, and electromagnetic loads. This can be achieved in the following ways: designing plasma torches with a reduced negative effect on the staff, using protective means and devices, and regulating the operating conditions under increased harmful effects. Note the actual absence of narrow-jet plasma torches from the domestic market. The narrow-jet plasma technology [1] significantly exceeds the conventional methods of plasma cutting of thin and medium-thickness metals in terms of efficiency and power consumption, and it is comparable with laser cutting of metals in terms of the width and quality of the cut [2].

Unfortunately, despite the existing theoretical developments of these technologies in Russian engineering [3], products of only foreign manufacturers (Hypertherm, Kjellberg, Messer Griesheim) are currently available on the market. In order to create a plasma torch implementing the narrow-jet technology, a multiannual experience in research and design of plasma torches for various applications was used [4]. As a result, several modifications of the PMVR-5 narrow-jet plasma torch with an upgraded system of gas vortex stabilization and plasma jet compression were developed. This system increases the efficiency of heat input and enhances the quality of the metal in the cutting area. The

IOP Conf. Series: Materials Science and Engineering **681** (2019) 012004 doi:10.1088/1757-899X/681/1/012004

quality of subsequent welding is affected by the thickness of the heat-affected zone (HAZ) after cutting, its microstructure and hardness, as well as by the cut surface roughness, the characteristics being regulated by GOST 14792-80.

The aim of this paper is to study the roughness, microstructure, and microhardness of the edge surface, which characterize the quality of structural low-alloy steel cuts made with the use of the PMVR-5.2 plasma torch.

2. Materials and research methods

Plasma cutting was performed with the use of a PMVR-5.2 plasma torch (Fig. 1a), produced by Poligon LLC, at different angles of plasma torch inclination, as shown in Fig. 1b. The design of the gas-air path (GAP) of the PMVR-5.2 plasma torch is based on the principle of stepwise equalization of plasma forming gas (PFG) velocity across the duct section (i.e., the gas comes from duct 1 to duct 2, then from 2 to 4, etc.). The main swirler is the final element in this chain. The design feature of this swirler is that the PFG flow passing through the swirler is supplied tangentially to the cathode surface in the plane perpendicular to the cathode. A commercially available A-141 cathode with a series of 1.4 to 1.8 mm outlet diameter nozzles is used in the plasma torch design. Due to the concepts used in the design of the gas-air path of the plasma torch and its cooling system, it was possible to increase by 15–20% the maximum permissible cutting current for the A-141 cathode and nozzle used in the design. The cutting technology was chosen in view of the possible further use of thin-sheet steel (up to 20–25 mm) for the production of butt welds (plasma torch kept at an angle of 90°), or for the welding of medium-thickness (25 to 40 mm) steels (edge preparation with a plasma torch inclined at an angle of 30°), Fig. 1b. The cutting process conditions are given in Table 1.

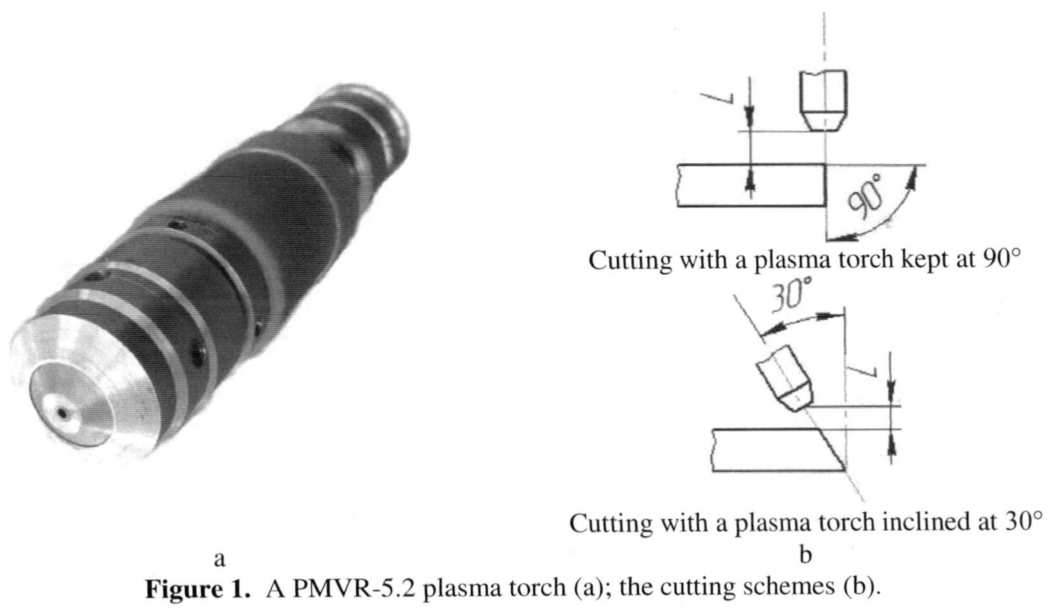

Figure 1. A PMVR-5.2 plasma torch (a); the cutting schemes (b).

Table 1. The cutting conditions

Inclination angle	Arc current [A]	Arc voltage [V]	Cutting speed [m/min]	Nozzle diameter [mm]	Distance from the nozzle section *L* [mm]	Plasma-forming gas pressure [MPa]
30°	115	200	1.25	1.7	5	0.5
90°	88	180	0.75	1.7	5	0.5

Plates of the 09G2S sheet steel, which is one of the main materials in the production of welded blanks, were subjected to plasma cutting. The plate thickness was 10 mm. It should be noted that 10 mm thick metal is cut by a plasma torch with upgraded individual gas vortex stabilization and plasma jet compression at a much higher speed (0.75 to 1.1 m/min) than by conventional single-tube gas plasma torches (0.54 m/min). As a result, in addition to the expected improvement of the cutting quality owing to reduced heat input into the cutting zone, a significant increase in productivity, energy efficiency, and process safety is achieved [5].

The sample surface profile after plasma cutting was examined with a Veeco optical interferometer in the center of the cut. The metallographic structural analysis of the samples etched in a reagent based on a 4% solution of nitric acid in ethyl alcohol was performed on a Neophot 21 microscope at a magnification of 100×. The steel microstructure was identified by comparison with the reference scales according to GOST 8233-56. The hardness of the samples was measured on a Leica device under loads of 0.245 (only the near-surface layer with a thickness of at most 20 μm) and 9.8 N on the indenter. The imprinting scheme is shown in Fig. 2.

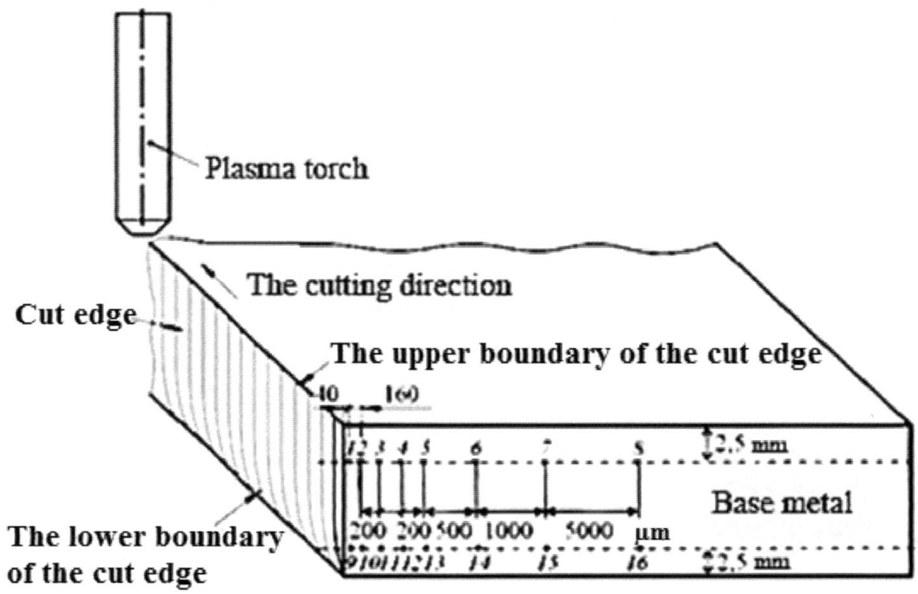

Figure 2. Cut edge hardness measurement scheme.

3. Results and discussion

The examination of the cut surface (Fig. 3) has shown that the average roughness value ranges between 7 and 9 μm (Table 2), and this corresponds to the 1st class of quality according to GOST 14792-80. The maximum surface roughness values are also within the permissible limits (at most 50 μm) for the 1st class of quality.

The base metal of the samples, which did not undergo heating and structural changes in the process of cutting, has a typical ferrite-pearlite structure, the perlite/ferrite (P/F) ratio being 20/80 on scale 7 according to GOST 8233-56 (Fig. 4). During plasma cutting, the metal melts in the zone of plasma jet action, whereby the plate is divided into two fragments. Near the cut surfaces, the steel is heated to different temperatures depending on the length deep into the plate. The material of the edge is the heat affected zone structurally consisting of 3 subzones. The first near-surface subzone of the maximum steel overheating consists of structureless martensite (the bright stripe in Fig. 4a) with a hardness of 900 HV 0.025, where the cooling rate has maximum values due to the summation of surface cooling

IOP Conf. Series: Materials Science and Engineering **681** (2019) 012004 doi:10.1088/1757-899X/681/1/012004

and heat dissipation through the plate material. The thickness of the martensitic zone ranges from 15 to 30 μm at a plasma torch inclination angle of 30° and does not exceed 5 μm at a cutting angle of 90°. The second subzone consists of pearlite, whose dispersion varies with the plasma torch inclination angle. At an angle of 30°, 5-grade fine lamellar perlite is formed on scale 1 according to GOST 8233-56, with an interlamellar distance of 0.80 μm (Fig. 4a), the subzone width being 185 μm. At a cutting angle of 90°, 6-grade medium lamellar pearlite with an interlamellar distance of 1.00 μm is formed, the subzone width being 100 μm (Fig. 4b).

The third subzone consists of pearlite and ferritic grains. At an inclination angle of 30°, P/F=65/35, the subzone width being 280 μm. When the angle is 90°, P/F=50/50, the subzone width being 250 μm. The total HAZ thickness for the plates cut at an angle of 30° is about 375 μm, and it is 500 μm for the plates cut at an angle of 90°. This corresponds to the 2nd class of quality according to GOST 14792-80.

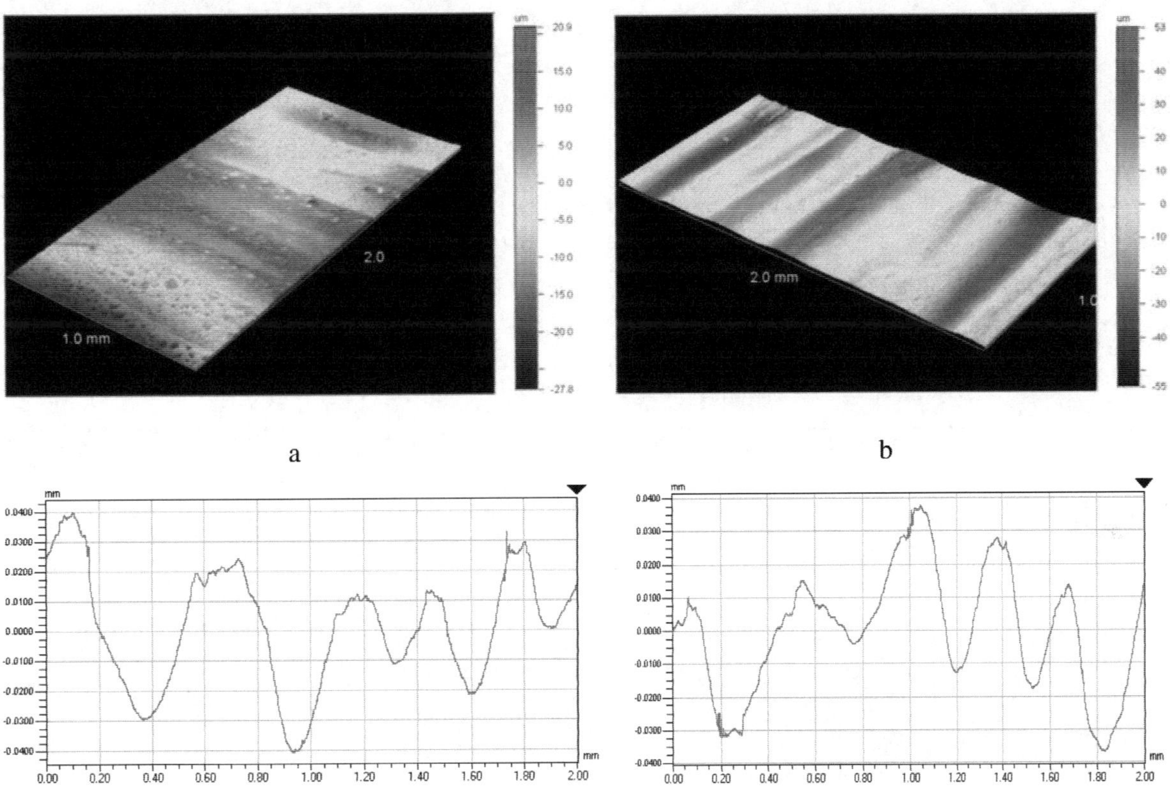

Figure 3. Cutting surface profile (a, b) and profilograms (c, d); a, c – plasma torch inclined at 30°; b, d – plasma torch kept at 90°.

Table 2. The results of surface roughness examination

Plasma torch inclination angle	Average surface roughness (Ra) [μm]	Maximum profile height (Rt) [μm]
30°	7.3	38.7
90°	8.0	48.2

A large proportion of ferrite in the HAZ structure for the plates cut at a plasma torch inclination angle of 90° causes smaller changes in hardness, as shown in Table 3 and Fig. 5. According to the

IOP Conf. Series: Materials Science and Engineering **681** (2019) 012004 doi:10.1088/1757-899X/681/1/012004

requirements of the STO Gazprom 2-2.4-083 application standard, the HAZ hardness values after plasma cutting must not exceed 300 HV. This requirement is fully met at a cutting angle of 90°, and this allows one to avoid additional machining of the cut surface during subsequent welding of the cut plates. Cutting at an angle of 30° requires the removal of a 40 μm thick surface layer before welding.

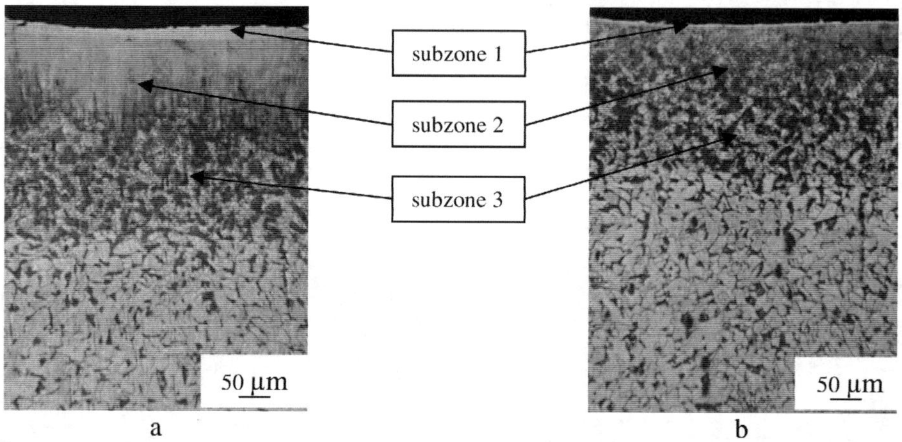

Figure 4. The structure of the sample surface in the cutting area: a – plasma torch inclined at 30°; b – plasma torch kept at 90°.

Table 3. Hardness variation in the surface layer on the side of the cut

Distance from the surface [μm]	Plasma torch inclined at 30°			Plasma torch kept at 90°		
	Input	Center	Output	Input	Center	Output
40	347	356	356	311	273	263
160	261	289	289	314	262	269
200	235	274	231	278	227	258
200	211	217	198	226	216	236
200	203	203	196	198	199	230
500	195	189	178	190	217	203
1000	180	179	174	174	180	194
5000	170	168	162	167	176	175

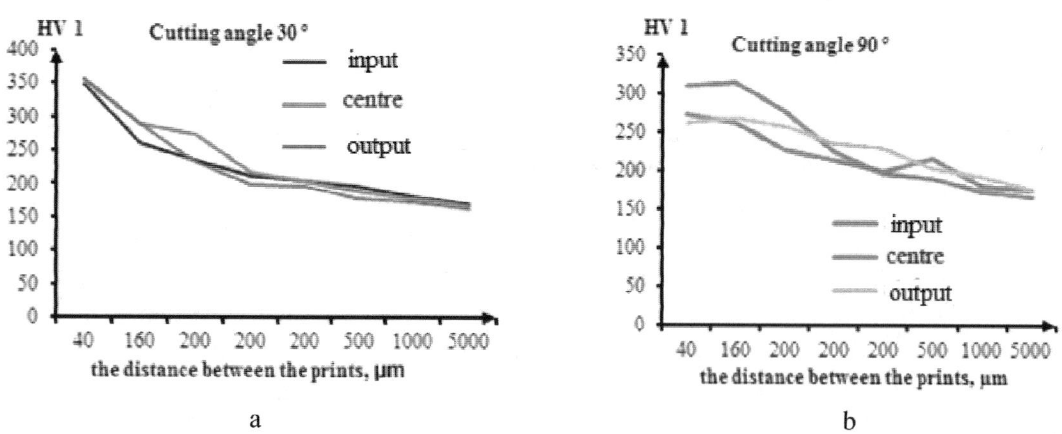

Figure 5. Hardness distribution in the sample: a – plasma torch inclined at 30°, b – plasma torch kept at 90°.

4. Conclusion

The use of the new plasma torch allows one to produce edges with medium (7 to 9 μm) and maximum (48 μm) surface roughness, corresponding to quality class 1 according to GOST 14792-80, on 09G2S steel plates. The heat-affected zones have a gradient structure with a proportion of pearlite smoothly increasing to 100% of its content near the cutting surface. High cooling rates during cutting at an angle of 30° induce structureless martensite on the surface, the thickness of this zone not exceeding 30 μm.

The requirement of the STO Gazprom 2-2.4-083 application standard for hardness values (≤ 300 HV) is fully feasible for plasma cutting of 09G2S steel plates at a plasma torch inclination angle of 90°; this enables edges on sheet steel to be used for welding without removing heat-affected zones. After edge preparation by cutting at an angle of 30°, typical for welding of medium-thickness steels, it is required to remove mechanically the surface layer no thicker than 40 μm.

Acknowledgments
The work was performed on the equipment installed in the Plastometriya collective use center of the IES UB RAS within the state assignment, themes Nos. AAAA-A18-118020790145-0 and AAAA-A18-118110790009-3.

References

[1] Krink V., Simler H., Laurisch F. 2006 Plasmaschneidtechnologie – Erweiterung wirtschaftlicher Anwendungsgebiete. Proc. Int. Schneidtechnische Tagung (International Conference on Cutting Technology – ICCT 2006) S.18–25.
[2] Kaydalov A. A. Modern technologies of thermal and remote cutting of structural materials, Kiev, Ecotechnology, 2007.
[3] Anakhov S. V. Principles and methods of plasma torch designing, RSVPU, Ekaterinburg, 2018.
[4] Anakhov S. V, Pykin Yu. A., Matushkin A. V. Narrow-Jet Plasma as the Energy Efficient and Safe Technology for Metal Cutting // Solid State Phenomena (Material Science Forum). 2016, Volume 870, p. 523–527
[5] Anakhov S. V. Qualitative study of steel cutting with the use of the narrow-jet plasma technology/ S. V. Anakhov, N. B. Pugacheva, Yu. A. Pykin, A. V. Matushkin, T. M. Bykova / Inorganic Materials: Applied Research, 2017, Vol. 8, No. 3, pp. 400–405
[6] Krink V., Simler H., Laurisch F. 2006 Plasmaschneidtechnologie – Erweiterung wirtschaftlicher Anwendungsgebiete. Proc. Int. Schneidtechnische Tagung (International Conference on Cutting Technology – ICCT 2006) S.18–25.
[7] Kaydalov A. A. Modern technologies of thermal and remote cutting of structural materials, Kiev, Ecotechnology, 2007.
[8] Anakhov S. V. Principles and methods of plasma torch designing, RSVPU, Ekaterinburg, 2018.
[9] Anakhov S. V, Pykin Yu. A., Matushkin A. V. Narrow-Jet Plasma as the Energy Efficient and Safe Technology for Metal Cutting // Solid State Phenomena (Material Science Forum). 2016, Volume 870, p. 523–527
[10] Anakhov S. V. Qualitative study of steel cutting with the use of the narrow-jet plasma technology/ S. V. Anakhov, N. B. Pugacheva, Yu. A. Pykin, A. V. Matushkin, T. M. Bykova / Inorganic Materials: Applied Research, 2017, Vol. 8, No. 3, pp. 400–405

IOP Conf. Series: Materials Science and Engineering **681** (2019) 012005 doi:10.1088/1757-899X/681/1/012005

Impact of heating conditions on the synthesis dynamics of composites in a cylindrical mold

N V Bukrina [1], A G Knyazeva [1,2]

[1]Institute of Strength Physics and Materials Science of SB RAS, Tomsk, ,634055 Russia
[2]National Research Tomsk Polytechnic University, 634050 Tomsk Russia

E-mail: bookr81@mail.ru

Abstract. The paper suggests a model of powder compact heating in a cylindrical mold. The model accounts for various conditions of the reactor wall heating by different devices available for a laboratory experiment. Different heating conditions are set as different boundary conditions. The effective thermophysical properties of the mixture in the reactor are considered to be constant. The treatise demonstrates the examples of numerical modeling for three heating conditions, which lead to different respective conditions of reaction initiation due to nonuniform temperature propagation in the compact.

1. Introduction

Heat transfer is one of the main processes in the formation of the materials' structure and surface during their synthesis and processing. Any high-temperature process is aimed at attaining a new type of material or composition that possesses the highest properties and low cost. Different conditions of the synthesis cause different composition of the material. Therefore, the mechanical, physicochemical and operational characteristics of materials and parts are mainly determined by such thermal parameters as temperature field, thermal cycle, heating and cooling rates.

Indeed, in [1] a compacted specimen was subjected to electric-field treatment. As a result, the reaction initiated inside the compact. In [2], the authors initiated the reaction through heating a steel part—touching one of the ends of the compacted specimens—using an induction coil. The authors of [3] ignited cylindrical compacts using a tungsten coil. In all the cases, the deliverable is a high-density product. The selection of the initiation source should also take into account the reaction type and particle size of the powder mixture. In some cases, low exothermicity of the reaction between powder compact components requires preliminary heating of the system, similarly to [3]. Sometimes, selecting a more suitable heating method is required [2].

Thus, the classification of heating sources becomes of high importance. As a rule, literature sources contain classification of sources into distributed and pointed; surface and volumetric; immovable and moving; continuous and pulsed ones.

The traditional approach to synthesis modeling introduces a concept of effective heat source for different processing types, which allows distinguishing in them a general regularity and specificity. Effective heat source parameters include thermophysical and optical characteristics of a material. In this case, the temperature field is the only independent characteristic through which those remaining are determined: motion of phase boundaries, chemical reaction rates, mass transfer of alloying additives, etc.

Current work is aimed at investigating the effect of heating conditions on the dynamics of a compact heating in a cylindrical reactor.

2. Mathematical statement of the problem

The mathematical statement of the problem of powder composite heating in a cylindrical reactor (Fig. 1) includes a two-dimensional heat conductivity equation for a powder mixture:

$$(c_A\rho_A)^0\frac{\partial T_A}{\partial t}=\frac{1}{r}\frac{\partial}{\partial r}\left(\lambda_A^0 r\frac{\partial T_A}{\partial r}\right)+\frac{\partial}{\partial z}\left(\lambda_A^0\frac{\partial T_A}{\partial z}\right)+W_H \quad (1)$$

The equation includes the term accounting for the heating through electric fields. For the reactor's walls, bottom and punch, equations similar to (1) are composed, but without source terms. It is assumed that the heat contact between the materials is ideal. In the center, a condition of symmetry is established.

Function W_H in eq. (1) characterizes heating due to current $I(t)$. This value is considered to be volumetric. In this case, function W_H changes in time following a certain law:

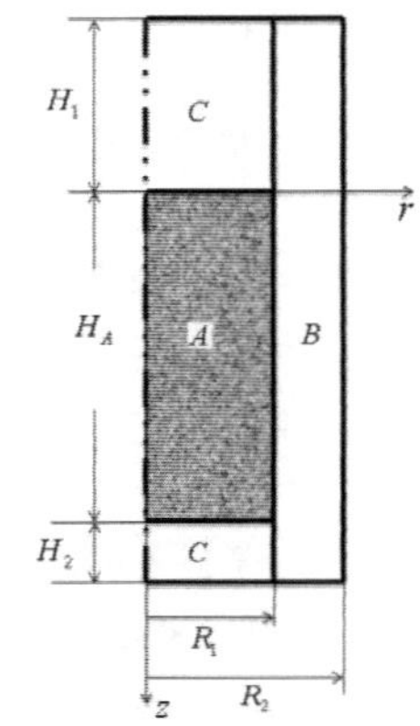

Figure 1. Scheme of the composite synthesis in the regime of thermal explosion after heating the powder compact in a steel reaction mold

$$W_H=R_e I^2(t). \quad (2)$$

In the case of heating by high-frequency currents, from the inductor at interface R_2 the heat flux will be

$$r=R_2:\quad -\lambda_B\frac{\partial T_B}{\partial r}=\sigma\varepsilon_B\left(T_W^4-T_B^4\right) \quad (3)$$

where T_W is the change in the temperature touching the external reactor wall obtained from the solution of a separate electromagnetic problem [4].

If the powder compact is heated through heat conductivity due to heat of bottom $z=H_A+H_2$ and punch $z=-H_1$, the heat source is set on the following boundaries: $T|_{z=-H_1}=T_C$ or $-\lambda_c\cdot\partial T/\partial z|_{z=-H_1}=q_C$; similarly, on the other surface $T|_{z=H_A+H_2}=-T_C$ or $\lambda_c\cdot\partial T/\partial z|_{z=H_A+H_2}=q_C$.

At the initial moment of time, the temperature in all regions is set: $t=0$: $T_A=T_{A0}$, $T_B=T_{B0}$, $T_C=T_{C0}$.

In general case, the problem is a set of equations that allows deriving the temperature distribution at any time for different composition of powder mixtures and different materials that may comprise the cylindrical reactor walls. The result will be also affected by the reactor wall thickness and porosity of the powder mixture, because by accumulating the heat, the reactor walls become an additional heat source for the mixture.

The problem was solved numerically using implicit difference scheme of second-order approximation in space and first-order approximation in time using splitting in coordinates and the linear sweep method. The calculations yielded temperature fields along the whole studied region for different heating variants.

In the case of electric heating (2), the electrical resistance, in general case, depends on such parameters as structure, medium composition, temperature and porosity, i.e. it changes in time. Volumetric heat sources feature relatively large heating time up to the working temperature. The energy concentration in such sources is usually small.

In the case of heating by high-frequency currents by inductor (3), the inner reactor surface heats up and then the heat propagates through heat conductivity inside the system. An important role is played by the current density in this case. An electromagnetic problem [4] was solved separately. Its solution gave the dependence of the heater temperature on the magnetic field strength. The body that is characterized by this changing temperature are reactor walls heated by radiation. Then, the powder mixture is heated by the walls through heat conductivity. The processed results showed that the temperature rise can be described by a straight line $T_w(t,H)=at+b$, where the coefficients depend on H_0.

In case (4), the calculations used constant temperature on the upper and lower boundaries of the reactor.

As an example, let us consider a compact comprised of Ni and Al powder mixture. The effective properties of the mixture obey the following rule:

$$(c_A\rho_A)^0 = c_{Ni}\rho_{Ni}\xi + c_{Al}\rho_{Al}(1-\xi);\ \lambda_A^0 = \lambda_{Ni}\xi + \lambda_{Al}(1-\xi);$$

ξ is the mass fraction of nickel in the initial mixture, $\rho_{Ni}, \rho_{Al}, c_{Ni}, c_{Al}$ are density and heat capacity of pure elements in the mixture. Hence, $(c_A\rho_A)^0 = 4.04$; $\lambda_A^0 = 1.024$.

In general case, the material of the reactor walls, bottom and punch may be different. The material of the reactor walls was steel E235 with the following properties: $c_B = c_C = 0.462$ [*J/(g·K)*]; $\rho_B = \rho_C = 7.87$ [*g/cm*^{3}]; $\lambda_B = \lambda_C = 0.46$ [*W/(cm·K)*].

The calculations were made with the following parameters [6]: $R_1 = 3$, $R_2 = 4.5$ [*cm*], $\sigma = 5.67\cdot10^{-12}$ [*W/cm*^{2}*·K*^{4}], $\varepsilon = 0.4$, $c_{Ni} = 0.482$; $c_{Al} = 0.951$ [*J/(g·K)*]; $\rho_{Ni} = 8.9$; $\rho_{Al} = 2.7$ [*g/cm*^{3}]; $\lambda_{Ni} = 0.8$; $\lambda_{Al} = 2.4$ [*W/(cm·K)*].

3. Results and Discussion

Below are the calculation examples for different heating conditions (Fig. 2–4).

In the case of powder compact heating, the heat propagates from upper and lower reactor parts (bottom and punch) through the heat conductivity of the materials. The reaction mixture gradually heats up. Since the side walls of the reactor, in this case, are the heat sinks (they should also heat up), the temperature distribution at the initial stage is nonuniform..

In the figures, the reactor bottom is at the top, while the punch is at the bottom.

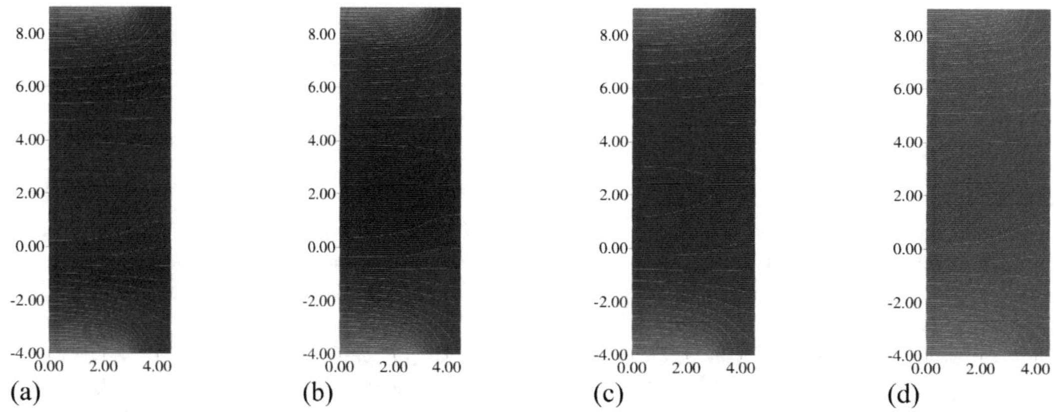

Figure 2. Temperature fields corresponding to heating of bottom and punch at different moments of time: a) 40 s; b) 70 s; c) 100 s; d) 200 s.

In the case of electric heating inside the powder compact, all reactor walls are heat sinks. Uniform heating of the powder compact is evident. However, since the reactor bottom thickness is smaller than that of the punch, it will require less heat for the heat-up. Therefore, more uniform temperature is expected to be closer to the reactor's bottom.

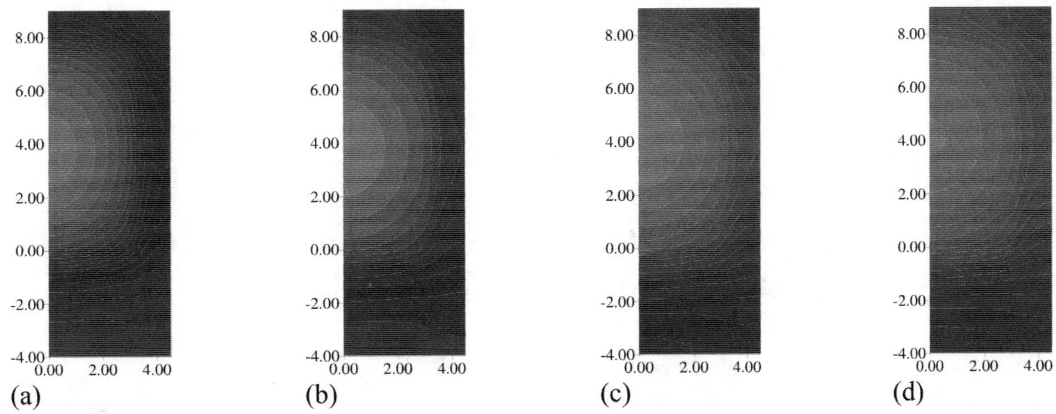

Figure 3. Temperature fields at different moments of time during heating by volumetric source (through Joule heating): a) 40 s; b) 70 s; c) 100 s; d) 200 s.

Heating by high-frequency currents by an inductor provides uniform heating from side surfaces. However, evidently from Fig. 4, the heat-up is nonuniform along the specimen.

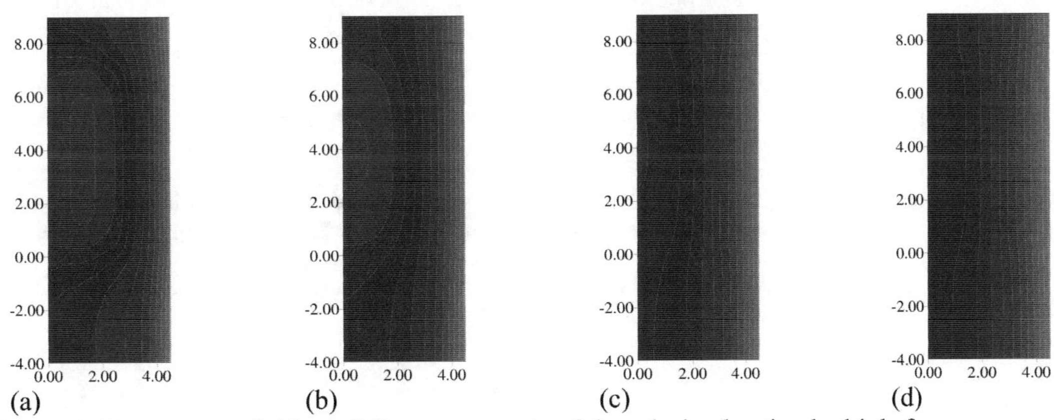

Figure 4. Temperature fields at different moments of time during heating by high-frequency currents from inductor a) 40 s; b) 70 s; c) 100 s; d) 200 s.

4. Conclusion

Thus, different heating conditions create different conditions for reaction initiation, which should be accounted when selecting the synthesis conditions and interpreting experimental data.

Acknowledgments

This work has been performed under the program of fundamental research project III. 23.2.2.

References

[1] Qiaodan Hu, Peng Luo, YouweiYan 2007 *J. Alloys Compd.* **439** (1-2), 132-136.
[2] Shekari M., Adeli M., Khobzi A. et. al. 2017 *Adv. Powder Technol.* **28** (11), 2974-2979.
[3] Yeh C.L. and Sung W.Y. 2004 *J. Alloys Compd.* **384,** 181–191.
[4] Bukrina N.V., Knyazeva A.G. and Ovcharenko V.E. 2018 *Interdisciplinary problems in additive technologies*, 2-9.
[5] Knyazeva A.G., Sorokova S.N., Pobol A.I. et.al. 2015 *Chemical Physics and Mesoscopy*, **17** (2), 239–252.
[6] Grigorev I.S., Meilikhov E.Z. 1991 *Physical values: Book of reference*. (Moscow: Energoatomizdat).

IOP Conf. Series: Materials Science and Engineering **681** (2019) 012006 doi:10.1088/1757-899X/681/1/012006

The issue of quality assurance of materials in additive manufacturing of metal products

V Burov[1], **T Ogneva**[1] **and O Dul'beeva**[1]

[1] Department of Materials Science, Novosibirsk State Technical University, Karl Marx ave., 20, Novosibirsk, 630073, Russia

E-mail: v.burov@corp.nstu.ru

Abstract. The article addresses the quality issues of additive manufactured products from the perspective of materials science. Additive manufacturing is multilayered step-by-step deposition of a material in micro-volumes following the trajectory controlled by a CNC program. Such heating sources as electric arc, laser, plasma and electron beam can be used to melt the metallic powder. This article investigates the structure and properties of titanium alloy Ti-6Al-4V ELI produced by selective laser melting (SLM). The samples were produced in pure argon environment in a system with a 3D printer EOS M290 (Electro Optical System GmbH, Germany) and then heat-treated in vacuum at 750...800 °C during 3 hours. It was revealed that the main structural defects reducing mechanical properties of material are pores and stress concentrators in the points of contact between titanium alloy particles and the newly built surface of the previous layer. The ways aimed to avoid the structural defects which reduce fracture toughness were suggested.

1. Introduction

Additive technology provides the ability to produce different complex geometric shapes without necessity to create any complex special equipment. It makes this method one of the most promising and commercially viable for single-piece or small-scale manufacturing of metallic products. In this area, a great practical experience has been gained with the use of laser and electron-beam sources [1]. The technological process can include one-step or multistep processes, when after forming a geometric shape the mechanical properties of materials can be modified. Heat treatment, as the second step of technological process, usually is applying for residual stress relief, and in some cases for additional structural homogenization [1].

Achievements in SLM showed the possibility to reduce the costs for single-piece manufacturing and to decrease the number of details in assembly units due to the complex design of the basic parts. All the materials (pure metals, alloys and composite materials with metallic matrix) applying in machine building theoretically can be used as the "building material" for additive manufacturing details for machines, mechanisms, devices and instruments. According to the world experience, even the most advanced SLM of powders can't always provide the proper quality of material, which is required for some products working as machine details and structural elements for important exploitation. Protective gas environment significantly expands technological capabilities and simplifies the process of 3D-printing, but can't eliminate porosity even when using such high-concentrated energy sources as laser [2], and electron beam [3]. Recent ten years in scientific publications much attention is paid to issues of quality of titanium alloy Ti-6Al-4V as the widely spread material in different areas of manufacturing. Instead of wide use of additive manufactured Ti-

IOP Conf. Series: Materials Science and Engineering **681** (2019) 012006 doi:10.1088/1757-899X/681/1/012006

6Al-4V products one of the urgent tasks is eliminating of structure defects as pores in the formed material. Another problem mostly concerns the stability of sintering process and anisotropy control of the structure, which appears in welded layers in local volumes of melted material during electron-beam [3] and laser heating [4].

2. Materials and methods

The structure and mechanical properties of materials obtained by SLM were studied on samples made of Ti-6Al-4V ELI powder alloy. The chemical composition of powder was determined using an ARL X-ray fluorescence spectrometer Optim X. The particle size distribution of the original powder was studied using a Nikon MM-400 instrumental microscope; spherical particles were ranging in size from 20 to 46 μm.

Table 1 - The chemical composition of the initial powder titanium alloy

The content of chemical elements in powder titanium alloy [wt%]									
V	Al	Fe	W	Mg	Ni	Si	Mn	S	Ti
3.76 ±0.18	5.54 ± 0.11	0.394 ± 0.02	0.098 ±0.011	0.052 ±0.004	0.051 ±0.003	0.012 ±0.002	0.011 ±0.004	0.006 ±0.001	Balance

The SLM was carried out in argon environment on a EOS M290 3D-printer (Electro Optical System GmbH, Germany), equipped with an ytterbium fiber-optic laser with a wavelength of 1.065 μm with an output power 400 W. The samples were built up on auxiliary supports on Ti-6Al-4V thick plate (Figure 1), which was attached to the desktop of the printer. Using the massive plate as a substrate prevents the deformation of the samples during their manufacture. Then the plate with the samples was annealed in vacuum at 750 °C for 3 hours with the heating rate not higher than 1 ° per second. Heat treatment of the samples with the plate allows relieving residual stresses and avoiding uncontrolled deformation of the samples after cutting them off the plate.

Samples for tensile tests with the size 150 × 30 × 3 mm and samples for impact bending tests with the size 60 × 12 × 2 mm were printed with an allowance to cut them off from the slab using a wire-spark machine (Fig. 2 a, b). Then the all sides of the samples for testing were grinded. The thickness of the sintered powder layer, the scanning direction of each layer and the scanning speed were controlled using software, which takes into account the product geometry and minimizes the value of residual stresses in the formed product.

To evaluate the level of stresses that appears after sintering, a flat sample 60 × 12 × 2.5 mm was sintered which was not heat treated. After breaking off the supports from the plate, the long side surfaces of the sample lost their flatness (Fig. 2c). To find out the cause of the curvature of the surfaces, the analysis of the stresses by the criterion of the maximum von Mises was carried out using the software package SYSWELD.

Figure 1. SLS samples of Ti-6Al-4V ELI alloy made for mechanical testing and grid-shaped products on supports (upper left part of the plate) built up on titanium alloy Ti-6Al-4V table plate.

IOP Conf. Series: Materials Science and Engineering **681** (2019) 012006 doi:10.1088/1757-899X/681/1/012006

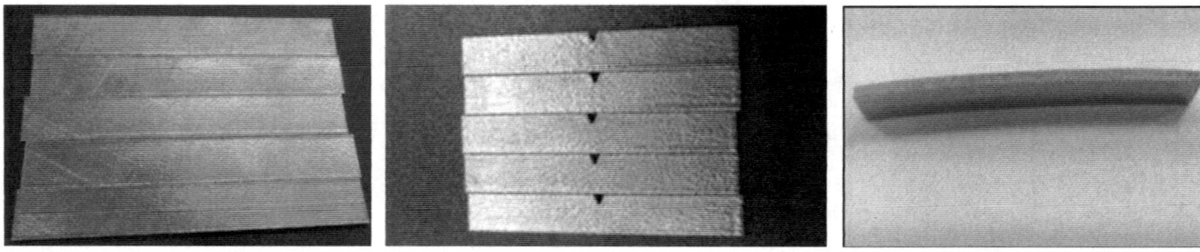

a b c

Figure 2. Samples for mechanical testing, made of titanium alloy: a - flat samples for tensile testing; b - flat samples for impact strength testing; c - not heat treated sample after sintering.

Tensile tests were conducted on an Instron 3369 testing machine, impact strength testing was carried out on a pendulum copra Metro Com 06103300. The structure was investigated using a Carl Zeiss AxioObserver Alm optical microscope.

3. Results and discussion

The stresses leading to the distortion of the samples and products appear while crystallization of the melted powder layers. Heat treatment of the samples with the plate was preformed to avoid the possible distortion when stress release after cutting the samples from the plate. Not heat-treated sample has a dome-shaped 1.5 mm deflection along the length of the sample and 0.2 mm along the width (Figure 2c). The curvature occurs because of the sequential multilayered crystallization of local volumes when the beam is scanning on the powder layer along the path on the already crystallized (previous) layer.

While modelling, the following assumptions were made to simulate the level of the stresses: the stresses in the sample were formed only during the surfacing of the last layer 0.1 mm thick, and the physic and mechanical properties of the underlying layers corresponded to the properties of Ti-6Al-4V ELI alloy after high-temperature annealing. The laser beam moved from the upper corner of the sample along its short side with the scanning speed 7 m $\times$ s^{-1} and the scanning step 140 μm. The largest calculated tensile stress in the sample with sizes 60 $\times$ 12 $\times$ 2.5 mm^3 was 314 MPa, which is 40% of the yield strength of the Ti-6Al-4V ELI alloy. The propeller-like deformation behavior of the sample matches with the calculated deformation, which indicates that the sequence formation of local volumes of material influences on the generated stresses, on the deviations from the specified surface shape, and on the deviations of their relative position during Bed Deposition. The samples annealed with the massive plate had flat parallel surfaces with the deviation from the flatness within the roughness, which is determined by the fused particles on the surface (Figure 3a) and the surface tension of the melt on the upper deposited layer (Figure 3b).

The surface roughness of the samples was estimated on ZYGO 7300 microinterferometer. The measurements showed that the roughness had a clear periodic profile. The highest Ra parameter (average deviation of the profile) was 5.167 μm and the Rms parameter (standard deviation of the profile) was 6.433 μm. Before mechanical testing the samples with the size 150 $\times$ 30 $\times$ 3 mm^3 were grinded, and the fillets were made. Thus, size of the testing parts was 10 mm. The tensile strength of the formed material was 1182 ±15 MPa, which is slightly higher than the strength values for deformed and annealed alloy (1103 MPa) and significantly higher than the values of the cast alloy (895 MPa). Comparison with the results of tensile strength tests given in [5] showed that the tensile strength of tested samples is at the same level [5] (1115 MPa). Comparison of the tensile test results carried out by the authors of the present study with the results of [5] is quite correct, since they were obtained at similar modes of scanning, heat treatment, and subsequent mechanical treatment to remove welded powder particles that works like stress concentrators. However, in the compared cases, the parameters of the process significantly differed: laser power, the rate of scanning beam relatively to the sintered layer, the location of the sample being formed relative to the object table (in [5] the samples were formed vertically). Two-stage heat treatment allows to increase the values of tensile strength, for instance, the authors of [6, 7] achieved the values of 1269 ± 10 and 1271 ± 8 MPa.

The tensile strength of produced samples has deviations less than ±1.2%, nevertheless, the values of yield strength, relative elongation and relative narrowing of the simultaneously sintered samples (sintered under the same conditions) have a significant variation (σ_T = 844 ± 261 MPa; ψ = 13.1 ... 20.3%; δ = 8.9 ... 13.3%).

Metallographic studies of the samples showed the presence of pores in the samples and grid-shaped products made by SLM, both before and after heat treatment (Figure 4). The presence of pores after the sintering process has been noted by the most studies dedicated to SLM of Ti-6Al-4V ELI powder alloy [8–12].

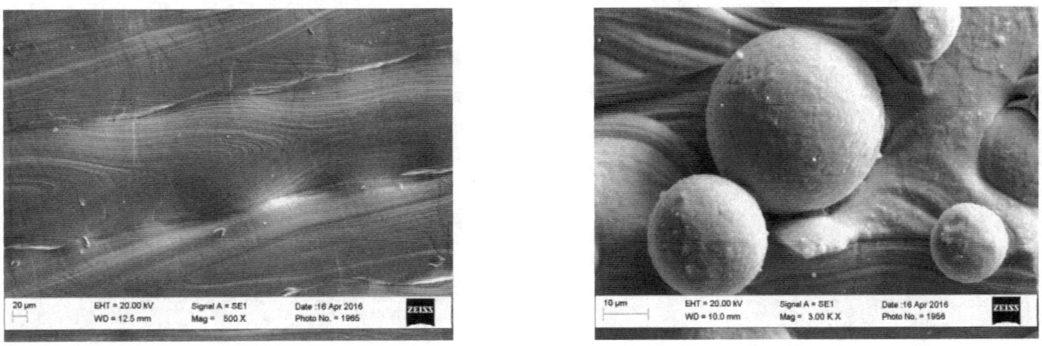

a b

Figure 3. The side surface (a) and the surface of the upper deposited layer (b) of samples for mechanical testing after laser layer-by-layer deposition and heat treatment.

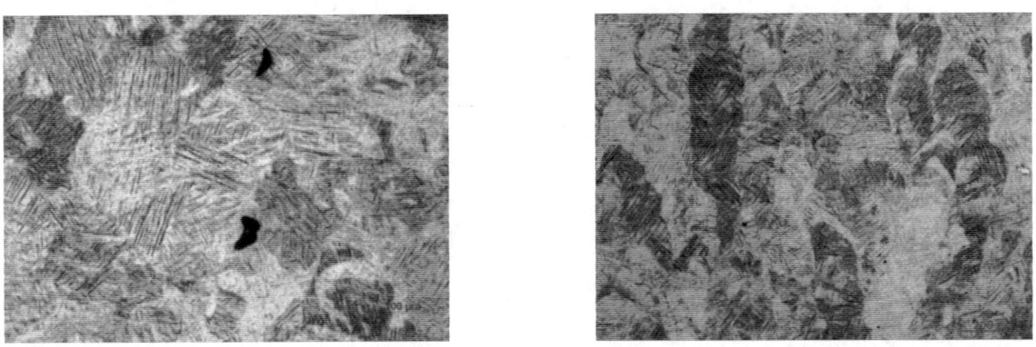

a b

Figure 4. Structure of the titanium alloy Ti-6Al-4V after laser layer-by-layer deposition and heat treatment: a –cross section of the sample; b - longitudinal section of the sample.

Pores have a negative effect on impact resistance. The impact strength (KCV) was 32 ... 36 J × cm^{-2}, while the impact toughness values for cast alloys of similar composition is 50 ... 100 J × cm^{-2}. The formation of pores in the formed material is associated with the capture of inert gas atoms and uneven heating of the particles of the powder material by a laser beam. The pores work as stress concentrators under impact action and their shape is well observed in the fracture surface (Figure 5).

The structure of the titanium alloy Ti-6Al-4V directly after SLM, as a rule, is a martensitic α'-phase [13]. The cooling rate can reach 10^4 °C × s^{-1}, which promotes the formation of a fine needle-shaped morphology. Annealing at 730-780 °C after SLM allows eliminating stresses in the material.

According to the literature, annealing of Ti-6Al-4V ELI at above mentioned temperatures for 2...3 hours leads to a partial decomposition of α'-martensite into more stable α and β phases. In the studied material, annealed at 750 °C for 3 hours, only a trace of the most intense β-phase reflex was detected by X-ray phase analysis, which indicates that only a small amount of the phase can possibly exist.

Metallographic study of the annealed structure clearly reveals the boundaries of the initial β-grains. Due to the fact that the initial martensitic structure of the Ti-6Al-4V ELI alloy is characterized by a high temperature stability [13, 14], the boundaries of the initial β-grains do not undergo changes after annealing. It should be noted that in different directions the morphology and dimensions of the primary β-phase grains are not the same. In the vertical cross-section, the grains have an elongated shape (Figure 4 a), while at the same time in the horizontal cross-section the material has an equiaxial structure (Figure 4 b). This structure feature can be explained by their epitaxial growth during the layer-by-layer melting of the powder material [13].

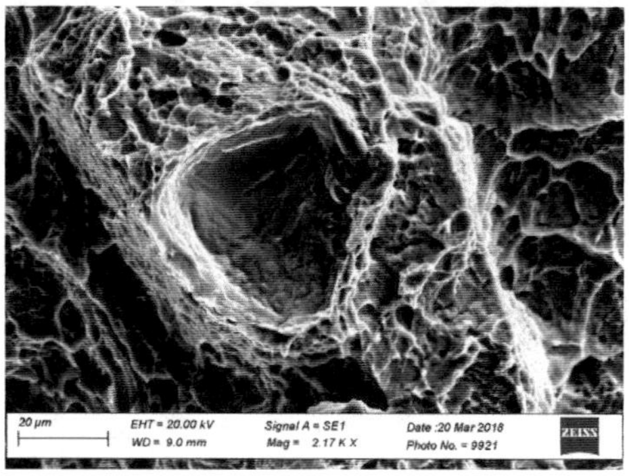

a

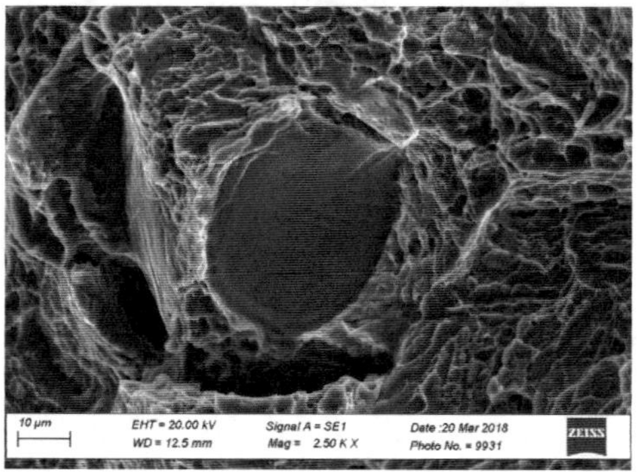

b

Figure 5. Fracture surface of the alloy Ti-6Al-4V, obtained by laser layer-by-layer deposition and heat treatment (of pores in the center of the images).

Due to the presence of gaps between particles the laser beam can heat and melt the particles not only by directly hitting on the particle but also because of the reflection of the beam from other particles [9]. The distribution of the power density and the incident angles of the laser beam on the particles lead to uneven heating within the beam spot from temperatures that are quite below the melting point to temperatures of the evaporation. Degassing of the deposited layers can be provided either by additional thermomechanical treatment or by implementing the SLM process in vacuum. Vacuum allows not only to conduct degassing the melt, which has a positive effect on the mechanical properties of the material, but also to apply electron beam as the source of heating. Electron-beam heating makes it possible to obtain products using the "Direct Deposition" method, in which the cladding of the layers is provided by melting filaments of metallic materials. Electron-beam layered surfacing in vacuum has its own benefits and drawbacks. In comparison to "Bed deposition" the main drawback of vacuum electron-beam surfacing is difficulty with the formation of smooth surfaces, thus, it's hard to control the shape of the products been formed. The main benefit of electron-beam methods of metal additive manufacturing are quality of the formed material, high performance of the process, and the low cost of raw materials, since the filaments are much cheaper than powder materials. Another advantage of electron-beam methods is that the maximum temperature of a metal material bombarded by an electron beam is achieved in the thin surface layer at a distance from nanometers to hundreds of micrometers from the surface. This feature of electron-beam heating allows avoiding uncontrollable transition of the molten metal to gaseous state. Besides, from 75 to 95% of electron beam energy spends into heating, which makes the process more energy efficient in comparison with laser heating. Electron beam heating allows to use metal filaments from a tenth of a millimeter to 5 mm thick when focusing the beam diameter from 200 nm to tens of millimeters, which potentially allows for both surfacing and heat treatment of any metallic material: from low-melting metal alloys to tungsten.

4. Conclusion
Selective laser melting of powder materials doesn't allow avoiding the formation of structural defects like pores. Heat treatment of products formed by additive welding makes possible reducing the residual stress that appear while layer-by-layer depositing, but is not sufficient to eliminate pores.

Different sequence of the laser beam moving direction on the deposited powder layer affects the geometry of the resulting surfaces. Heat treatment with alternating layer-by-layer depositing makes possible to reduce the shape deviations of the formed products.

The elimination of such defects in structure as porosity and the improvement of physical and mechanical characteristics of additive manufactured materials can be achieved using vacuum electron-beam heating and filaments as the row material.

Acknowledgments
The authors are grateful to the Ministry of Education and Science of the Russian Federation for the financial support of research (Agreement No. 14.610.21.0013, project identifier RFMEFI61017X0013).

References
[1] Herzog D Seyda V Wycisk E Emmelmann C 2016, Acta Mater. **117** 371
[2] Vilaro T Colin C Bartout J D 2011 *Metall. Mater. Trans.* A **42** (10) 3190
[3] Galarraga H Lados D A Dehoff R R Kirka M M Nandwana2016 *Addit. Manuf.* **10** 47
[4] Li P Warner D H Fatemi A Phan N 2016 *Int. J. Fatigue* **85** 130–143
[5] Kasperovich G Hausmann J 2015 *J. of Mater. Proc. Technol.* **220** 202
[6] Van Hooreweder B Boonen R Moens D Kruth J P Sas P 2012 *Proc.* 53rd *Conf* AIAA/ASME/ASCE/AHS/ASC Structures, Structural Dynamics and Materials 1173
[7] Rafi H K Starr T L Stucker B E 2013 *Int. J. Adv. Manuf. Technol.* **69(5-8)** 1299
[8] Herzog D Seyda V Wycisk E Emmelmann C 2016 *Acta Mater.* **117** 371
[9] Wysocki B Maj P Sitek R Buhagiar J Kurzydłowski K Święszkowski W 2017 *Appl. Sci.* **7(7) 657** 1
[10] Vilaro T Colin C Bartout J D 2011 *Metall. Mater. Trans. A* **42 (10)** 3190
[11] Vrancken B Thijs L Kruth J P Humbeeck J V 2012 *J. Alloys Compd* **541** 177
[12] Qiu C Adkins N J E Attallah 2013 M M *Mater. Sci. Eng. A* **578** 230
[13] Vrancken B Thijs L Kruth J P Van Humbeeck J 2012 *J. Alloys Compd.* **541** 177
[14] Facchini L Magalini E Robotti P Molinari A Hoges S Wissenbach K 2010 Rapid Prototyping Journal **16** 450

Precondition for the use of hybrid additive-thermomechanical technology

A V Chumaevskii[1], A V Gusarova[1], D A Gurianov[1], K N Kalashnikov[1], T A Kalashnikova[1]

[1] Institute of Strength Physics and Materials Science of SB RAS, Tomsk, Russia

E-mail: tch7av@gmail.ru, gusarova@ispms.ru, desa-93@mail.ru, sso.spektr.asu@gmail.com, gelombang@ispms.tsc.ru

Abstract. Structural-phase state of gradient structures obtained in the course of additive electron-beam production, friction-stir treatment and hybrid additive-thermomechanical technology was investigated by optical and scanning electron microscopy. Characteristic features of defect formation in materials obtained by additive method are revealed. The peculiarities of defects "healing" in additive materials after friction-stir treatment are determined. In tensile tests the mechanical properties of the samples before and after further post-treatment in aluminum alloys are measured.

1. Introduction

At present, the production processes of parts made of heterogeneous metals and alloys are mainly based on the additive method of forming products from different metals [1-4]. In turn, such techniques are based on the melting of metal powders or wire filament and the deposition of molten material on the substrate. Such a process can be carried out in an air environment as well as in a shielding gas or vacuum. Both the electron beam and the laser beam are used as heat sources for melting the material.

But, despite the sufficient versatility of additive production technologies of parts from metal materials, the resulting products have a number of significant drawbacks. The most significant drawback of products obtained from homogeneous metals and alloys is the microstructure of the samples after printing, represented mainly by the cast dendritic structure. This provides quite low mechanical properties of the obtained materials and in some cases requires additional heat treatment after the process of 3d-printing of samples.

At the building of samples from heterogeneous materials, at the expense of fusion forming intermetallic compounds, for example in pair Cu-Al, Al-Ti, strength properties of samples essentially decrease. In this case, the destruction occurs along the boundary of intermetallic-solid solution. Also, when printing heterogeneous materials, for example, the Cu-Al system, due to the higher melting point of copper, its deposition on the surface of the aluminum alloy is limited. Its' occur due to the excessive melting of aluminum and a high degree of diffusion of aluminum alloy in copper, forming a large proportion of brittle intermetallic compounds in the volume. From the experience of friction-stir welding and friction-stir treatment it is known that intermetallic compounds in the above mentioned metal vapors are formed and under these effects, but, unlike the processes based on the melting of metals, intermetallic layers have a smaller thickness and do not have such a significant impact on the mechanical properties of the resulting products [5-8].

The study of physical processes occurring in the manufacture of samples by technology of additive-thermomechanical influence will allow, in addition to the elimination of a number of negative structural effects in samples with a homogeneous and heterogeneous structure, to form composite metal materials with hierarchical structures due to the alternation of metal layers with processed ultrafine disperse structure and cast undeformed structure. Such composites can be made of homogeneous and heterogeneous metal materials, with a layered or complex three-dimensional structure, with areas with different not only structural-phase composition, but also grain-subgrain structure.

At present, there is no data in the literature on the structural response of materials in the conditions of additive production with thermomechanical treatment of heterogeneous materials, although there is a number of studies on structural changes in the materials obtained by the additive method of friction stir welding and processing [9,10]. For this reason, it is impossible to establish the relationship between the parameters of the process of obtaining the structure and properties of the resulting products by this method. Thus, practical expediency of carrying out works in this direction and fundamental scientific task determine one purpose of the present work. Obtaining these data on the structure and phase composition of the samples and the relationship between structure and properties of heterogeneous materials formed by the hybrid method will allow to obtain products with known properties in advance.

Proceeding from the aforesaid, the present work is directed on studying of structural-phase features of components interaction at additive electron-beam manufacturing and revealing of influence on structure and mechanical properties of the subsequent frictional mixing processing.

2. Materials and methods

The paper studies the processes of materials formation with gradient structure by the method of additive electron-beam production, friction mixing treatment and hybrid method. The samples were obtained on experimental equipment at the Institute of Strength Physics and Materials Science.

Metal wire from Ti-6Al-4V, AA5056, C11000 alloys with a diameter of 1-1.2 mm was used for obtaining samples by the method of electron-beam additive production. Samples were produced by layer-by-layer deposition on the surface of the substrate made of the corresponding alloy at different parameters of beam current, beam voltage, feed rate of the sample or wire. Production of polymetallic samples was carried out with the help of two wire feeders in the chamber, with successive deposition of materials of one filament and another.

Friction-stir treatment of rolled sheet was performed on C11000 copper and polymetallic material from AA5056 and copper alloy sheets, with copper interfering into aluminum alloy to obtain gradient material in the mixing zone. The thickness of the processed sheets was from 2 to 5 mm. For reception of polymetallic system Cu-Al sheets in the thickness of 2 and 3 mm are used.

Hybrid obtaining of materials was made by sequential additive obtaining of materials of aluminum alloy AA5056 by the method of electron-beam production with the following frictional mixing treatment at a depth of 1.8 to 5 mm. The processing was carried out with a standard tool for friction-stir welding.

After printing, samples for optical and scanning metallography were cut out using the electrospark method. The research was carried out using optical and scanning electron microscopy on the Olympus LEXT 4100 laser scanning microscope and Zeiss LEO EVO 50 scanning electron microscope.

3. Results and discussion

The materials produced by the electron-beam additive production method have a structure similar to that observed in a standard metallurgical foundry. Metal in the printing zone is represented by a typical dendritic structure with elongated in the direction of heat dissipation. In the process of products formation there is a formation of rough wavy morphology on the surface, correlating with the layering of the material structure. In the internal volume of the material pores and heterogeneity of the structure

IOP Conf. Series: Materials Science and Engineering **681** (2019) 012007 doi:10.1088/1757-899X/681/1/012007

of different sizes are formed. The elimination of such defects is possible only in case of different post-treatments after manufacturing of parts.

At additive obtaining of details from heterogeneous metals and alloys with the limited solubility and propensity to formation of intermetallic phases in structure the complex multiphase structure, and special distribution of chemical elements and specific kinds of defects is formed (Figs. 1 and 2). At non-uniform cooling and heterogeneous distribution of chemical elements in the structure of the sample cracks of different structural and large-scale levels are formed (Fig. 1). In the case of mismatch of the parameters of the electron beam leading to non-fusing of the wire, it is possible to introduce wire into the material of the melted substrate. In this case, a defect is formed, for example (when printing copper-titanium samples), in the form of introduction of pure copper in the resulting material, with the formation of large areas of formation of intermetallic materials of different compositions around it (Figs. 1 and 2). Also, the formation of heterogeneous distribution of intermetallic phases in terms of size and volume fraction in the sample material (Fig. 2) can be referred to as defects.

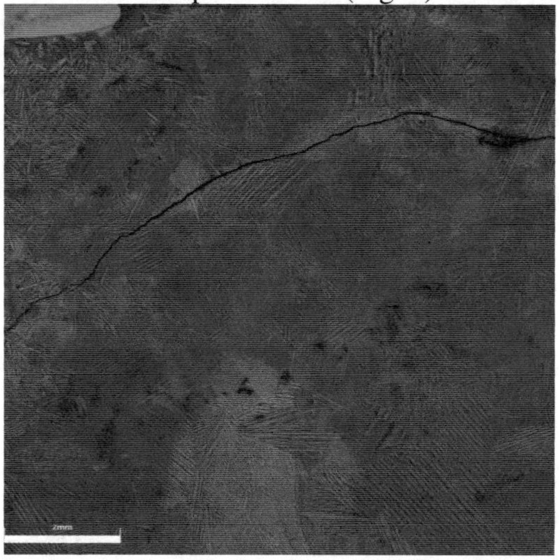

Figure 1. Structure of the polymetallic sample of the Cu-Ti system with defects obtained by the method of electron-beam additive production.

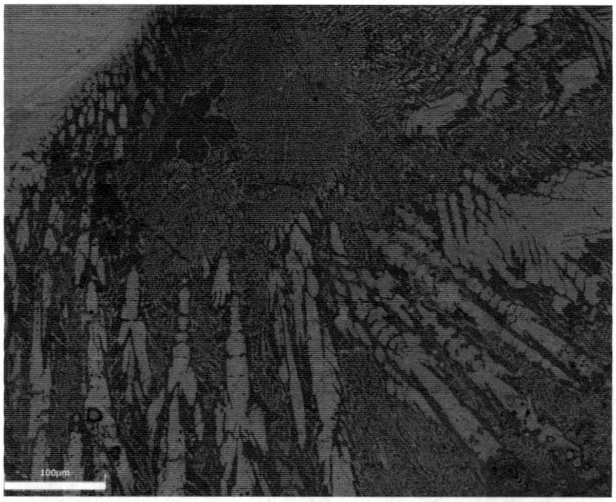

Figure 2. Inhomogeneity of the polymetallic sample structure of the Cu-Ti system, obtained by the method of electron-beam additive production.

IOP Conf. Series: Materials Science and Engineering **681** (2019) 012007 doi:10.1088/1757-899X/681/1/012007

In polymetallic materials, which are not prone to the formation of intermetallic phases and do not have mutual solubility in the solid state, for example, as copper and austenitic steel, a mechanical mixture of system components with different concentrations is formed in the boundary zone. Pore formation and inhomogeneities of both grain structure and phase composition are possible in the structure. The use of two gradient wires from two different feeders in the same melt bath at the same time allows the structure to be aligned and the resulting material structures to be controlled during printing.

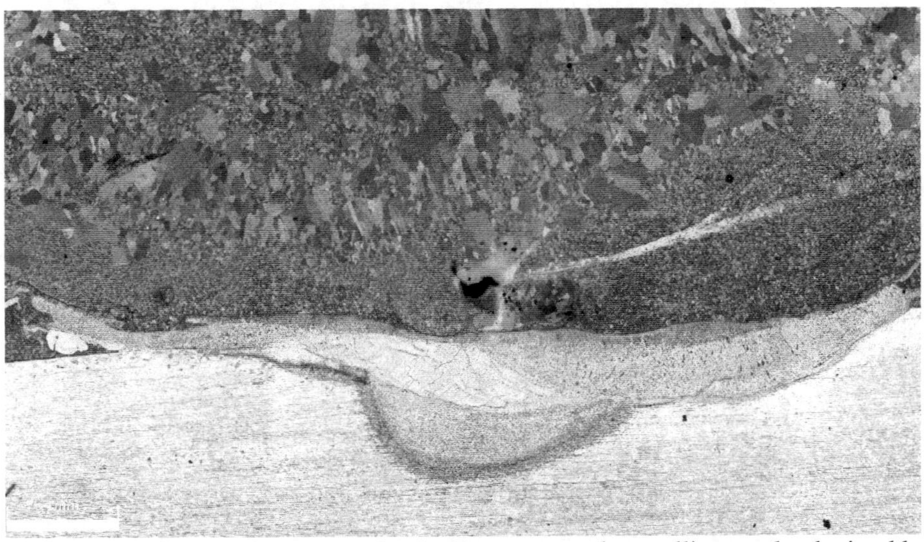

Figure 3. Structure of the boundary zone of a Cu-Fe system polymetallic sample obtained by electron-beam additive production.

From the conducted works on manufacturing of samples by the method of additive electron-beam production it is possible to draw a conclusion about heterogeneity of structure, formation of defects of various type, the form, the size and irregularity of a surface of received samples. Part of the above mentioned defects can be removed by normalization of the parameters of the sample production process. The other part cannot be eliminated at normalization of parameters, for example, low durability of the received materials in comparison with sheet rolling, heterogeneity of structure, large-crystal structure, etc. For this reason, the formation of a homogeneous structure of the surface layer or internal volumes in the material obtained by the additive electron-beam technology, it is necessary to use different post-treatment.

The formation of ultrafine disperse materials by friction-stir has been actively developed over the past decades. This method is based on the friction-stir process, which is also used for friction stir welding of different metals and alloys. At frictional processing the tool under the influence of axial force at the certain frequency of rotation at the expense of frictional heating is introduced into the sample, after that there is a progressive movement of the tool in the warmed up material and formation of the recrystallized finely dispersed structure in a mixing zone (Fig. 4). Despite the heterogeneity of the processed material structure in the thermomechanical affected zone, this technology allows to produce materials with high mechanical properties and homogeneous structure. The heterogeneity of the structure of the thermomechanical affected zone can be removed by multi-pass treatment.

At formation of materials by a method of frictional mixing processing from heterogeneous metals and the alloys on the basis of the metals testing at mixing the intensive phase interactions including mutual dissolution of components of system, formation of secondary phases, solid solutions or intermetallic compounds, the structure in a zone of mixing is formed more difficult type. In the structure there is a pronounced layering with alternating layers of initial components and intermetallic compounds formed on their boundaries (Fig. 5).

Figure 4. Structure of copper sample C11000 after friction-stir treatment.

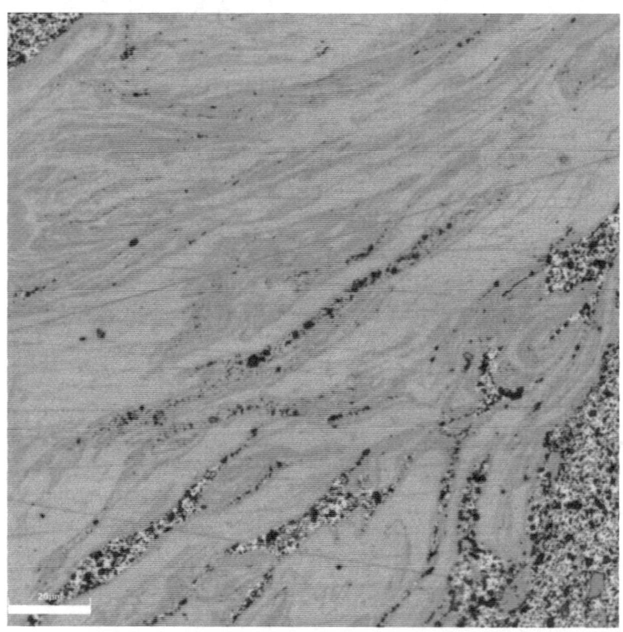

Figure 5. Structure of a polymetallic sample of the Cu-Al system obtained by friction-stir treatment.

At intensive diffusion of system components, it is possible to form intermetallic phases with complex structure and high heterogeneity of structure in different directions (Fig. 6). In a number of cases, during the formation of materials of such systems as Cu-Al, the material overheated with the destruction of the sample structure and the melting of aluminum alloy material. At the same time, by regulating the parameters of the machining process or by carrying out multiple passages with the tool along the machining line, it is possible to eliminate such defects with the formation of a more homogeneous structure in the machining zone. Such materials with homogeneous or heterogeneous structure have rather high mechanical properties, significantly exceeding the properties of the samples obtained by the method of additive electron-beam production and being at the level of properties of rolled sheets, which makes it expedient to combine these methods. Also relevant is the use of friction treatment to improve the properties of additive materials, which makes the absence of pores and other defects, typical for additive production.

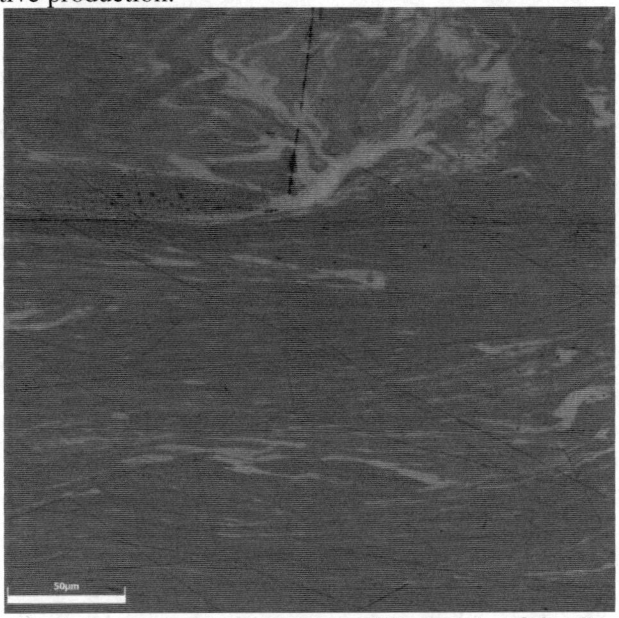

Figure 6. Inhomogeneity of the structure of a polymetallic sample of the Cu-Al system obtained by friction-stir treatment.

Formation of materials by combined additive-thermomechanical method was tested on AA5056 alloy. The obtained additive structures with defects and low mechanical strength from 190 to 240 MPa were treated with friction-stir treatment. After the treatment mechanical properties of the samples increased up to 300-330 MPa. Mechanical properties were determined on the samples cut along the processing line in such a way that only the treated area was located in the stretching zone. In the structure of samples (Fig. 7) in the region without defects the structure of the material is presented by the standard distribution of the mixing zone, the thermomechanical affected zone and the thermal affected zone. A recrystallized ultrafine disperse structure is formed in the area of the mixing zone, which is favorable for mechanical properties.

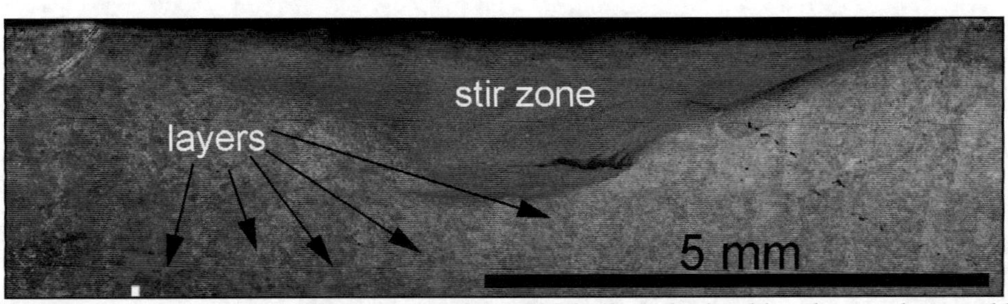

Figure 7. Structure of defect-free sample of AA5056 alloy obtained by electron beam additive technology and subsequent friction stir processing.

In a sample with a large number of defects in the structure in the form of pores, the structure in the mixing zone has traces of defects through which the processing is carried out (Fig. 8). Pores in the area of treatment at friction mixing influence almost completely collapse and are removed at the expense of material mixing. The pores in the thermomechanical affected zone are squeezed due to material heating and tool pressure, but the structure remains defective in the form of irregularities. Mechanical properties of such samples decrease by a small value and are about 280-300 MPa. Thus, frictional stirring treatment of additively obtained material allows not only to obtain ultrafine disperse structure of surface or internal volume of the material, but also to eliminate defects that occur during printing.

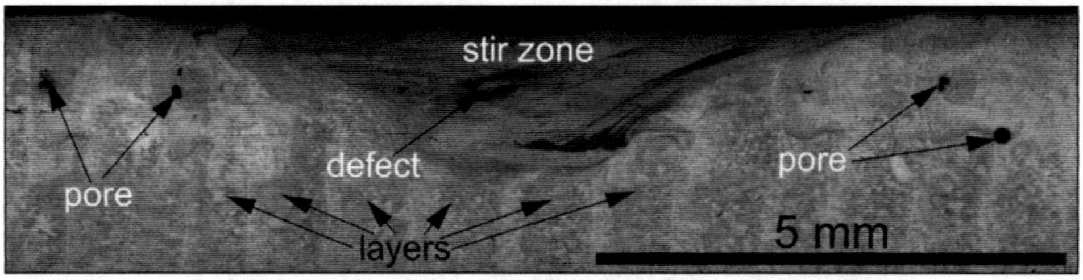

Figure 8. Structure of defective sample of AA5056 alloy obtained by electron beam additive technology and subsequent friction stir processing.

4. Conclusion

The research shows that the process of additive reception of materials by a method of electron-beam manufacture with wire filament possesses a number of the lacks leading to formation in structure of defects connected with inhomogeneity of process or discrepancy of parameters to demanded values. The obtained materials have rather low mechanical properties, which is unacceptable in many cases. Friction mixing treatment with individual structure defects allows to obtain materials with higher mechanical properties and more homogeneous and favorable metal structure. Defects resulting from the manufacture of materials by additive manufacturing are successfully eliminated by friction treatment, and their impact on the mechanical properties is leveled. However, it is impossible to obtain complex shaped parts by frictional mixing treatment, and without additive technology it is extremely ineffective for these purposes. Thus, it is shown that the hybrid additive-thermomechanical process of obtaining parts is actual and feasible and combines the positive aspects of additive obtaining and friction treatment.

Acknowledgement

The work on obtaining and studying the structure of samples by the method of additive production was performed within the frame of the Fundamental Research Program of the State Academies of Sciences for 2013-2020, line of research III.23. The friction mixing treatment of sheet metal and additive-derived materials was carried out by a grant from the Russian Scientific Foundation (Project No. 19-79-00136).

References

[1] Debroy T *et al* 2018 *Prog. Mater Sci.* **92** 112-224
[2] Gorsse S *et al* 2017 *AM&T* **18(1)** 1-27
[3] Liu W P and DuPont J N 2003 *Scripta Mater* **48(9)** 1337-42
[4] Fessler J *et al* 1997 Proceedings of the solid freeform fabrication Symp. (Austin (TX): University of Texas at Austin) pp. 521-8
[5] Tan C W *et al* 2013 *Mater. Des.* **51** 466-473
[6] Allah-Zadeh A, Saeid T and Sazgari B 2008 *J. Alloys Compd.* **460** 535-8
[7] Avettand M N, Racineux G, Debeugny L and Taillard R 2016 *Mater. Des.* **98** 305-18
[8] Bhattacharya T K, Das H, Jana S S and Pal T K 2016 *Int. J. Adv. Manuf. Technol* **88(1-4)** 847-61
[9] Maamoun A H, Veldhuis S C and Elbestawi M 2019 *J. Mater. Process. Technol.* **263** 308-20
[10] Rubinoa F *et al* 2019 *JMFP* **37** 392-401

IOP Conf. Series: Materials Science and Engineering **681** (2019) 012008 doi:10.1088/1757-899X/681/1/012008

Investigation of the optimal modes of electron-beam wire deposition

V K Dragunov[1], **M V Goryachkina**[1], **A V Gudenko**[1], **A P Sliva**[1] **and A V Shcherbakov**[1]

[1] National Research University "Moscow Power Engineering Institute", 14, Krasnokazarmennaya str., Moscow, 111250, Russia

E-mail: alexsguddy@gmail.com

Abstract. The paper is devoted to the definition of the minimum electron beam power that is required to obtain stable layer formations during the additive electron beam wire deposition. Correlation between the minimum energy that is needed for stable deposition and the deposition speed has been established. Additionally, the influence of the electron beam oscillation type on geometrical parameters of single layers have been investigated.

1. Introduction

Current research in the field of electron beam metal wire technologies pays special attention to the structure and properties of the products made of titanium [1, 2], nickel [3, 4] aluminum alloys [5], and stainless steel [6, 7]. Also, heat and mass transfer models of wire deposition process were investigated [8]. However, in these studies the technological aspects of the process are not considered. They lack attention to the selection of the main parameters of additive electron beam wire deposition such as the electron beam power, feed rate, deposition rate, oscillation type and its parameters and their influence on the parameters of deposited layers and products.

The development of the technological basis for manufacturing of components with complicated shape includes determining relations between deposition process parameters, layers geometry, their structure and properties for obtaining the required wall thickness and metal structure. It is necessary to define the criteria that limit the value range of deposition process parameters and allow to obtain the product with the needed shape and properties. To limit the range of parameters it is advisable to determine the minimal values of the electron beam power under which the stability of deposition is saved.

The purpose of this research was to determine the electron beam power required to ensure the stable formation of beads at different feed rates and deposition speeds, and to investigate the influence of the beam oscillation type on the geometrical parameters of beads.

2. Research methods

The process of electron beam wire deposition was carried out using an electron-beam machine with an accelerating voltage of 60 kV that was equipped with an automatic wire feeder.

The determination of the minimal electron beam power required to ensure the stable formation of a single layer at different wire feed rates and surfacing rates was divided into two stages.

The first stage was to establish the minimum electron beam power required for melting the feeding wire. The investigation was conducted using wires of various materials: stainless austenitic steel AISI

IOP Conf. Series: Materials Science and Engineering **681** (2019) 012008 doi:10.1088/1757-899X/681/1/012008

316L, Fe-Ni alloy 36NKhTU (36% Ni, 12% Cr, 3% Ti, 1% Al, Fe balanced) and titanium alloy 2V (2% Al, 1.5 % V, Ti balanced). The feed rate varied in the range of 1.0–3.0 m/min with a step of 0.5 m/min. The beam oscillation represented a circle with the diameter equal to half of the wire diameter. The frequency of the oscillation was 500 Hz.

The second stage involved the determination of the minimum beam power that ensured the stable formation of the layer. The investigation consisted of deposition of the 316L steel wire with the diameter of 1.2 mm on a substrate of the same material at different feed and deposition rates (Table 1). The wire was fed into the melting front of a liquid pool (Figure 1). The electron beam oscillation of type "sawtooth to crystallization front" was used with the length of 1 mm during these experiments.

The selection of the oscillation type is based on the hypothesis that this type of oscillation can lead to height increase of deposited layer. The electron beam movement cycle during the oscillation consists of two parts: the translation along the longitudinal axis and the rapid displacement to the beginning of oscillation trajectory (Figure 2). It is known that the use of the sawtooth oscillation in the process of electron beam welding results in intensification of liquid metal transfer to a rear part of the liquid pool and in metal displacement from the root to the top of a joint [9]. This effect can be used in additive electron beam wire deposition.

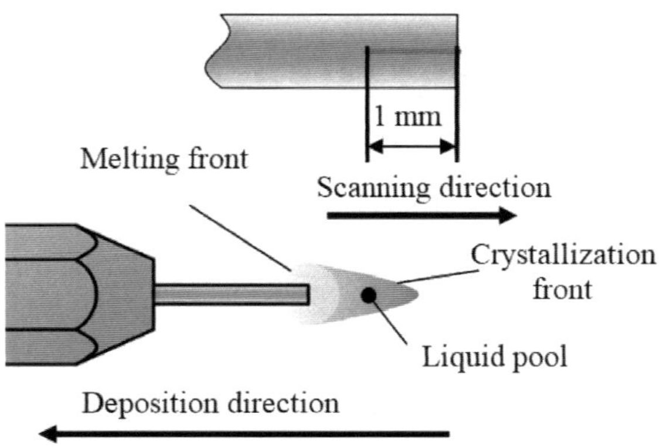

Figure 1. Feeding the wire in the deposition process.

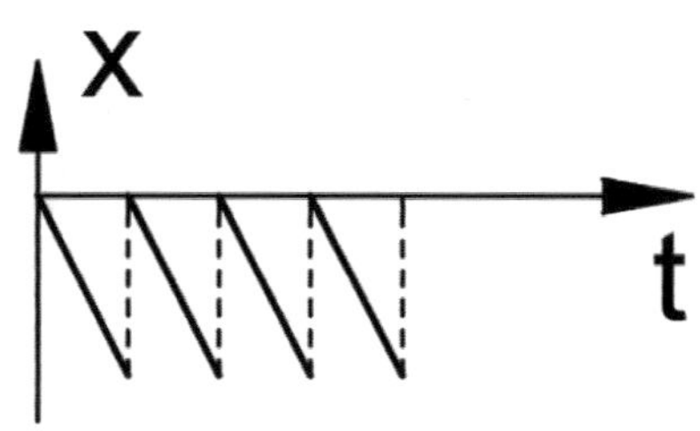

Figure 2. Movement of the electron beam with the use of saw tooth oscillation $x(t)$. x direction is identical to the deposition direction.

Table 1. Combination of wire feed rate and deposition rate

Wire feed rate V_f (m/min)	Deposition rate V_d (m/h)		
	15	30	45
1.0	×	×	×
2.0	×	×	×
3.0	×	×	×

The influence of the application of different types of electron beam oscillation on the single layer beads geometry was investigated via deposition using the wire with the diameter of 1.2 mm which was made of steel 316L. The deposition process was carried out with the feed rate of 2 m/min and the deposition rate of 30 m/h. The beam current was 16 mA. The oscillation types "sawtooth to crystallization front", "sinusoidal", "zigzag", and "circle" were used (Figure 3). Also, the comparison of deposition with and without the oscillation was done. X axis of the oscillation was oriented along the wire feed direction. The oscillation parameters are shown in Table 2.

In order to compare geometrical parameters of the beads cross-sections were made. The analysis of the geometry was conducted using an optical microscope Zeiss Observer Z1.

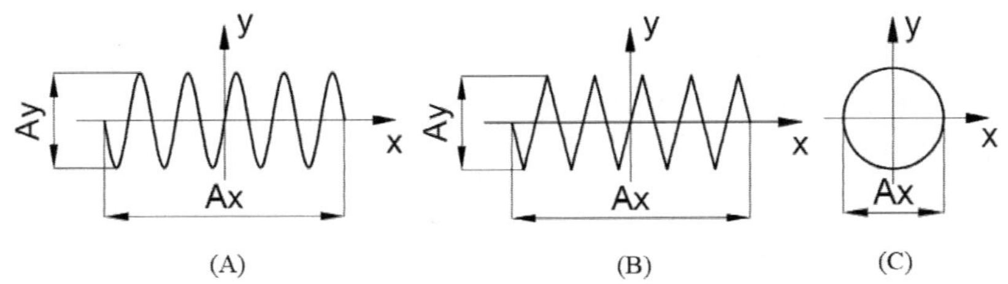

Figure 3. Trajectory of the electron beam movement using different types of oscillation: (A) sinusoidal, (B) zigzag, (C) circle.

Table 2. Geometric parameters of the oscillations

Oscillation type	Dimensions		Frequency f (Hz)
	Length A_x (mm)	Width A_y (mm)	
Sawtooth	1	0	100
Zigzag	1	0.6	100
Sinusoidal	1	0.6	100
Circle	0.6	0.6	500

3. Results

It was shown that the minimal electron beam power required to melt the wire that is fed with different rates is linearly related to the feed rate (Figure 4).

The minimum power required to melt a wire is the power below which the wire is melting incompletely. The minimum power could be presented as follows:

$$q = V_f \cdot F \cdot S_m \tag{1}$$

where V_f – wire feed rate, m/s, F – wire cross-section area, m^2, S_m – enthalpy of liquid metal at the melting point, J/m^3.

Thus, the tilt angle of the graph equals numerically to energy per volume required to melt the wire and depends on the thermophysical properties of any material. Energy per volume equals the enthalpy of liquid metal in this case. The effective electron beam power can be determined as:

$$q_{ef} = n_{ef} \cdot U \cdot I \tag{2}$$

where n_{ef} – beam heating efficiency equal to 0.95, U – accelerating voltage, V, I – beam current, A.

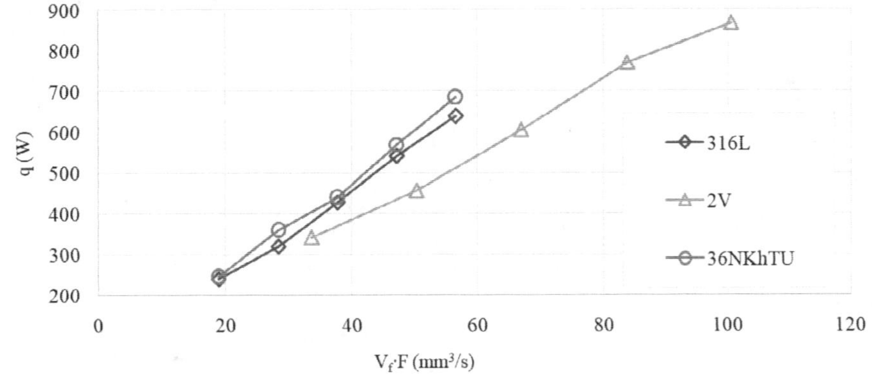

Figure 4. Relation between electron beam energy that is needed to melt the wires of different materials and wire volume that is fed per unit of time. Diameter of the wires of 316L and 36NKhTU – 1.2 mm, and 2V – 1.6 mm.

IOP Conf. Series: Materials Science and Engineering **681** (2019) 012008 doi:10.1088/1757-899X/681/1/012008

During the process of wire melting some part of electron beam power is spent on losses which connected with the overheating of liquid metal and radiation. The thermal efficiency of melting a wire $\eta_t$is defined as the ratio of spent electron beam energy (2) to the power needed to turn solid metal of wire to liquid metal (1).

Considering the dependence of thermophysical properties of stainless steel 316L on temperature [10], the thermal efficiency η_t of deposition process is close to the value of 1 (table 3). At the feed rate of 1 m/min the melting of the wire took place with the formation of big metal droplets. Decreasing η_t is connected with beam energy input both into wire melting and overheating the droplet.

Table 3. Thermal efficiency η_t of melting the wire of 316L, minimum beam power q vor various values of feed rate V_f

V_f (mm/min)	η_t	q (W)
1.0	0.867	239
1.5	0.975	319
2.0	0.971	427
2.5	0.958	542
3.0	0.975	638

During the process of deposition, electron beam energy is distributed between melting the wire and melting the substrate. To ensure the stable layer formation it is required to sustain the melting of the substrate for creating a joint liquid pool. Decreasing the electron beam power in the process of deposition has resulted at some point in the change of the stable deposition into the formation of droplets (figure 5). This effect relates to the lack of energy input to melt the substrate. The electron beam power below which occurs a shift from the layer formation to the droplets is the minimum for obtaining stable layers.

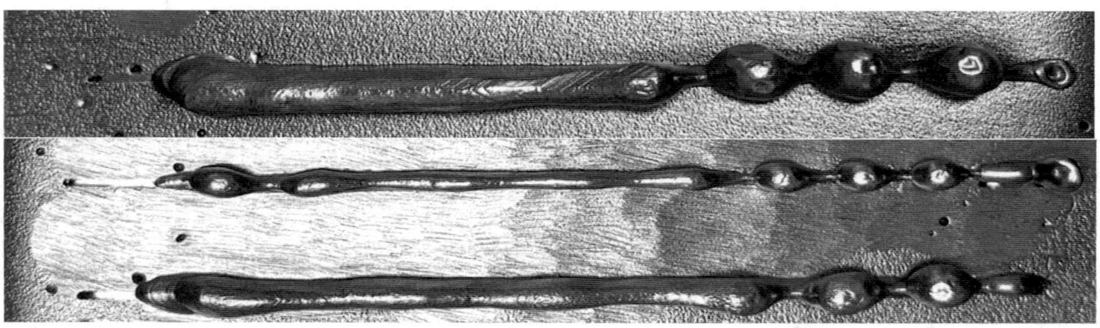

Figure 5. Changing the type of deposition as a result of decreasing the electron beam power. Deposition parameters: 1 - V_f = 3 m/min, V_d = 15 m/h; 2 - V_f = 2 m/min, V_d = 45 m/h; 3 - V_f = 3 m/min, V_d = 30 m/h.

As a result of layer deposition investigation with the decline of the beam power it was determined that the minimum energy per wire volume $q/V_f F$ which is required for sustaining stable process relates linearly to the deposition rate (Figure 6). This relation makes it possible to define the limit value of energy per volume below which layer formation will be unstable or interrupted. The horizontal dotted line corresponds to energy per volume that is required to melt the wire of 316L without deposition. The line divides the area below the graph into two parts. The upper part between the graph and the dotted line corresponds to the energy per volume required to melt the substrate, and the lower part to wire melting.

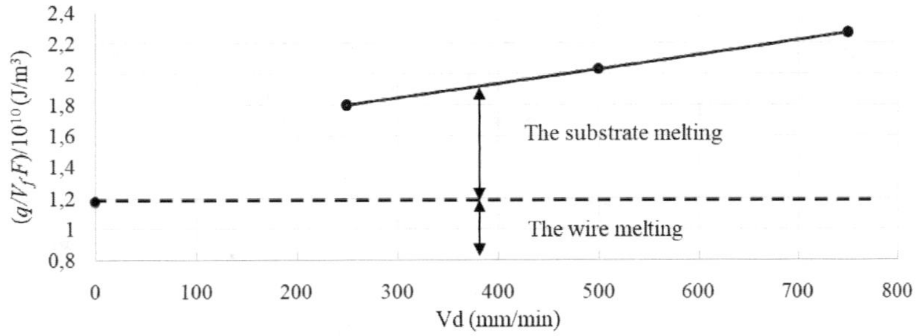

Figure 6. Relation between the minimum energy per volume ($q/V_f F$) required to stable deposition and deposition rate V_d.

It is possible to adjust geometrical parameters of the layers in different ways. The main method consists of regulating feed and deposition rates. In this case the layer size changes without changing the form. It is possible to change the layers form using electron beam oscillation. The comparison of geometrical parameters of deposited layers, obtained with the same electron beam power, feed and deposition rates, but different types of oscillation (Figure 7), made it possible to establish the impact of the oscillation type on the layer height.

The oscillations can be divided into 2 types. The first type includes the oscillations types that do not impact on the movement of liquid metal from the melting front to the crystallization front. This includes oscillation types like circle, concentric circles, and ellipses. The second type includes oscillations that intensify the transfer of liquid metal to the crystallization area. Oscillation types "sawtooth", "sinusoidal" and "zigzag" belong to this group. The movement of electron beam with the use of the second type oscillations leads to the displacement of liquid metal on to crystallized layer. This effect results in the increase of the layer height.

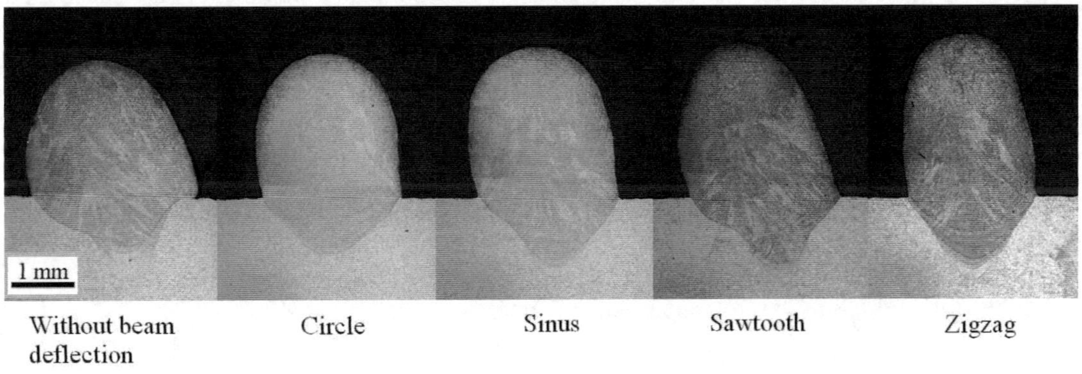

Figure 7. Cross-section of the single layers, deposited with different type of oscillation.

The height of the beads which were deposited with the use of "sawtooth", "sinusoidal", and "zigzag" oscillations is greater than the height of beads deposited with the use of circle or without oscillation at all due to the above-mentioned effect (Figure 8). The layer that was made with "zigzag" is the highest and has the biggest relation of height to average width ($B_a = F_{dep}/H$), probably due to the beam power density distribution that was proportional to the width of the wire. The beam power is mostly concentrated in axis of the wire and decreases at the edges.

The application of oscillations allows to redistribute the electron beam power density on the surface of the substrate. It is possible to adjust the width of the layers in a certain range due to the regulation the width of the oscillation A_y. To obtain the thinnest layers it is advisable to decrease the width of the oscillation to 0. Thus, the transverse size of heating source becomes equal to the size of the beam focused on the surface of the substrate.

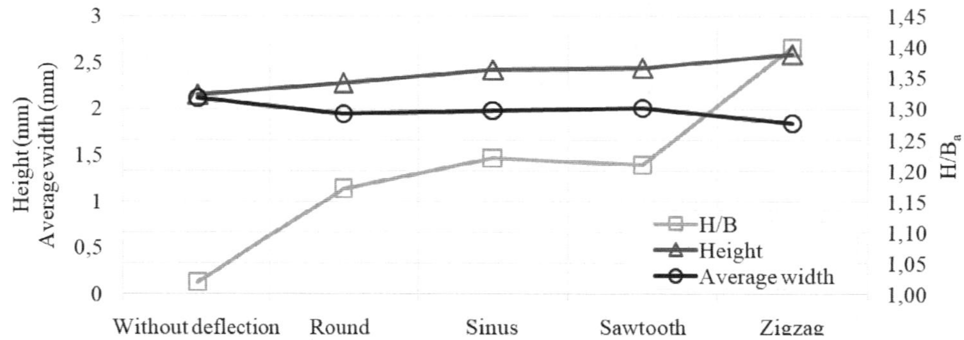

Figure 8. Changing of height, average width, and relation of height to average width (H/B_a) depending on oscillation type.

However, it is shown that the relation of the average width to the layer's height allowed to establish that the oscillations with the width equal to the half of wire diameter or lower do not impact on the width of the layer. In this case, the main parameters leading to the change of the layer width are feed and deposition rates.

4. Conclusions

The investigation of wire melting with the use of the electron beam showed that the thermal efficiency of the wire melting equals to 0.867 with the feed rate of 1 m/min. If the feed rate is more than 1.5 m/min then the thermal efficiency is constant and does not drop below 0.95. It is not advisable to carry out the process using low feed rates because a large proportion of beam energy is wasted on overheating the liquid droplets. The minimum energy per volume q/V_fF that is needed to sustain stable deposition process grows proportionally to the deposition rate and does not depend on the feed rate. The application of oscillations like "sinusoidal", "sawtooth" and "zigzag" that intensify liquid metal transfer in the liquid pool, leads to the increase of the layers height as the beam movement during the oscillations displaces liquid metal to crystallized bead. As a result of the effect, the bead with the biggest relation of height to average width is obtained, as compared to the beads that are deposited with other types of oscillation or without it. Presumably, the effect should be more visible in case of higher feed rates. The application of combinations of oscillations and their parameters may allow to ensure constancy of layers width during the process of obtaining products with a high wall.

Acknowledgments

The research was carried out in National Research University "Moscow Power Engineering Institute" with the financial support of the Russian Science Foundation (project No. 17-79-20015).

References

[1] Huang Z, Suo H, Gong S, Liu J, Yang G and YangY 2016 *Mater. Techn.* **31(2)** 115-20
[2] Brandl E *et al* 2011 *IOP Conf. Ser.: Mater. Sci. Eng.* **26** 012004
[3] Tayon W A, Shenoy R N, Redding M R *et al.* 2014 *J. Manuf. Sci. Eng.* **136** 61005
[4] Waters B R 2018 *Mechanical Properties of Inconel 718 Processed Using Electron Beam Free Form Fabrication (EBF3)* (Brigham Young University: Master of Science Thesis) p 65
[5] Brice C A and Hofmeister W H 2013 *Metal and Mat Trans A* **44** 5147
[6] Tarasov S Y, Filippov A V, Savchenko N L, Fortuna S V, Rubtsov V E, Kolubaev E A *et al.* 2018 *Int. J. Adv. Manuf. Technol.* **99 (9-12)** 2353-2363
[7] Tang Q, Pang S, Chen B, Suo H and Zhou J 2014 *Int. J. Heat and Mass Transfer* **78** 203-15
[8] Hu R *et. al.* 2019 *Sci. and Technology of Welding and Joining* **24(5)** 401-11
[9] Sliva A P 2017 *Proc. 2nd Int. Conf. "Electron Beam Welding and Related Technologies"* (Moscow: MPEI) pp 506-20
[10] Kim C S 1975 *Thermophysical Properties of Stainless Steels* (USA: Argonne Nat. Lab.) p 24

IOP Conf. Series: Materials Science and Engineering **681** (2019) 012009 doi:10.1088/1757-899X/681/1/012009

Features of sustainable localization of deformation zones forming in construction steels welded joints

V I Danilov[1], N V Ababkov[2], E A Ozhiganov[2]

[1] Institute of Strength Physics and Materials Science SB RAS, Academic Avenue, 2/4, Tomsk
[2] T.F. Gorbachev Kuzbass State Technical University, 28 str. Vesennyaya, Kemerovo,

E-mail: n.ababkov@rambler.ru

Abstract. The paper discusses the features of elastic-plastic transition in welded joints of Mn-Si steel samples made by stationary arc welding consumable electrodes to assess the possibility of exploitation of such compounds after the stress tests. In welded joints made by a stationary arc, the elastoplastic transition is realized by initiating the Chernov-Lüders band in the weld metal, followed by the formation of moving fronts in heat-affected zones. Local internal stresses in the heat-affected zones of welded joints made by a stationary arc are at the level of the material strength, therefore, the generation of the Chernov-Lüders bands occurs as relaxation processes.

1. Introduction

Conducting non-destructive testing of welded joints, as the most damaged elements, is one of the most important tasks of technical diagnostics at present. Therefore, the solution of the issue of quality control of welding work is an urgent task. Carbon or low-alloy structural steels are used to manufacture a large number of metal structures of potentially hazardous fuel and energy equipment. Welding in installation or repair conditions is performed mainly by manual arc welding. New methods of arc welding are being developed to reduce the impact of the human factor on the quality of welding and increase productivity. An initial assessment of the quality of repair or installation work is performed by testing equipment with increased loads (increased pressure). Different test methods and different loads are used depending on the type and purpose of the equipment. Test requirements are regulated by the regulatory documents of Federal Service for Ecological, Technological and Nuclear Supervision. However, to date, in the technical literature, there is practically no information about the physical processes occurring in the base metal and in the welded joints when performing such tests. The effect of preload (deformation) on the structural-phase state and internal stress fields in welded joints and, as a result, on the further trouble-free operation of the tested equipment has not been evaluated.

For this purpose, in the present work, a study was conducted of welded joints obtained by manual arc welding with a melting electrode in the initial state and in the process of an elastoplastic transition. This work is a continuation of [1].

2. Materials and methods

Welded joints of Mn-Si steel, which were made by welding with a stationary arc and current-modulated electrodes TsU-5, under conditions that are automatically tuned depending on the welding conditions, were investigated.

IOP Conf. Series: Materials Science and Engineering **681** (2019) 012009 doi:10.1088/1757-899X/681/1/012009

Mechanical tests and preliminary deformation of double-blade samples with a working part of 2×6×40 mm were carried out on a Walter + Bai AG LFM-125 machine (Switzerland) with a moving speed of 0.2 mm / min, which ensured the speed of deformation $8.3 \cdot 10^{-5}$ s^{-1}. The specified testing machine not only registers the stress diagram automatically, but also records the load-elongation-time primary data file, which can be used to analyze and highlight the stages of the deformation curve. This procedure was carried out according to the Trefilov method with employees [2]. First, the diagram from the conditional coordinates (σ - ε) was rearranged to the true coordinates (s - e). Moreover, s = σ (1 + ε), and e = ln (1 + ε). According to the Kholomon-Ludvik model [3], the deformation curve can be represented as

$$s = s_0 + Ke^n . \quad (1)$$

Then the stage of hardening is determined by the value of the index n. When n = 0, the stress is constant and does not depend on the deformation, that is, it is the stage of the yield pad. If n = 1, then the stress is proportional to the strain, a linear hardening stage is observed. The stages of parabolic hardening of Taylor correspond to n < ½ [3]. Finally, before the formation of the fracture neck begins at n <½, a pre-fracture stage is observed. This stage was first noted by the authors [4]. The stages of strain hardening are easily distinguished when the graph is linearized in coordinates $\ln(s - s_0) - \ln e$, such as the tangents of the angles of inclination of the corresponding sections [2].

Simultaneously with the registration of the loading diagram, the pictures of the localization of plastic deformation were recorded by the method of digital statistical speckle photography [5] developed on the basis of traditional speckle photography recorded on a photo carrier [6]. To implement it, the stretchable sample was illuminated with coherent light of a semiconductor laser with a wavelength of 635 nm and a power of 15 mW. Images of a deformable sample obtained with such illumination with speckle patterns superimposed on them were recorded with a PixeLink PL-B781 digital video camera with a frequency of 10 Hz, digitized and stored. For each point of the image, a sequence of samples was formed, characterizing the time course of its brightness, the variance and the expectation were calculated, the ratios of which were used to display the deformation localization zones. Using such a method, it is possible to record practically in situ the regions in which, at a given increase in the total elongation of the sample, the deformation of the material is localized.

The structure of welded joints was analyzed by optical and transmission electron microscopy (TEM) in the initial state and after deformations corresponding to different stages of the loading curve. After reaching the required degree of deformation, the sample was unloaded and removed from the grips of the machine. Metallographic studies were carried out on thin sections made on the working part of the sample. The weld metal zone (weld metal), the heat-affected zone and the base metal were studied in detail. The choice of these sites for research is due to the fact that it is in the weld metal and the heat-affected zone (HAZ) that there is the greatest probability of cracking both after welding and during long-term operation [7]. Foils for electron-microscopic analysis of the stressed and structural-phase states of these zones were made of appropriate thin sections.

Metallographic structural studies were carried out using a Neophot-21 microscope from a Genius VileaCam digital camera. In the manufacture of thin sections used mechanical grinding, mechanical polishing on the diamond paste ACM 10/7 NVL and chemical etching with a four percent alcoholic solution of nitric acid. The morphology and size of the phase and structural components were determined at magnifications of 100 and 500 times.

The TEM method was carried out on an EM-125 electron microscope with an accelerating voltage of 125 kV. The working magnification in the microscope column was 25,000 times. Electron microscopic images, confirmed by microdiffraction patterns and dark-field images obtained in the corresponding reflexes, were subjected to phase analysis. The following parameters of the material structure were measured: the scalar (ρ) and excess ($\rho_{\pm}$) dislocation densities, and the amplitude of the internal stresses σ_d).

3. Results and discussions

At the first stage, the analysis of the staging of the deformation curves of samples of welded joints was carried out. In fig. 1 shows the corresponding deformation curves, the form of which is qualitatively similar. An analysis of the deformation curves, performed according to the method, showed that the latter are multistage and contain, in addition to the obvious yield area, the linear hardening stage, Taylor parabolic hardening and the prefracture stage. The yield area is not smooth, fluctuations of the deforming stress and slight hardening are detected. Tooth turnover is not pronounced. A short microplastic area stands out in front of the tooth. It begins at the limit of proportionality, which corresponds to a relative deformation εpts = 0.006. Before the beginning of the hardening stages, there is always a drop in voltage, which we called the "reverse tooth".

Table 1. The boundaries of the stages and the corresponding indicators of strain hardening.

№	Stage name	Beginning of the stage e_1 (ε_1)	End of stage e_2 (ε_2)	Strain hardening rate n
1	Yield area	0.010 (0.010)	0.033 (0.034)	0
2	Linear Hardening Stage	0.037 (0.038)	0.067 (0.069)	1.01
3	Taylor Parabolic Hardening Stage	0.067 (0.069)	0.126 (0.136)	0.50
4	Pre-failure stage	0.126 (0.136)	0.188 (0.207)	0.38

The hardening rate at the yield area varies in a complex way, so the value n = 0 should be considered conditional. Between the yield point and the linear hardening stage, there is a short transitional stage, which includes a "reverse tooth" and a section where the hardening rate is constantly decreasing. There are such gaps between other stages, but their duration in deformation is less than 0.001. It should be noted that the pre-destruction stage is large and even exceeds the Taylor parabolic hardening stage in duration. According to the autowave ideas of the authors [8], each of the named stages has its own type of autowave of localized deformation. The yield point corresponds to a switching autowave, a linear hardening stage — a phase autowave, a Taylor stage — a stationary dissipative structure, and a collapse of a localized deformation wave develops at the pre-fracture stage. Types of autowaves differ from each other in the behavior of foci of localization of deformation. So at the stage of linear hardening, the foci are located equidistantly and synchronously move along the loading axis. At the Taylor stage, they are also equidistant, but still. At the pre-fracture stage, there is a stable zone with a high amplitude of the autowave, to which the other localization centers move. The end of the pre-fracture stage corresponds to the achievement of the highest stress σB, that is, the final design of the macroscopic cervical fracture. In [9], it was shown that the position of the specified high-amplitude zone of strain localization determines the place of future destruction. Indicated in table indicators of hardening at the described stages coincide with the values given in [8, 9].

Analysis of the staging and determining the type of autowave makes sense to use when developing modes of processing materials by pressure, as shown in [10]. In order to diagnose the state of the material in structures and parts, it is important to know its behavior during the elastoplastic transition, that is, before the hardening stages. This transition in the studied steel consists of a stage of microplasticity and a stage of a tooth and a yield point; therefore, it takes a considerable period of time. It accumulates up to 3.5% strain. In previous experiments [11], it was found that the stage of microplasticity, as well as the ascending and falling branches of the tooth yield corresponds to the process of germination through the cross section of the embryo of the Chernov-Lüders band (PCL). According to generally accepted concepts [11], the formed strip expands and transfers the sample material from an elastically stressed to a plastically deformed state. In this case, at each instant of time, the deformation processes are localized on the moving edges (fronts) of the PCL, so when using the digital statistical speckle-photography method, the fronts are distinguished as bright light bands (Fig. 2). The expansion of the strip proceeds at the stage of the yield pad. The transition ends when the entire sample workspace is "swept up" by the

PCL fronts. Thus, the time of the elastoplastic transition is determined by the germination rate of the PCL nucleus and the velocities of the strip fronts.

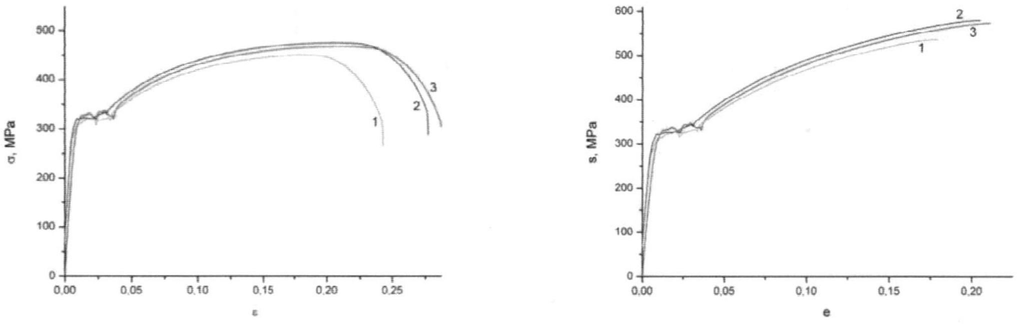

a b

Figure 1. Diagrams of loading samples with welds:
a – conditional stresses – conditional deformations, b – true stresses – true deformations.

It should be noted that in the flat specimens two PCLs are usually formed near the grips of the loading device (Fig. 2). The process of nucleation is stochastic and attempts to initiate it in a uniformly cross-sectioned object by creating a precise mechanical stress concentrator (identification) or by irradiating a small region of the sample with a force beam have not been successful.

Figure 2. The origin and distribution of PCL in a uniform sample. The time interval between images is 35 s.

The weld and heat-affected zone represent a macroscopic region of structural-phase heterogeneity, the grain structure of which does not depend on the welding method at the level of optical microscopy. The seam (weld metal) is represented by columnar dendrites characteristic of the cast state (Fig. 3a). The plates of the Widmannstatt ferrite reach a length of 1 mm and a width of 20 microns. The interplacial space is occupied by smaller dendrites. Other phases and structural components other than ferrite are not detected, which corresponds to the composition of the electrode rod. The width of the seam zone was ≈ 5 mm. The transition from the weld metal to the heat-affected zone and then to the structure of the base metal occurs smoothly without abrupt changes. The entire heat-affected zone is represented by polyhedral ferrite grains. Pearlite inclusions are practically absent, which apparently was due to the diffusion of carbon into the decarburized weld metal. A complete recrystallization occurred near the fusion line and, as a result, the growth of ferritic grain to an average size of 18±8 μm (Fig. 3b). In this area, the grains are unequal. The area of recrystallization is clearly not allocated. The total width of the heat-affected zone is small and does not exceed 1.5 mm. The structure of the base metal is ferritic-pearlitic (Fig. 3c). The volume fraction of perlite at the level of 10–12%, which corresponds to the chemical composition of Mn-Si steel. Grains of ferrite are polyhedral with well-defined, clean borders and an average size of 7±5 μm.

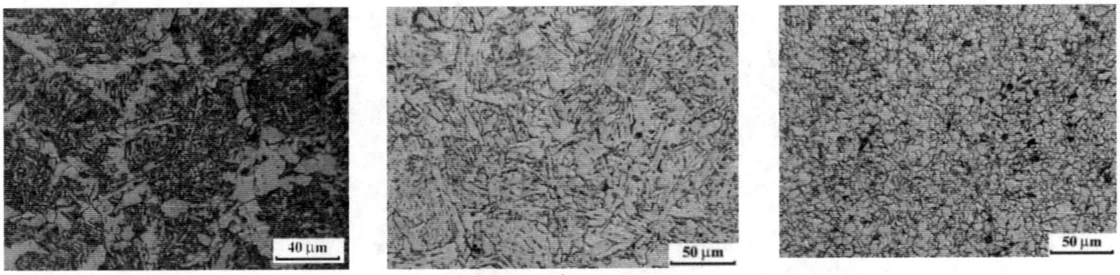

a b c

Figure 3. Microstructure of welded joint:
a – weld metal, b – heat-affected zone, c – base metal

Precision studies using transmission electron microscopy revealed the influence of the welding method on the defect structure and stress state of the heat-affected zone. Research foils were cut at a distance of <1 mm from the fusion line.

When using manual arc welding, ferritic grains are completely fragmented. The degree of fragmentation is higher, the smaller the distance to the fusion line. The dislocation structure is polarized (bending extinction contours are present inside the fragments, Fig. 4). The average scalar dislocation density is $2.1 \cdot 10^{10}$ cm^{-2}, and the average excess dislocation density is significantly less than меньше $0.9 \cdot 10^{10}$ cm^{-2}. The amplitude of local internal stresses reaches 450 MPa, which is commensurate with the temporary resistance of the material under study (see Fig. 1).

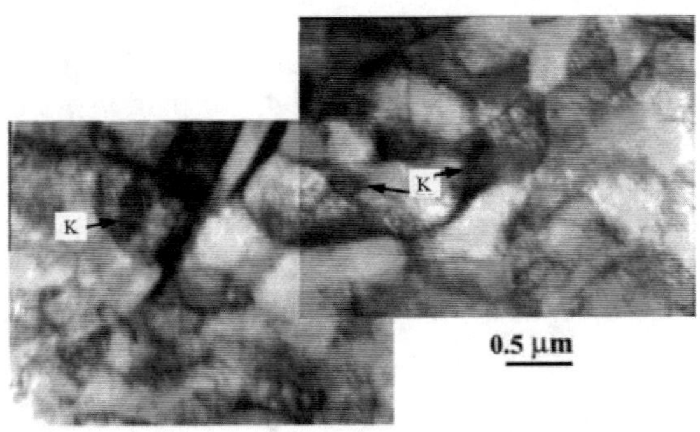

Figure 4. Fragmented structure of ferrite grains in the heat-affected zone. K – bending extinction contours

As mentioned above, the front PCL separates the material in an elastically stressed and plastically deformed state [11]. In this work, we investigated the effect of PCL propagation on the microstructure, structural phase state, and internal stress fields in a welded joint. To do this, after passing the PCL front through the deposited metal and heat affected zones, the test sample was unloaded, a thin section was made on its working surface, and electron microscopic studies were performed on foils prepared from the heat affected zone. The total strain of the sample was 2%

Optical microscopy studies have shown that the grain structure of the material in all three investigated zones did not undergo changes after the passage of PCL fronts. The morphology of the borders, the average grain sizes, and the intragranular structure remained unchanged.

Figure 5. The origin of the Chernov-Lüders band in the weld metal

The influence of the passage of PCL fronts was discovered during the study of the dislocation substructure and the stress state by the TEM method. Scalar and excess dislocation densities grow at the same rate and slightly. After the passage of the PCL front in the heat-affected zone of the weld obtained by the stationary arc, the scalar dislocation density reached $2.6{\cdot}10^{10}$ cm^{-2}, and the excess one — $1.7{\cdot}10^{10}$ cm^{-2}. At the same time, the level of internal stresses in ferritic grains of the heat-affected zone decreased almost twice from 450 MPa to 280 MPa. This confirms the above-stated assumption that the formation of PCL fronts in the heat-affected zone is a relaxation process and is accompanied by changes in the dislocation structure.

Conclusions

1. In welded joints made by a stationary arc, the elastoplastic transition is realized by initiating the Chernov-Lüders band in the weld metal, followed by the formation of moving fronts in heat-affected zones.
2. Local internal stresses in the heat-affected zones of welded joints made by a stationary arc are at the level of the material strength, therefore, the generation of the Chernov-Lüders bands occurs as relaxation processes.

References

[1] Smirnov A. N., Ozhiganov E. A., Danilov V. I., Gorbatenko V. V. and Murav'ev V.V. 2015. *Russ. J. Nondestr. Test.* **51** 705.

[2] Trefilov V. I., Moiseev V. F., Pechkovsky E. P. 1987 *Strain hardening and destruction of polyc rystalline metals* (Kiev: Naukova Dumka).

[3] Honicomb R. 1972 *Plastic deformation of metals* (Moscow: Mir).

[4] Koneva N. A., Trishkina L. I. and Kozlov E. V. 1998 *Bulletin of the Russian Academy of Scie nces. Physics* **62** 1352.

[5] Zuev L. B., Gorbatenko V. V. and Pavlichev K. V. 2010 *Measur. Science Technol.* **21** 054014-18.

[6] Dzhouns R. and Uayks K. 1986 *Holographic and speckle interferometry* (Moscow: Mir) p 327.

[7] Livshits L. S. and Khakimov A. N. 1989 *Welding metal science and heat treatment of welded joints* (Moscow: Mashinostroyeniye) p 336.

[8] Zuev L. B., Danilov V. I. and Barannikova S. A. 2008 *Physics of Macrolocalization of Plastic Flow* (Novosibirsk: Nauka) p 327

[9] Danilov V. I., Orlova D. V., Zuev L. B. 2011 *J. Appl. Phys* **81** 51.

[10] Zuev L. B. 2014 *Bulletin of the Russian Academy of Sciences. Physics* **78** 957.

[11] Pelleg J. 2013 *Mechanical properties of materials* (Heidelberg, New York, London: Springer) p 634.

IOP Conf. Series: Materials Science and Engineering **681** (2019) 012010 doi:10.1088/1757-899X/681/1/012010

New approaches to the rational manufacturing of combined constructions by EBW

V K Dragunov[1], **A L Goncharov**[1]

[1] National Research University Moscow Power Engineering Institute, ul. Krasnokazarmennaya, 14, Moscow, 111250

E-mail: DragunovVK@mpei.ru.

Annotation. The paper derives a relation describing the shape of an electron beam axis affected by thermoelectric currents during welding of dissimilar materials. The treatise suggests a new method for determining the adjustment actions for preventing lack of penetration in the weld root and derives an expression for determining the critical degree of weld penetration during welding of dissimilar austenitic and pearlitic steels. The received results can be used to make high-quality welding connections of thick dissimilar materials by eliminating incomplete penetration and ensuring necessary penetration degree of welded edges along the whole part thickness.

1. Introduction

In the welding seam of dissimilar alloys made by different welding methods, including electron-beam one, the melting boundary undergoes formation of regions with transitional composition (crystallization layers), which properties differ from those of welded alloys and cladded metal. The reason is virtually absent mixing of the liquid metal in the melting pool near the interface of solid and liquid phases [1]. The dimensions of these sections depend on the penetration degree, intensity of liquid metal mixing in the melting pool and its life duration, i.e. on the welding technology. It is be believed that the width of crystallization layers during manual arc welding is 0.4–0.6 mm, in the case of automatic submerged arc welding it is 0.4–0.6 mm, while in the case of electron-beam welding it does not exceed 0.1 mm [2]. In the latter case, the dimensions of the layer with variable chemical composition turns out to be smaller as compared to arc welding, because of more intense mixing of metal in the melting pool and its reduced life.

If the welded connection is formed by steels of the same structural class and different alloying, the properties of the crystallization layers practically have no effect on the operation characteristics of the parts. Unlike such situation, the layers of welded connections of steels from different structural classes, for instance austenitic and pearlitic, can be expected to include brittle martensitic regions and for cracks during welding or operation. In the first place, the width of the regions will depend—under given welding conditions—on the weld seam composition. Reducing content of Ni in the weld seam metal increases the width of the martensitic layer in the fusion zone, while its exceedingly low content yields martensitic structure of the seam. Such a case appreciably increases the affinity to form cold cracks that, as a rule, arrange along the seam in its middle. Therefore, the structural nonuniformity, conditioned by the presence of layers with reduced ductility in the fusion zone, reduces the performance of combined welded structures from dissimilar metals and alloys.

In this connection, one of the important challenges in welding dissimilar metals and alloys is obtaining the required chemical composition of the seam metal determining the properties of the welded connection as a whole. The most widely spread methods for adjusting the chemical composition in welding dissimilar metals are based on the implementation of intermediate inserts between welded parts, which main purpose is alloying the seam metal. In addition, the chemical composition of the seam metal can be adjusted by beam displacement towards one of the parts.

However, welding of dissimilar steels gives rise to magnetic fields, which sources are residual magnetization of ferromagnetics and thermoelectric currents [3]. These fields interact with the electron beam and deflect it from the joint region. Hence, the required degree of edge penetration and seam composition are determined by the preset accuracy of beam positioning relatively to the joint.

The work is aimed at scientifically justifying the technological methods for stabilizing the spatial parameters of the beam relatively to the joint during electron-beam welding of dissimilar steels and alloys providing necessary chemical composition, structure and properties of welded connections in combined constructions.

2. Investigation Methods

Currently, different methods are used to reduce the action of magnetic fields on electron beam: demagnetizing the parts before welding, implementation of screens from high-permeability materials for transporting the electron beam from the gun anode to the part's surface, field compensation in the welding zone, introduction of additional materials into the welding zone or orientation of the beam oppositely to its deflection [4–9]. These methods can also be combined; for instance, demagnetizing with application of ferromagnetic shields considerably increase the EBW conditions. The effective implementation of such technological techniques is possible only in cases with known distribution of magnetic field induction in the region of electron beam propagation in the penetration channel [10].

In the absence of external high-current sources, for preliminarily demagnetized materials, the thermoelectric currents are the main source of magnetic field. The field lateral component distribution $B_{xm}(z)$ is practically linear along the depth of the gas-vapor channel ($0 \leq z \leq \delta$) and reaches extremums at its opposite ends. The longitudinal field induction component over the surface of welded parts in the electron drift space ($z \leq 0$) decreases almost exponentially. The direction of the thermoelectric current field magnetic induction vector, in this case, is determined by the sign of relative thermal EMF of the welded pair of materials. The magnetic field of thermoelectric currents $B_{xm}(z)$—while switching its sign along the thickness of the welded connection, deflects the electron beam up to the semi-thickness of the connection towards the material with positive thermoelectric potential, and then towards the opposite side (Fig. 1). In this case, the deflection angles of the beam axial electron relatively to the initial direction in the top and root of the penetration channel are equal ($\theta(0)=\theta(\delta)$), while the inflection point of its trajectory is at the middle of the channel depth. Such spatial parameters of the beam usually fail to provide necessary penetration degree and cause lack of penetration along the part thickness. Thus, the adjustment of the parameters is needed. The adjustment means providing such beam inclination angle to the joint $\varphi(0)$ (Fig. 2), at which there is no beam deflection in the weld top and root, i.e. $\psi(0) = \psi(\delta) \approx 0$. This condition is satisfied when turning line CD — connecting the point of entrance of a beam electron into the part and its exit point — around point C by angle Θ and displacing it by distance $\psi(0)$ up to the alignment with the joint.

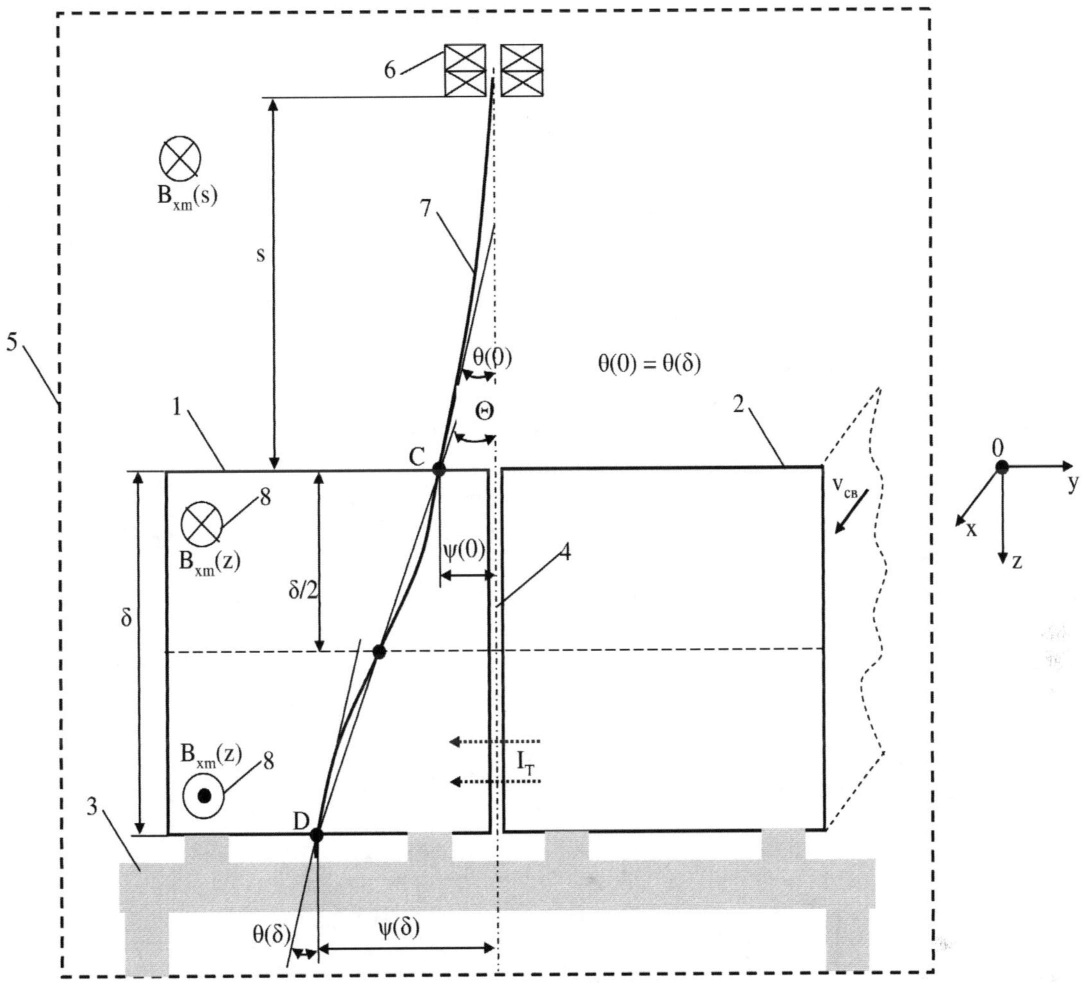

Figure 1. Deflection of beam axial electron in magnetic field of thermoelectric currents: 1) welded part with positive thermoelectric potential; 2) welded part with negative thermoelectric potential; 3) welding bench of the electron-beam setup; 4) joint of welded parts; 5) vacuum chamber of the electron-beam setup; 6) electron-beam gun; 7) electron beam axis; 8) direction of the magnetic field induction $B_{xm}(z)$.

IOP Conf. Series: Materials Science and Engineering **681** (2019) 012010 doi:10.1088/1757-899X/681/1/012010

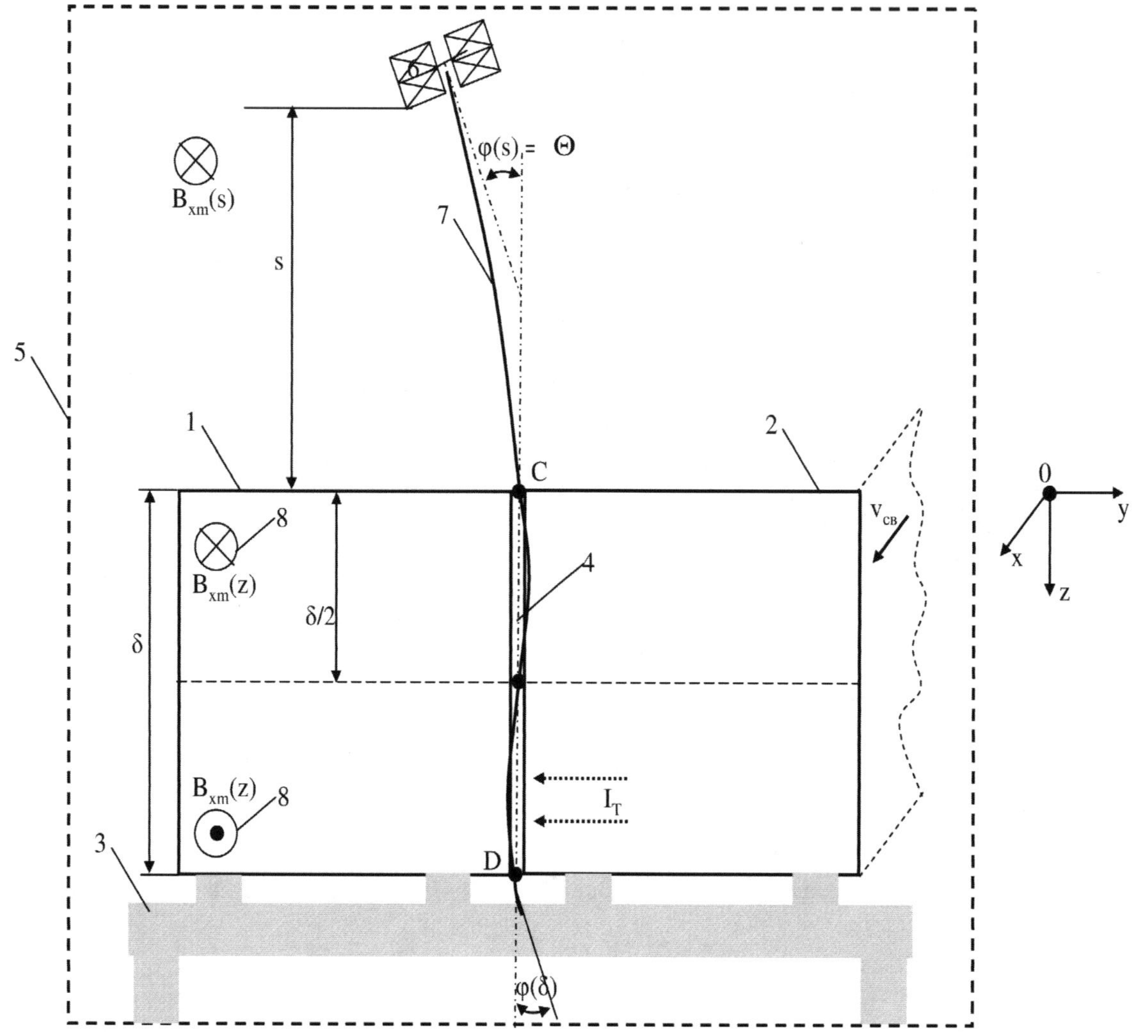

Figure 2. Adjustment of spatial position of the electron beam in magnetic field of thermoelectric currents: 1) welded part with positive thermoelectric potential; 2) welded part with negative thermoelectric potential; 3) welding bench of the electron-beam setup; 4) joint of welded parts; 5) vacuum chamber of the electron-beam setup; 6) electron-beam gun; 7) electron beam axis; 8) direction of the magnetic field induction $B_{xm}(z)$.

In this case, angle Θ equals

$$\Theta \cong tg\Theta = \frac{\psi(\delta)-\psi(0)}{\delta} \quad (1)$$

In the absence of residual magnetization field, the following expression was derived for determining the values of ψ(δ) and ψ(0):

$$\psi(z) = \sqrt{\frac{e}{2m_e U}} B_{xm}(0)\left[\frac{1-(1+\alpha s)\exp(-\alpha s)}{\alpha^2} + \frac{1-\exp(-\alpha s)}{\alpha} z + \frac{z^2}{2} - \frac{z^3}{3\delta}\right], \quad (2)$$

IOP Conf. Series: Materials Science and Engineering **681** (2019) 012010 doi:10.1088/1757-899X/681/1/012010

where e is the electron charge, C; m_e is the electron mass, kg; U is the accelerating voltage, V; α is the empirical coefficient characterizing the plasma density above the part, m^{-1}; s is the distance between the gun and part, m; δ is the thickness of welded parts, m.

The inclination angle between the electron-beam gun to the joint φ(s)—with due consideration of the beam deflection in the magnetic field above the part—will equal Θ. Neglecting the minuscule change in the magnetic field induction in X0Y plane (1) with due regard to (2) gives the following:

$$\Theta \cong \sqrt{\frac{e}{2m_e U}} B_{xm}(0) \left[\frac{1-\exp(-\alpha s)}{\alpha} + \frac{\delta}{6} \right] \quad (3)$$

Thus, depending on the welding regimes, the adjustment of the beam axis inclination to the joint can be used to set up the spatial beam position during EBW of thick parts, which ensures necessary penetration degree and reduces the possibility of lack of penetration.

Below is the experimental method for determining the deflection angle Θ between the beam and joint. The ferromagnetic specimens were demagnetized before welding by alternating magnetic field at industrial frequency with gradual reduction of its strength. The induction of the residual magnetization field was ensured to be less than 0.05–0.06 mT. Then, trial welding was performed for the dissimilar materials to measure the beam deflection from the joint. The electron gun axis (electron beam) was aligned with the joint, and then the penetration was carried out. The welding process induced thermoelectric currents IT which magnetic field $B_{xm}(z)$ was deflecting the electron beam from the joint by angle θ(0) towards the metal with positive thermoelectric potential. To compensate such beam deflection, the electron-beam gun axis was revolved around the seam top by angle Θ that was determined from eq. (3) and then, the gun was displaced towards metal 1 with negative thermoelectric potential so the beam axis (axial electron beam trajectory) was pointing to metal 2 with negative thermoelectric potential. The gun was displaced up to the alignment of the electron beam on the part's surface with the joint. This provided the angle between the beam axis and joint at the entrance and exit from the part to be φ(0)= φ(δ)≈θ.

3. Results

The study established that if the angle of electron beam entrance into the part φ(0) is larger than θ(0), then the electron beam axis at the exit from the part deflects towards the material 2 with negative thermoelectric potential. On the contrary, if φ(0)<θ(0), the beam deflects towards the material 1 with positive thermoelectric potential. In both the cases, one of the materials is predominantly melted. In addition, the welding seam features defects of lack of penetration along the thickness.

4. Discussion

In the presence of the magnetic field of thermoelectric currents and with due account for the seam shape, the beam adjustment leads to melting of different alloys in the top and in the root of the seam, the penetration areas being not equal to each other. The suggested method for adjusting the spatial position of the beam in this case can increase the penetration degree of the edges of non-austenitic steel, which may negatively affect the properties of weld connections. Thus, this method is effective for welding different steels with austenitic steels, if the penetration degree for the austenitic steel is

$$\gamma_1 = \frac{F_1}{F_1 + F_2} \geq \frac{13Cr_{э2} + 16Ni_{э2} - 416\%}{13(Cr_{э2} - Cr_{э1}) + 16(Ni_{э2} - Ni_{э1})}, \quad (4)$$

where $Cr_э = \%Cr + \%Mo + 1{,}5\%Si + 0{,}5\%Nb + \%V$ and $Ni_э = \%Ni + 30\%C + 0{,}5\%Mn + 30\%N_2$ are equivalent contents of chrome and nickel in the seam metal.

IOP Conf. Series: Materials Science and Engineering **681** (2019) 012010 doi:10.1088/1757-899X/681/1/012010

The specificities of the technological process presented above can be used to make high-quality welding connections of thick dissimilar materials by eliminating incomplete penetration and ensuring necessary penetration degree of welded edges along the whole thickness, thus forming necessary structure and properties of welded connections.

Acknowledgements
The research was performed at the National Research University Moscow Power Engineering Institute under financial support of the Russian Science Foundation (project No. 18-19-00652).

References
[1] Ryabov V R, Rabkin D M, Kurochko R S and Strizhevskaya L G 1984 *Welding of Dissimilar Metals and Alloys* (Moscow: Mechanical Engineering) p 239
[2] Dragunov V K, Muravyova T P and Rodionov Yu A 1990 *Welding production* **4** 2–4
[3] Dragunov V K and Chepurin M V 2001 *Welding Production* **12** 8–16
[4] Nazarenko O K 1982 *Automatic welding* **1** 33–9
[5] Leskov G I and Zhivaga L I 1981 *Automatic welding* **11** 38–40
[6] Akopyants K S, Levitsky S M, Nazarenko O K, Neporozhny V Yu and Shilov G A 1989 *Let. to J. Tech. Phys*. **15(22)** 20–3
[7] Shilov G A 1977 *Automatic welding* **6** 71–2
[8] Novokreshchenov V V, Rodyakina R V and Karimbekov M A 2017 *Solid State Phenomena* **265** 237–45
[9] Shilov G A and Kovbasenko S N 1982 *Automatic welding* **8** 58–9
[10] Dragunov V K 2003 *Welding Production* **5** 15–20

IOP Conf. Series: Materials Science and Engineering **681** (2019) 012011 doi:10.1088/1757-899X/681/1/012011

Analytical methods of electron beam power evaluation for electron-beam welding with deep penetration

V K Dragunov[1], E V Terentyev[1], A P Sliva[1], A L Goncharov[1] and D A Zhgut[1]

[1] National Research University "Moscow Power Engineering Institute", 14, Krasnokazarmennaya str., Moscow, 111250, Russia

E-mail: terentyevyv@mpei.ru

Abstract. Existing analytical methods for calculating electron-beam welding modes, which provide welds with required penetration depth, are described, and main disadvantages of these methods are shown. A new calculation method based on a model of moving cylindrical heat source, which involves moving of a cylinder with constant surface temperature in an infinite plate, is proposed. The proposed calculation method takes into account energy losses during metal melting and its partial evaporation. A comparison of experimental and calculated data obtained using the developed method and existing methods, is presented. Plates with a thickness of 32 mm from 5V titanium alloy, 27 mm from 40 Kh13 steel and 20 mm from D16 duralumin were used for experiments. It was shown that the developed method application made it possible to calculate with appropriate accuracy a required beam power during EBW of various materials.

1. Introduction

Beam welding methods for thick-walled structures joining are used in almost all industries [1]. Preliminary calculations of welding modes significantly reduce labor and material costs during welding technology development. Specifically, the main parameter, which is usually used for weld penetration regulation during electron beam welding (EBW), is beam current. There are a large number of methods for establishing the interrelation between beam energy characteristics and weld penetration. Some methods were determined empirically. Such methods usually contain correction factors, due to the complexity of the processes taking place during electron-beam welding. The main disadvantage of empirical methods is that they can be applied correctly only in a quite narrow range of mode parameters and materials being welded. As regards analytical methods, they produce very approximate results; therefore, they are rarely used in practice. Using of numerical methods or complex models of electron beam interaction with a material yields accurate results after verification, but considerable computational power, high cost of software and large process duration severely limit their applicability during production process. At the same time, it is often necessary to determine quickly the required beam current at the stage of EBW technology development or improvement during production process. For example, the transition from EBW on a backing plate to EBW with free formation of weld root, as a rule, requires significant increase in welding speeds, but existing methods cannot achieve acceptable accuracy during recalculation of required beam current. Thus, the problem of appropriate beam current evaluation with the use of minimum time and software resources remains highly relevant. The aim of this paper is the analysis of existing calculation methods and development of the universal analytical method for calculating electron beam power during EBW with deep penetration.

IOP Conf. Series: Materials Science and Engineering **681** (2019) 012011 doi:10.1088/1757-899X/681/1/012011

2. Methods of required beam current calculation

During the review several specific calculating analytical methods of establishment interrelation between beam energy parameters and weld dimensions were identified. Several methods are described below. For comfort comparison, in each case, beam current required for penetration of a certain depth was calculated depending on welding speed, and all the remaining parameters were set. Table 1 shows parameters, used for beam current calculation.

Table 1. Parameters for beam current calculation

Parameter	5V titanium alloy	40Kh13 steel	Duralumin D16
Specific heat capacity [J·(kg·K)$^{-1}$]	530.8	452	879
Melting temperature [K]	1944	1773	850
Channel surface temperature [K]	2519	2508	1421
Metal's density [kg·m^{-3}]	4505	7650	2780
Thermal conductivity [W·(m·K)$^{-1}$]	18.85	25	119
Specific melting heat [J·kg^{-1}]	470	270	393
Specific boiling heat [J·kg^{-1}]	9800	6300	9210
Channel diameter·10^3 [m]	1.1	1.1	0.7
Weld width mean value·10^3 [m]	2	2	2
Nominal penetration depth·10^3 [m]	30	27	20

Method 1. The calculation method proposed by I.V. Zuev is based on energy balance in melting zone. For a continuous weld this method is reduced to the equation for beam current calculating:

$$I = \frac{\pi \cdot B_{\text{mean}}^2 H \cdot S_{\text{m}} \cdot v}{4\eta_{\text{ef}} \cdot \eta_{\text{T}} \cdot U \cdot d}, \tag{1}$$

where B_{mean} is value of weld width; H is penetration depth; S_{m} is heat content of liquid metal at melting temperature; it is determined as $S_{\text{m}} = \rho(cT_{\text{m}}+L_{\text{m}})$, where ρ is metal's density; c is metal's specific heat capacity (J·kg^{-1}·K^{-1}); T_{m} is melting temperature, K; L_{m} is specific melting heat; v is welding speed; η_{ef} is effective efficiency; η_{T} is thermal efficiency of welding process, which depends on physical properties of welded material and welding speed; U is accelerating voltage; d is diameter of heat source.

The main disadvantage of this method is a necessity to know in advance mean weld width and thermal efficiency of welding processes. Additional experiments to determine these parameters are very time-consuming, thus weld width is set on the basis of experimental data, and thermal efficiency is assumed to be 0.484 as in the case of fast-moving linear heat source. Obviously, a change in welding speed will significantly affect both weld width and thermal efficiency. Therefore, practical use of this method for theoretical determination of welding modes with acceptable accuracy without proper experience is difficult.

Method 2. T. Hashimoto at the start of electron-beam technologies development proposed an equation for calculating penetration depth depending on welding mode parameters and physical properties of the material [2]. This equation is based on a simplified weld formation model. Beam current in this method can be determined by following equation:

$$I = \frac{H \cdot r_e \cdot v}{U} \cdot (c \cdot \rho \cdot T_{mean} + \rho \cdot L_m + \frac{5 \cdot \lambda \cdot T_{mean}}{2 \cdot a} + \frac{5 \cdot \lambda \cdot T_{mean}}{2 \cdot v \cdot r_e}), \tag{2}$$

where r_{e} is beam effective radius, which was taken as $d/2$, T_{mean} is mean temperature of weld pool channel, which was taken as $T_{\text{m}} + 300$ (K), λ is thermal conductivity, and a is thermal diffusivity.

Method 3. The equation below was obtained by A.A. Lopatko and colleagues by formation of calculated ratios using Similarity and Dimensional Methods [3]. Then beam current can be written as follows:

$$I = \frac{\lambda \cdot H \cdot T_m}{0,132 \cdot U} \cdot \left(\frac{2 \cdot v \cdot r_e}{a}\right)^{0,815} \cdot \left(1 + \frac{L_m}{c \cdot T_m}\right)^{0,815}. \tag{3}$$

Method 4. Under these conditions, a mathematical model of a moving cylindrical heat source, which involves moving of a cylinder with constant surface temperature in an infinite plate [4], is interesting. The cylindrical heat source simulates the movement of vapor-gas channel during electron beam or laser welding. In this case channel surface temperature can be taken constant. By integrating heat flux through the surface of cylindrical source, it is possible to obtain full power of heat source moving at a speed v. It can be written as follows:

$$q = 2\pi H\lambda(T_s - T_0)\sum_{n=0}^{\infty}(-1)^n \varepsilon_n \frac{I_n\left(\frac{vd}{2a}\right)}{K_n\left(\frac{vd}{2a}\right)}, \tag{4}$$

where λ is metal thermal conductivity, T_0 is temperature at infinite distance from heat source; T_s is surface temperature of cylindrical source; v is speed of heat source movement; a is thermal diffusivity of the material being welded; d is diameter of cylindrical heat source; I_n and K_n are modified Bessel functions of nth order; $\varepsilon_0 = 1$ for $n = 0$, $\varepsilon_n = 2$, for $n \geq 1$.

Using full power of moving cylindrical source to determine beam power is possible if vapor-gas channel diameter is specified as cylindrical heat source and channel surface temperature is calculated from the condition of forces balance in the channel [5]. Using this method, change in the nature of heat transfer depending on welding speed and thermophysical properties of the material is considered, but costs on metal's melting and its partial evaporation are not taken into account.

To eliminate this drawback, a new method was developed on the base of this method. To determine heat source total power, in the equation (4) one component was added that takes into account heat consumption for metal melting and its partial evaporation. Then the equation (4) can be written as follows:

$$q = 2\pi H\lambda(T_s - T_0)\sum_{n=0}^{\infty}(-1)^n \varepsilon_n \frac{I_n\left(\frac{vd}{2a}\right)}{K_n\left(\frac{vd}{2a}\right)} + Bhv\rho\left(L_{\text{m}} + \xi L_{\text{boil}}\right), \tag{5}$$

where ξ is evaporated metal part (in the calculations it was assumed to be equal 2%).

To determine energy costs for melting it is necessary to know weld width. It was explored using the same model of moving cylindrical heat source. For this purpose, distribution of temperature fields from such source with a constant surface temperature in polar coordinates was considered [6].

The isotherm $T(r,\theta) = T_{\text{m}}$ is actually the boundary of weld pool (figure 1). The maximum width of weld pool is also weld width. Solving equation $T(r,\theta) = T_{\text{m}}$ for r, the dependence $r(\theta)$ in polar coordinates was obtained. Then weld width can be determined by simultaneous solution of following equations:

$$T_{\text{m}} = T_0 + (T_s - T_0)e^{\left(-\frac{vr\cos\theta}{2a}\right)}\sum_{n=0}^{\infty}\varepsilon_n I_n\left(\frac{vd}{4a}\right)\frac{K_n\left(\frac{vr}{2a}\right)}{K_n\left(\frac{vd}{4a}\right)}\cos(n\theta), \tag{6}$$

$$\frac{dr}{d\theta \cdot r} = ctg(\pi - \theta), \tag{7}$$

$$B = 2 \cdot r(\theta) \cdot \sin(\pi - \theta). \quad (8)$$

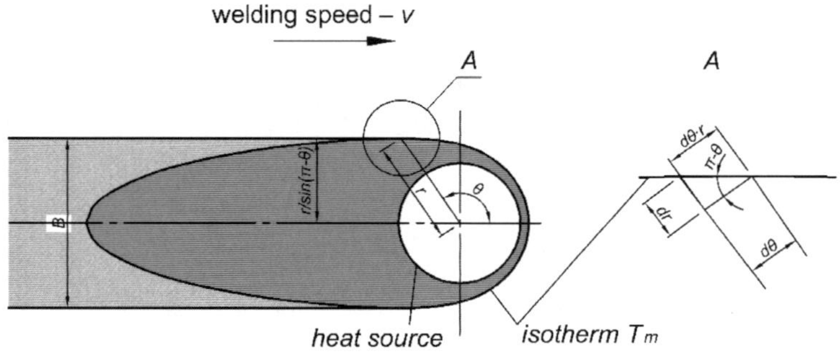

Figure 1. Scheme for weld width determination in a case of cylindrical heat source moving.

For comparison calculated and experimental data, several tests were carried out on electron-beam welding plates. Plates with a thickness of 32 mm from 5V titanium alloy (Al = 4.7–6.3%, Mo = 0.7–2%, V = 1.–1.9%, Ti balanced), 27 mm from 40 Kh13 steel (C = 0.35–0.44%, Cr = 12–14%, Fe balanced) and 20 mm from D16 duralumin (Mn = 0.3–0.9%, Cu = 3.8–4.9%, Mg = 1.2–1.8%, Al balanced) were used for experiments. Welding speed was changed in the range of 20…120 m/h. During complete fusion part of energy is lost on the screen installed on the back side of weld root and it is difficult to take into account its quantity. Therefore, during experiments absence of complete fusion was guaranteed and possibility of penetration depth measuring was maintained. For this purpose, beam current was selected experimentally, providing point penetration during sharp focusing, i.e. focusing current, which provides maximum penetration depth [7].

Then beam current was reduced by 5% and plates were welded with different welding speeds. The ELA-15I power complex with accelerating voltage of 60 kV and maximum beam current of 650 mA was used for welding. The pressure in vacuum chamber was 10^{-3} mm of Hg.

3. Results of studies

Welding modes, on which samples from different alloys were welded, are presented in Table 2. Macrostructure of welded joints from 5V titanium alloy is shown in figure 2.

Figure 3 shows comparative results of experimental and calculated data, obtained using the developed method and existing methods. Since penetration depth is slightly varied at different welding speeds, for objective comparison of calculated and experimental data, the actual beam current was normalized to the penetration depth of 30 mm by a proportional change relative to the actual penetration depth. Using the proposed method, it is possible to calculate required electron beam power with satisfactory accuracy for various materials, whereas using existing methods it is possible to obtain results with acceptable accuracy only in a rather narrow range of welding speeds or only for some materials.

4. Discussion of studies results

Results of calculations according to method 1 are in poor agreement with experimental data, because in the calculations the same weld width and thermal efficiency set for all welding speeds. The accuracy of this method will increase significantly if these parameters known previously for various conditions. Acceptable results can be obtained using method 2 for titanium alloy and steel, especially at high speeds, but for duralumin calculated beam current exceeded real one by about 2 times. Method 3 can be considered unsuitable for titanium alloy and steel. An approximate regularity has been obtained for duralumin, but it is hardly suitable for practical use, especially at high speeds. The calculation results by method 4 show stably underestimated values of beam current, because energy costs for metal's melting and its evaporation are not taken into account.

Table 2. EBW modes (I_{fakt} – actual beam current, H_{fakt} – actual penetration depth, I – corrected beam current value, calculated by proposed method)

Material	U [kV]	v [m/h]	I_{fakt} [mA]	H_{fakt} [mm]	B_{mean} [mm]	I [mA]	H [mm]
5V titanium alloy	60	20	100	29.4	2.68	102	30
		40	135	26.8	2.16	151	
		60	170	28.1	1.79	181	
		90	215	26.4	1.63	244	
		120	250	27.7	1.43	270	
40Kh13 steel		20	115	25.4	2.10	122	27
		60	200	24.8	1.52	217	
		120	350	26.6	1.37	355	
D16 duralumin		20	65	17.1	2.34	76	20
		60	100	19.4	1.77	103	
		120	140	16.0	1.90	175	

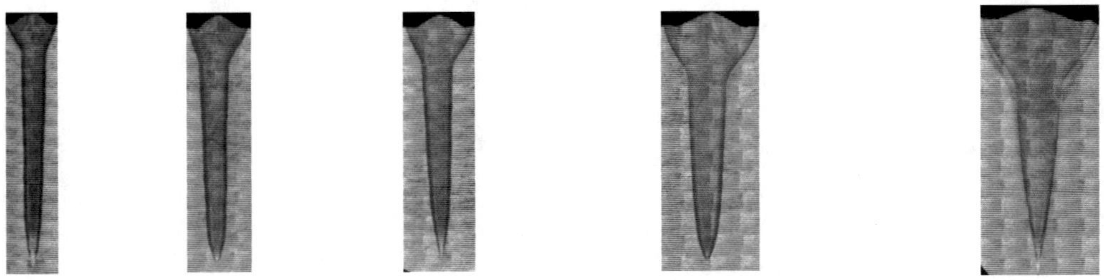

a b c d e

Figure 2. Macrostructure of welded joints from 5V titanium alloy, obtained at different speeds: 120 (a), 90 (b), 60 (c), 40 (d) and 20 (e) m/h.

Using the developed method for determination required beam current and a program for mathematical calculations (for example, Mathcad), it is possible to obtain almost instantly a result with acceptable accuracy for practical use. However, this fact does not preclude the need of welding mode experimental verification, but it reduces the number of preliminary experiments up to one.

It is obvious that the developed method can be valid only for welding with deep penetration, when heat transfer along beam axis can be neglected. The disadvantages of this method also include the fact that characteristic widening of weld top, which in some cases can reach considerable sizes, is not taken into account. In addition, welding of metals with high thermal conductivity, such as aluminium and copper alloys, is characterized by formation of a weld with smooth increase of weld width from its root to its top part. For such alloys, the accuracy of welding modes determination by the developed method can be significantly lower.

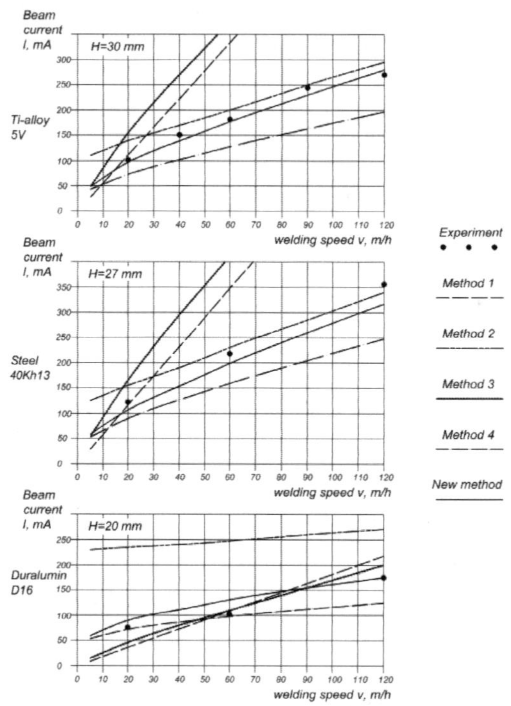

Figure 3. Comparative results of experimental and calculated data, obtained using the developed method and existing methods.

5. Conclusions

Using the developed calculation method, it is possible to obtain required electron beam power during EBW of various materials (including complex alloys), depending on welding speed, diameter of vapour-gas channel and thermophysical properties and taking into account energy costs on metal melting and its partial evaporation.

It is established that in order to maintain a constant penetration depth, it is necessary to change electron beam power almost linear to welding speed.

Acknowledgments

The research was carried out in National Research University "Moscow Power Engineering Institute" with the financial support of the Russian Science Foundation (project No. 18-79-00335).

References

[1] Zuev I V 1998 *Materials Processing by Concentrated Energy Flows* (Moscow: MPEI) p 162
[2] Hashimoto T and Matsuda E 1964 *J. Jap. Weld. Soc.* **33(9)** 38-46
[3] Lopatko A A, Kartashov G V, Tkachev L G *et al.* 1977 *5th All-Union Conf. on Electron Beam Welding* (Kiev: Naukova Dumka) p 16
[4] Arutyunyan R V, Bolshov L A, Vasilyev A D, Malyuta D D and Sebrant A Yu 1989 *Effect of Laser Radiation on Materials* (Moscow: Science) p 367
[5] Terentyev E V, Dragunov V K, Sliva A P and Scherbakov A V 2015 *Welding International* **29(2)** 150–4
[6] Dragunov V K, Terentyev E V, Sliva A P, Goncharov A L and Marchenkov A Yu 2017 *Welding International* **31(4)** 307–11
[7] Shcherbakov A V, Rodyakina R V, Kozhechenko A S, Gaponova D A, Goncharov A L, Dragunov V K 2017 *Techn. Phys. Let.* **43(11)** 958–60.

Wear resistance and characteristics of the friction surface of the coating metal with carbide-boride-nitride-intermetallic alloying

E N Eremin[1,a], A S Losev[1,b], S A Borodikhin[1,c], I A Ponomarev[1,d]

[1] Department of Engineering and Materials, Omsk State Technical University, Omsk, Russia

E-mail: [a]weld_techn@mail.ru, [b]alesha-los@yandex.ru, [c]frozze8@mail.ru, [d]don_13@bk.ru

Abstract. The metal deposited by flux-cored wire containing 15% chromium, 0.5% boron carbide, 0.5% boron nitride, 2.5% titanium diboride and 1.0% zirconium diboride was investigated. The hardness of such a metal reaches a maximum value of 58 HRC. The microhardness of the structural objects is 521-593 HV for the matrix, 829-978 HV for eutectics and 1262-1342 HV for reinforcing phases. It was established that the average value of the relative mass wear per test was 0.0034 g/m, and linear wear was 0.00862 mm/m, which is 2.2 times, respectively, and 2.7 times less than that of the coating metal without borides. The average value of the friction moment for the entire friction path of 452.16 m was 19.075 N·m. The coefficient of friction was 0.364. It is shown that the metal structure is an iron-chromium martensitic matrix with a eutectic component formed on the basis of boride $(Fe, Cr)_2B$, having a frame structure, dispersed inclusions of $(Fe, Cr)_7(C, B)_3$ carboronides and high strength nitrides ε –$(Fe, Cr)_{2-3}N$. High-chromium flux-cored wire alloyed by a complex of boron and nitrogen compounds provides a composite-type metal with high wear resistance and can be used for hardening the surfacing of parts operating under conditions of significant frictional loading.

1. Introduction

One of the rapidly developing areas in the field of hardening high-wear parts and equipment components is the surfacing of powder wire coatings. The longevity of a large group of equipment is determined by the resistance to corrosion. There is a great potential for creating corrosion coatings by surfacing with high-chromium cored wires. At the same time, most of these coatings have high corrosion resistance, but insufficient wear resistance under frictional loading conditions.

One of such effective methods of metal hardening is alloying it by boron compounds such as ferrobor, boron carbide, chromium diboride, titanium diboride [1-6]. Previously carried out studies of the authors established the positive effect of titanium and zirconium diborides on the wear resistance of nickel-chromium martensite-aging steels deposited by flux-cored wires [7, 8]. Nitrogen has found quite wide application for alloying corrosion-resistant steels [9-13]. At the same time, during the deposition, nitrogen is introduced into the metal through the use of nitrated chromium or ferrochrome in the cored wires. However, for these purposes it is possible to use boron nitride, which is due to the similarity of a number of properties as an electronic analogue of carbon [14]. However, boron nitride is extremely rarely used in cored wires.

The authors showed the promise of using filler cored wire alloyed by the B_4C-BN-TiB_2-ZrB_2 complex [15, 16].

At the same time, the current state of the problem of wear requires knowledge of the basic laws of hardening and destruction of the surface layers of the metal during its frictional loading.

Based on this, the purpose of this work is to study the wear resistance and characteristics of the friction surface of a metal deposited by chromium cored wire with a complex of boron and nitrogen compounds.

2. Objects and research methods

Studies were conducted on the effect of boride-nitride alloying on the wear resistance, structure, tribological and durometric properties of chromium steel, obtained by surfacing with flux-cored wire with the composition: 15% Cr; 0.5% B_4C; 0.5% BN; 2.5% TiB_2; 1.0% ZrB_2. For comparison, steel without borides was investigated. The surfacing was carried out on plates made of St3 steel 200×50×10 mm in size by experienced flux-cored wires with a diameter of 2.4 mm in argon in three layers.

Tests under conditions of adhesive wear of metal-to-metal pairs were carried out under sliding friction without lubrication according to the finger-rotating disk (counterbody) scheme, using a UMT-2168 friction machine. The ring surfacing was made on the disk by an OK Weartrode 55 HD (70Kh10GS) coated electrode, providing the hardness of the metal coating 56-59 HRC. After grinding, the deposited coating contacting with samples in the form of a finger, the butt-end of which was surfacing by the studied composition. The test was carried out at a load F=0.75 kN and a counterbody disk rotation speed of 100 rpm.

Electron microscopic studies were performed on a JEOL JSM-6610-LV raster electron microscope with an Inca-350 attachment for energy dispersive analysis.

3. Results of the experiment and discussion

The test results on the adhesive wear of the metal coatings without borides and alloyed with boride compounds are given in Table 1. The main tribological properties of the deposited metal of the investigated coatings are given in Table 2.

Table 1. Mass wear of metal samples (g).

Number of revolutions	Coating without borides				
	Sample 1	Sample 2	Sample 3	Average value	Relative wear, [g/m]
0-300	0.965	1.1382	0.9861	**1.0297**	0.0091
300-600	0.6443	0.7108	0.6670	**0.6740**	0.0060
0-600	1.6093	1.849	1.6531	**1.7038**	**0.0075**
	Coating with borides				
0-300	0.4292	0.3751	0.3429	**0.3824**	0.0034
300-600	0.1714	0.3587	0.3722	**0.3008**	0.0027
600-900	0.4173	0.4138	0.5101	**0.4471**	0.0040
900-1200	0.4715	0.3179	0.415	**0.4015**	0.0036
0-1200	1.4894	1.4655	1.6402	**1.5317**	**0.0034**

The metal coating without borides withstood only 600 revolutions with a friction path of 226 m. The average mass wear of samples after 300 revolutions is 1.029 g. After 600 revolutions, the mass wear amounted to 1.70 g, while the wear along the length reached 5.33 mm, the value of which is comparable to the length of the surfacing part of the finger. Therefore, the samples were removed from further testing.

The average value of the relative wear of coating's metal without borides for the test was 0.0075 g/m. Mass wear for the friction path in 113 m was 0.674 g. The average linear wear for the test was 0.0235 mm/m. As a result of tests, the friction torque decreased from 23.709 N·m (for a friction path of 113 m) to 19.601 N·m (for a friction path of 226 m). The average value of the friction torque for the friction path of 226 m was 21.634 N·m. The friction coefficient decreased from 0.504 (for the friction path of 113 m) to 0.423 (for the friction path of 226 m), and the average value for the test was 0.464.

Table 2. Tribological properties of metal

Number of revolutions	Coating without borides	
	The average value of the friction torque, [N•m]	Friction coefficient
0-300	23.709	0.504
300-600	19.601	0.423
0-600	**21.634**	**0.464**
	Coating with borides	
0-300	19.026	0.390
300-600	15.472	0.354
600-900	20.880	0.363
900-1200	20.910	0.348
0-1200	**19.075**	**0.364**

The boride coating metal withstood 1200 revolutions with a friction path of 452.16 m, and then the samples were removed from further tests. The average mass wear of samples after 300 revolutions was 0.382 g, after 600 revolutions it was 0.683 g, and after 900 revolutions it reached 1.13 g. After 1200 revolutions, wear was 1.531 g. Relative wear slightly increases from 0.0034 g/m at the beginning of the journey friction to 0.0036 g/m at the end. The average value of the relative mass wear per test was 0.0034 g/m, which is 2.2 times less than that of the coating metal without borides. At the same time, wear in length reached 3.9 mm. The average value of linear wear was 0.00862 mm/m, which is 2.7 times less than that of the coating metal without borides. At the same time, the friction torque increased from 19.026 N·m (for a friction path of 113.04 m) to 20.910 N·m (for a friction path of 452.16 m). The average value of the friction torque for the friction path of 452.16 m was 19.075 N·m. The friction coefficient fell from 0.390 (for a friction path 113.04 m) to 0.348 (for a friction path 452.16 m), and amounted to an average for wear tests of 0.364.

In order to compare the results of tribological tests with the microstructure of the surface layers under friction, electron microscopic studies of the topography of the specimens at characteristic loading sites were carried out (Fig. 1).

There is a high degree of fragmentation of the coating metal without borides due to the formation of slip bands (Fig. 1, a).

The microstructure of the formation of these slip bands is characterized by a regular arrangement of dipole dislocation clusters. The loading turns these bands into localized deformation foci in the form of mezholos. This process acquires an avalanche-like character and leads to selective emissions of destruction products from the frictional contact zone.

The local nature of the fracture with a chip along the boundaries of the substructure elements manifests itself. The destruction of the frictional contact zone is characterized as a global loss of shear stability of the surface layer after nucleation and movement of the mezholos by localized plastic deformation [17], which causes a significant wear of such material already on small friction paths.

IOP Conf. Series: Materials Science and Engineering **681** (2019) 012012 doi:10.1088/1757-899X/681/1/012012

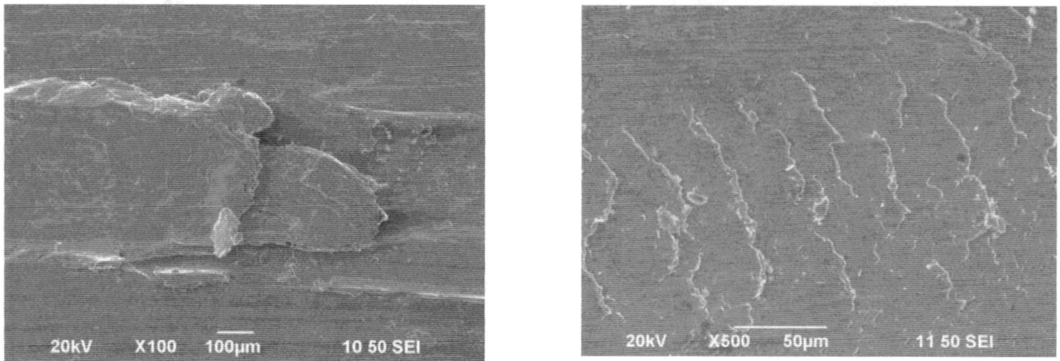

a) б)

Figure 1. General view of the surface of wear of the deposited coatings:
a) without borides; b) with borides

Microstructural studies of the coating metal with borides after frictional loading reveal the formation of a block structure that is quasi-evenly distributed over the volume of the surface layer (Fig. 1, b). In the surface layer, together with hardening, some plasticity of the material is realized due to the high length of the block boundaries and the boundary slippage. Simultaneous grinding of the block microstructure leads to an increase the limit of macroscopic elasticity and resistance to fatigue failure by reducing grain-boundary cracking [18]. There is a strong tendency to textured dislocation clusters in one direction and the development of microstrip localized plastic deformation.

In the microstructure, the presence of small microcracks is noted, which are combined into a main crack by destroying the material separating them, thus forming material wear lobes.

To identify the causes of differences in the hardness of coatings deposited by flux-cored wires of the studied compositions was performed a complex of metallographic and durometric studies.

The microstructure of the metal deposited by flux-cored wire without borides is a mixture of low carbon martensite with highly dispersed perlite. Allocation of strengthening phases is observed along the grain boundaries, and δ-ferrite inside the grains (Fig. 2, a). Such a mixed structure provides the hardness of the metal up to 40 HRC.

The deposited metal of the boride coating after surfacing has a complex composition structure with a martensitic matrix, a large number of eutectics and particles of reinforcing phases (Fig. 2, b). The hardness of such a metal reaches a maximum value of 58 HRC.

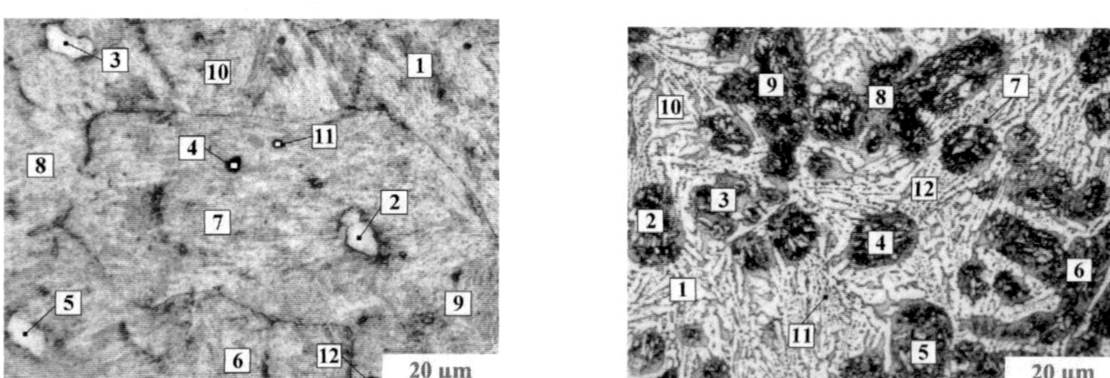

a) b)

Figure 2. The image of the microstructure with the distribution of locations for measuring the microhardness of the metal components of the coatings deposited by flux-cored wires:
a) without borides; b) with borides

Durometric studies showed significant differences in the microhardness of the structural components of the metal of the investigated coatings (Table 3).

Table 3. Microhardness $HV_{0.01}$* and $HV_{0.05}$ structural components of the metal of the coating deposited by flux-cored wires

Puncture no.	1	2*	3*	4*	5*	6	7	8	9	10	11*	12*
without borides	408	376	358	551	384	424	393	429	389	411	609	572
№ puncture	1	2	3	4	5	6	7*	8	9	10	11*	12
with borides	978	587	540	552	575	546	1342	521	593	874	1262	829

As can be seen, the microhardness of the metal matrix without borides varies from 390 to 430 HV, δ-ferrite from 358 to 384 HV, and reinforcing phases 550-610 HV.

The microhardness of the structural components of the metal deposited by flux-cored wire with borides is for the matrix 521-593 HV, the eutectic 829-978 HV and the reinforcing phases 1262-1342 HV.

To identify the mechanism of wear of the deposited metal of the coating with borides, electron microscopic studies were carried out.

The results of a scanning electron microscopic analysis of a metal with borides, the structure of which is shown in Fig. 3 are summarized in Table 4. Energy dispersive analysis (EDA) showed that the metal structure is an iron-chromium martensitic matrix with a eutectic component formed on the basis of boride $(Fe, Cr)_2B$, having a frame structure, dispersed inclusions of carborides type $(Fe, Cr)_7(C, B)_3$ and high-strength nitrides ε –$(Fe, Cr)_{2\text{-}3}N$.

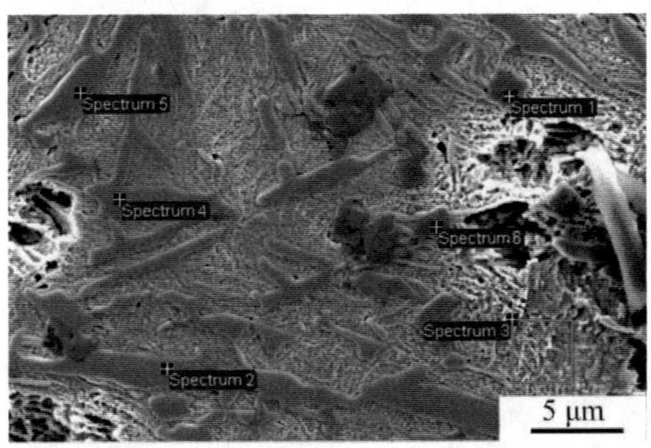

Figure 3. Research by the method of scanning electron microscopy of the structure metal of the coating deposited by flux cored wire

Table 4. The chemical composition of the areas of the coating metal, deposited by cored wire with borides

Spectrum	B [%]	N [%]	C [%]	Cr [%]	Fe [%]	Ti [%]	Zr [%]
1	1.95	23.37	7.59	11.40	19.13	29.47	7.09
2	24.32	0	5.32	24.92	45.22	0.17	0.05
3	8.78	0.66	10.52	17.39	62.55	0.07	0.04
4	5.65	1.37	7.77	15.08	70.03	0.06	0.04
5	7.73	0.27	6.32	15.50	70.09	0.06	0.02
6	2.96	20.75	9.16	16.48	10.45	31.16	9.04

The study of the distribution maps of the main elements showed that the composite structure of the coating is characterized by the presence of a significant amount of dispersed inclusions by the majority of titanium nitrides and carbides and a smaller part by zirconium nitrides ranging in size from 0.2 to 3.4 microns.

The results indicate that the mechanism of wear of the metal of the coating with borides is associated with its increased hardness caused not only by dispersion hardening by particles of boride compounds, but also by the presence of boride eutectics, which are effective obstacles to dislocation slip under conditions of plastic deformation of the surface during wear [18].

The higher wear resistance of the metal with borides can be explained by the combined effect of the framework of solid eutectic carboborides and boride-nitride- and intermetallic particles, which, under abrasive wear conditions, effectively protect the coating matrix. As a result, weaker traces of abrasive, adhesive and deep fracture, as well as a very slight crushing of the material, are characteristic of the worn surface of such a metal.

4. Conclusion

The high wear resistance of metal deposited by cored wire PPKh15 + 0.5%B_4C + 0.5%BN + 2.5%TiB_2 + 1.0%ZrB_2 is explained by the presence of solid eutectic carboborides and particles of the carbide-boride-nitride-intermetallic compounds in its structure, which, under abrasive wear conditions, receive a part of contact interaction, which increases the resistance to tearing of a metal operating under abrasion conditions. Apparently, this is due to the fact that the dispersed high-strength phases are effective obstacles to the dislocation slip in the surface layers of the metal during its frictional interaction.

Thus, high-chromium cored wire alloyed by a complex of compounds of boron and nitrogen, provides a composite-type metal with high wear resistance and can be used for hardening the surfacing of parts operating under conditions of significant frictional loading.

Acknowledgments
This research was made possible through the Russian Science Foundation grant (project No. 17-19-01224).

References

[1] I. I. Iskol'dskij, *Surfacing boride hard alloys*, *Mashinostroenie*, Moscow, 1965
[2] I. N. Sheenko, et al., *Welding production*, 5, 27-28 (1969)
[3] N. P. Ljakishev, Ju. L. Pliner, S. I. Lappo, *Boron steels and alloys*, Metallurgija, Moscow, 1986
[4] V. Raghavan, *Journal of Phase Equlibria*, 24, 459-460 (2003)
[5] L. Zhong, et al., *Journal of Iron and Steel Research International*, 16, 37-42 (2009)
[6] A. A. Artem'ev, et al., *Proceedings of higher educational institutions. Powder Metallurgy and Functional Coatings*, 2, 44-48 (2011)
[7] E. N. Eremin, *Welding International*, 27, 144-146 (2013)
[8] Ye. N. Yeremin, et al., *Welding International*, 28, 465-468(2014)
[9] V. G. Gavriljuk, *ISIJ Inter*, 36, 840-845 (1996)
[10] M.O. Speidel, *Materialwissenschaft und Werkstofftechnik*, 37, 875-880 (2006)
[11] M. M. L. Saucedo, et al., *Materials Science Forum*, 561-565, 2275-2278 (2007)
[12] M.V. Kostina, et al., *Metal technology*, 10, 2-12 (2000)
[13] A.P. Shljamnev, *Problems of ferrous metallurgy and materials science*, 1, 53-60 (2007)
[14] A. V. Kurdjumov, A. N. *Piljankevich, Phase transformations in carbon and boron nitride, Naukova Dumka*, Kiev, 1979
[15] E. N. Eremin, et al., *Oil and Gas Engineering* (OGE-2017) AIP Conference Proceeding, 1876, 020071-1–020071-6 (2017)
[16] E. N. Eremin, et al., *Welding production*, 7, 3-8 (2018)
[17] V.E. Panin, et al., *Wear*, 249, 860-867 (2002)
[18] V. G. Pinchuk, et al., *Global Journal For Research Analysis*, 4, 255-257 (2015)

Structure and mechanical properties of 09Mn2Si steel surfaced layers and welds modified by adding titanium carbonitride nanoparticles

N K Galchenko[1], V P Samartsev[1], K A Kolesnikova[1], I V Vlasov[1], S V Panin[1,2], A V Yakovlev[2]

[1]Institute of Strength Physics and Materials Science SB RAS, 2/4, pr. Akademicheskii, Tomsk, 634055, Russia
[2]National Research Tomsk Polytechnic University, 30, Lenin Avenue, Tomsk, 634050, Russia

E-mail: viv@ispms.tsc.ru

Abstract. The study on structure and mechanical properties of surfaced coatings and welded joints of 09Mn2Si steel formed with the use of a standard electrode as well as one modified by adding titanium carbonitride nanoparticles was carried out. Optical and scanning electron microscopy, microhardness measurement were employed to investigate the structure at welding with a standard electrode, as well as to illustrate changes resulting from the electrode modification through adding titanium carbonitride nanoparticles. Evaluation of mechanical properties was performed by conducting tests on static, cyclic and impact loading. It is shown that welding with the use of the modified electrode gives rise to decreasing the size of ferrite grains and increasing impact toughness and fatigue life.

1. Introduction

Welding is one of the most efficient and widespread technologies for producing permanent joints [1]. Parts connected in this way are characterized by high strength, while the welding process itself is not costly. However, the weld itself represents a pronounced macroheterogeneity at the scale of entire structure (component) [2] comprising of a zone of increased brittleness (weld material) and a heat-affected zone (HAZ) located next to the weld and characterized by lower strength [3, 4]. The reduction of the mechanical properties of the weld as compared with the bulk material is observed under impact loading [5] and testing under negative temperatures [6]. To strengthen the welded joint some alloying elements are added into the electrodes which substantially modify the weld material [7]. In this concern the most relevant are nickel and molybdenum which contribute to both austenite stabilizing (in the first case) and the formation of molybdenum carbides during the secondary crystallization of the melt which can concentrate at $\gamma \rightarrow \alpha$-transformation boundary. The latter contributes to the refinement of ferrite grains.

However, increasing of strength and toughness may also be achieved due to synergistic effect of several structure–dependent factors. So, another promising way to improve properties of a welded joint might be adding of titanium carbonitride nanoparticles to the melting bath. This should reduce the size of the primary austenite grains and shift the transformation region into a zone of lower temperatures. In so doing, the reason of the latter is appearance of additional crystallization centers or supercooling zones.

Thus, the aim of the paper is studying structure and mechanical properties of surfaced coatings and welded joints formed with the use of electrodes modified through adding of nickel, molybdenum as well as titanium carbonitride nanoparticles. It is expected that this modification will increase mechanical properties of welded joints and protective coatings of low-carbon ductile steel operated at low temperature conditions as well. The low–alloyed structural steel 09Mn2Si used for manufacturing hot-rolled pipes and welded joints was the object of the studies.

2. Fabrication of experimental electrodes

Electrodes for manual arc welding/surfacing were made of a wire grade SV-08 with the diameter of 3 mm (for a root weld) and 4 mm (for coating surfacing) with the covering obtained on the basis of the charge from "OZS-12" electrodes (GOST-2246-60) as well as "MP-3" (GOST 9466-75, GOST 9467-75) whose chemical composition is presented in Table 1.

Self-propagating High-temperature Synthesis (SHS) composite powder of Fe-$TiC_{0.5}N_{0.5}$ system was loaded into the electrode covering at the production stage as an additional component with the content of 0.15 wt% $TiC_{0.5}N_{0.5}$ (for the electrode "MP-3") and 0.25 wt% $TiC_{0.5}N_{0.5}$ (for the electrode "OZS-12"). The content of $TIC_{0.5}N_{0.5}$ was selected on the basis of preliminary tests on impact toughness. Welding and surfacing of sheet samples were performed by manual arc method.

Table 1. Chemical composition of welding electrodes "OZS-12" and "MP-3"

Grade	Fraction of elements (wt%)				
	C	Si	Mn	S	P
"OZS-12"	0.09	0.15	0.6	0.017	0.026
"MP-3"	0.1	0.2	0.5–0.8	0.04	0.045

Composite powders ($TiC_{0.5}N_{0.5}$ – Fe) with the component ratio of 1/1 were prepared by the method of self-propagating high-temperature synthesis (SHS) in a flow reactor. The initial particle size of the composite mixture was less than 2 mm. Grinding was carried out in a continuous mode in a vibromill in argon atmosphere. The size of complex-content particles of $TiC_{0.5}N_{0.5}$–Fe composite after milling was 250–500 nm; the particle size of titanium carbonitride ($TiC_{0.5}N_{0.5}$) in the fused iron matrix was 80–120 nm. The phase composition of the SHS powder is presented in Table 2. The adding of nickel and molybdenum to the electrode covering was made in the following ratio: Ni (3 wt%), Mo (0.6 wt%).

Table 2. Phase composition of the synthesized composite powder $TiC_{0.5}N_{0.5}$-Fe

Phases	$TiC_{0.5}N_{0.5}$	Ti_2CN	Fe_3C	α-Fe
Content (vol.%)	71	9.25	5.72	13.46
Lattice parameter (nm)	a=0.4277	a=0.30367, c=1.49905	a = 0.50525, c =0.45429	a = 0.28696

For fabricating welded samples some plates with the size of 65×300 mm and the thickness of 16 mm with X and V - shaped groove edges were used. The angle between cutting edges was 35°, the blunting of edges made 1.8 ± 0.8 mm. Welding was conducted at the current – 200-220 A of a constant reverse polarity.

The strength properties of welded joints and surfaced coatings were evaluated under static stretching with the help of electromechanical testing machine "Instron 8801". Macro- and microstructure were analyzed with the used of optical microscope AXIOVERT-200MAT (Zeiss, Germany), microhardness tester PMT-3 (according to the RF State standard referred to as GOST 9450-76) with the load of 100 g. Optical microscope Neophot-21 (Zeiss, Germany) was used for metallographic analysis.

The fatigue tests were carried out with the specimens of 09Mn2Si steel cut out from surfaced layers with the thickness of 6 mm produced by "MP-3" and "OZS-12" electrodes. These layers were then subjected to symmetrical cyclic bending loading at the frequency of 20 Hz at a constant load of 490 MPa with the help of Biss Nano servo-hydraulic testing machine. Impact toughness tests were performed with the use of instrumented pendulum Amsler RKP450 (according to GOST 10708-82) and a cryogenic chamber (produced by Lauda). Specimens for impact bending tests were fabricated in accordance with GOST 6996-66. The specimen design is shown in Figure 1.

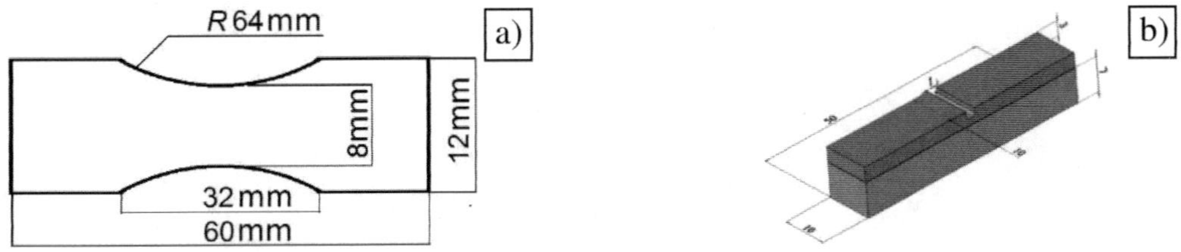

Figure 1. Schematic of specimens for fatigue (a) and impact bending tests with a surfaced coating (b)

3. Testing results

3.1. Coating surfacing

3.1.1. Structural studies. Metallographic analysis of all surfaced coatings deposited by the electrodes with the basic composition I) MP-3 and II) OZS-12 as well as experimental ones (with additives of titanium carbonitride nanoparticles) possess a ferrite-pearlite structure (Fig. 2). In contrast with the structures modified by adding titanium carbonitrides the basic composition is characterized by formation of numerous regions of Widmanstatten ferrite (Fig. 2a) as well as colonies oriented towards the direction of heat sink where dendrite structures represents axes of the first and second orders.

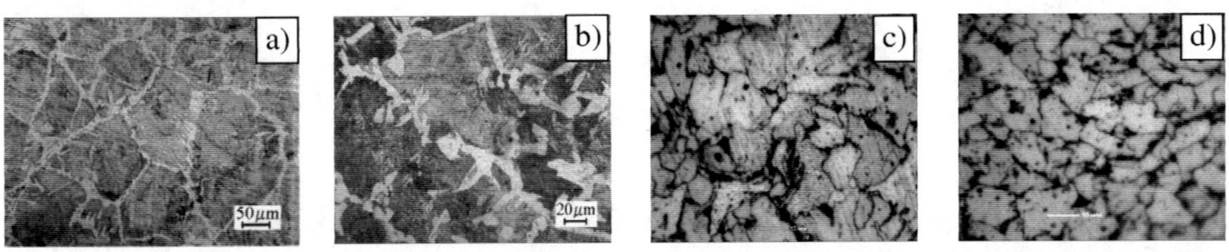

Figure 2. Optical images of the ferrite-pearlite coating structure obtained by surfacing with a standard electrode (MP-3) (a), (OZS-12) (c) and experimental ones: "MP-3 + 0.15 wt% TiCN" (b) and "OZS-12 + 0.25 wt% TiCN" (d)

The presence of Widmanstatten ferrite at the grain boundaries indicates on the overheating of the fusion zone during electric arc surfacing with the use of standard "MP-3" and "OZS-12" electrodes. Grain size of the coatings surfaced by the standard electrode "MP-3" makes $d \sim$ 100-200 μm (Fig. 2c). Surfacing with the electrode "MP-3 + 0.15 wt% TiCN" gave rise to refinement of the ferrite-pearlite structure down to a grain size of $d \sim$ 20-100 μm (Fig. 2d). The average grain size also decreased almost twice at coating surfacing with the modified electrodes "OZS-12 + titanium carbonitrides": from $d \sim$ 5–25 μm down to $d \sim$ 3–10 μm. In both cases, along with the refinement of the macrograins the columnar pattern of the grains as well as variation of grain size in the surfaced coating were also eliminated.

3.1.2. Mechanical properties. The microhardness of the surfaced coatings was evaluated for the use of the "MP-3" electrodes. It is shown that adding of the nanoparticles did not give rise to significant change of their values. The mechanical properties of the deposited coatings were estimated over the testing results on impact bending, static tension (Table 3) and cyclic bending (Table 4). The tests were carried out at room temperature. The structural changes occurred in the coatings modified with nanosized particles of titanium carbonitride did not give rise to any significant increase of their basic strength properties; however, the impact toughness (*KCV*) has increased by ~ 2.5 times (Table 3).

Table 3. Mechanical properties of surfaced coating specimens applied by the electrodes "MP-3" and "MP-3 + 0.15 wt% TiCN" at T = 20ºC.

Electrode composition	KCV (J/cm^2)	$\sigma_в$ (MPa)	$\sigma_{0,2}$ (MPa)	δ (%)
"MP-3"	42	679	375	11
"MP-3 + 0.15 wt% $TiC_{0.5}N_{0.5}$"	104	710	425	13

Table 4. Fatigue life of 09Mn2Si steel welds

№	Type of welding electrode	Load / frequency (MPa/Hz)	Cycles
1	"OZS-12"	490/20	76445
2	"OZS + 0.25 wt% TiCN"	490/20	106994
3	"MP-3"	490/20	186917
4	"MP-3 + 0.15 wt% TiCN"	490/20	519564

More than twofold increase of the impact toughness for the coated specimens formed with the use of "MP-3 + titanium carbonitride nanoparticles" electrode can be explained by I) structure refinement, II) significant decrease in the number and size of non-metallic inclusions and III) increase of the ductile structural component in the coatings by 70 % of. In addition, the increase of the impact toughness was stimulated by more dense and uniform grain structure of the crystallized metal.

The fatigue test results have shown that coating structure modification by nanosized titanium carbonitride particles introduced through covering of MP-3 and OZS-12 electrodes helps to slow down the accumulation of fatigue damage and increase cyclic durability: by 40% for specimens formed with the help of "OZS-12 + 0.25 wt% TiCN" electrode; by 278 % for the specimens formed with the use of "MP-3 + 0.15 wt% TiCN" one.

3.2. Weld

3.2.1. Structural studies. Additional alloying of the welding pool through the "OZS-12" electrode covering by nickel (3 wt%) and molybdenum (0.6 wt%) contributed to the enlargement of ferritic grains in comparison with the structure shown in Figure 3a.

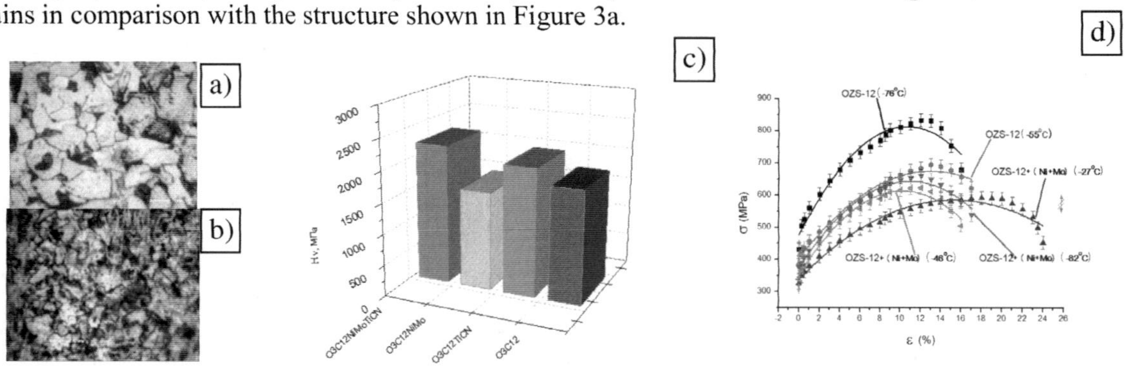

Figure 3. Microstructure of welds produced by "OZS-12 + (3 wt% Ni + 0.6 wt% Mo)" (a) and "OZS-12 + (3 wt% Ni + 0.6 wt% Mo) + 0.25 wt% TiCN" (b) electrodes; microhardness values of welds depending on the electrode composition (c); loading diagrams of welded joints at various testing temperatures (d)

The adding of nano-dispersed refractory titanium carbonitrides into the weld metal made it possible to significantly reduce the size of the primary austenitic grain from 30 μm down to 1 μm (Fig. 3b). Also, the transformation region was shifted towards low temperature zone that contributes to the formation of needle-shaped ferrite (up to 30%).

The complex alloying of the weld with nickel, molybdenum and TiCN nanoparticles (Fig. 3b) proved to be the most effective from the stand point of enhancement dispersion of structural components. Obviously, this is due to additional formation of crystallization centers in the molten bath as well as initiation of needle-shaped ferrites on them.

Thus, the analysis of the microstructures formed in the surfaced coating metal and weld metal has shown that depending on the alloying and modifying system with refractory nanosized particles the morphology and dimensions of the structural elements change. Also, additional amount of the large number of fine particles is stimulated which contributes to the formation of needle-shaped ferrite and bainite component improving the performance of the welded joint.

3.2.2. Mechanical properties. Comparative studies of the microhardness of welded joints were carried out for specimens prepared with the use of welding electrodes of 4 compositions (Fig. 3c):

1. standard electrode "OZS-12";
2. experimental – "OZS -12 + 0.25 wt% TiCN";
3. experimental – "OZS -12 + Ni (3 wt%) and Mo (0.6 wt%)".
4. experimental – "OZS -12 + (3 wt% Ni + 0.6 wt% Mo) + TiCN nanoparticles (0.25 wt%)".

The results obtained have shown at the use of "OZS-12 + titanium carbonitrides" electrodes regardless of the alloying system employed the weld structures with higher microhardness is formed. Welded joints were tested at static tension for specimens formed with the use of experimental electrodes "OZS-12 + Ni and Mo". Figure 3d presents their loading diagrams for testing under various temperatures. It is seen that with decreasing the temperature the values of the ultimate strength σ_U increases with a slight decrease of the relative elongation at failure. The values of the ultimate strength corresponding to the test temperatures of welds are following:

1. OZS -12 (σ_U) max (at T=-55° C) = 675 MPa, δ = 16%
2. OZS -12 (σ_U) max (at T=-76° C) = 800 MPa, δ = 18 %
3. "OZS -12 + (3 wt% Ni+ 0.6 wt% Mo)" (at T=-27 °C) (σ_U) max= 575 MPa, δ = 22 %
4. "OZS -12 + (3 wt% Ni+ 0.6 wt% Mo)" (at T=-46 °C) (σ_U) max= 600 MPa δ = 17 %
5. "OZS -12 + (3 wt% Ni+ 0.6 wt% Mo)" (at T=-82 °C) (σ_U) max= 650 MPa δ = 18 %

The condition for a brittle crack initiation is running of the preceding microplastic deformation process that gives rise to a local stress concentration and further cracking. The ability of a metal to develop microdeformation predetermines running of local stress relaxation processes and, therefore, the possibility of preventing brittle fracture.

One of the important features of the materials for structural applications including formation of welded joints is ability to operate at low temperatures. In so doing, crack propagation resistance under dynamic bending conditions is of key relevance.

Table 5. Toughness of welded joints produced by electrodes the "OZS -12 + (Mo+Ni)" and "OZS -12 + (Mo+Ni) + 0.25 wt% TiCN"

Electrode composition	T	Impact toughness (J/cm^2)
"OZS-12 + (Mo+Ni)"	+20 °C	82
"OZS -12 +(Mo+Ni) + 0.25 wt%TiCN"	+20 °C	92
"OZS-12 + (Mo+Ni)"	-20 °C	51
"OZS -12 +(Mo+Ni) + 0.25 wt% TiCN"	-20 °C	53
"OZS-12 + (Mo+Ni)"	-40 °C	23
"OZS -12 +(Mo+Ni) + 0.25 wt% TiCN"	-40 °C	32
"OZS-12 + (Mo+Ni)"	-60 °C	9
"OZS -12 +(Mo+Ni) + 0.25 wt% TiCN"	-60 °C	17

The obtained results have shown that adding of the titanium carbonitride nanoparticles is manifested only when specimens were tested at lowest temperatures (-40 and -70 ºC). Thus, the fracture resistance under dynamic loading of the weld metal with titanium carbonitrides exceeded by 24 % ones of the welds formed by a standard "OZS-12" electrode (at testing under similar conditions).

4. Conclusion

Adding of 0.25 wt% titanium carbonitride in the welding pool for the "OZS-12" electrode surfaced coating results in decreasing average grain size and increasing impact toughness by 24 % at T=-40 °C and -70 °C. Increased impact toughness is primarily associated with more effective arresting of the main crack development at low temperatures.

Complex alloying with nickel (3 wt%), molybdenum (0.6 wt%) and titanium carbonitride nanoparticles (0.25 wt%) turned out to be the most efficient for enhancement dispersion of structural components and microhardness of the welds.

The surfaced coatings deposited with "MP-3 + 0.15 wt% TiC_xN_y" electrode undergoes certain changes which are expressed through increasing uniformity, significant structure refinement and substantial increasing impact toughness from 42 J/cm^2 to 104.5 J/ cm^2. In doing so, during the crystallization process of the coating deposited by the experimental electrode "MP-3 + 0.15 wt% TiCN" the more uniform structures with a smooth transition zone were formed that possesses more dense arrangement throughout the entire cross section of the coating.

The titanium carbonitride nanoparticles presented in the composition of the electrode covering exert a noticeable effect on the crystallization processes in the molten bath. They primarily act as additional crystallization centers and contribute to the accelerated and more uniform solidification of the surfaced coating.

Acknowledgments

The study was performed in the framework of fundamental research Program of the Russian State Academies of Sciences for 2013–2020, line of research III.23 as well as a partial support of RFBR Grants № 18-38-00679 and 18-08-00516. Financial support of RF President grants i) Scholarship SP-2456.2019.1 and ii) RF President Council Grant for the support of leading research schools NSh-5875.2018.8 is also acknowledged.

References

[1] Cheng Z, Huang JH, Ye Z, Liu H, Yang J, Chen SH, Zhao XK *J. Manufac. Proc.* **36** 115-125
[2] Ramkumar KD, Prabu SS, Arivazhagan N *Materials Research Express* **6(8)** 086540
[3] Ramasamy N, Jeyasimman D, Kathiravan R, Raju N *Transactions of the Indian Institute of Metals* **72(5)** 1361-1369
[4] Deng D and Murakawa H *Computational Materials Science* **43(2)** 353-365
[5] An G, Woo W, Park J *Journal Of Nanoscience And Nanotechnology* **19(4)** 2323-2328
[6] Panin S, Byakov A, Vlasov I, Maruschak P, Berto F, Vinogradov A *Engineering Fracture Mechanics* **210(1)** 288-299.
[7] Kolooshani SH, Atrian A, Mohseni E *Materials Research Express* 5(11) 116511.

Structure and strength research of polymer pipe welded joints

A I Gerasimov[1], **E V Danzanova**[1] **and G V Botvin**[1]

[1] Institute of Oil and Gas Problems, the Siberian Branch of the Russian Academy of Science, Federal Research Center 'Yakut Scientific Center of the Siberian Branch of the Russian Academy of Sciences', Yakutsk, Russia, 677007

E-mail: dhv4071@mail.ru

Abstract. Research results of the effect of low ambient air temperatures during welding on structure and strength of polymer pipes welded joints are given. The connection of polymer pipes was made in two main ways: by hot element butt welding and socket welding. Structure studies of the material of welded joints have shown that the welding technologies developed at IOGP SB RAS provide formation of the same structure in the material of welded joints as in standard welding.

1. Introduction

Polymer pipes have numerous undeniable advantages over steel pipes, such as low weight and flexibility, simplicity and low cost of transportation and installation, absence of insulation, high durability, etc. However, during long-term operation, the weakest point is welded joints. Polymer pipes are usually connected by hot element welding, when welded surfaces are heated by their physical contact with heated tool. In this paper, two types of welding with heated tool are considered: butt welding of polyethylene pipes and hot element welding of polypropylene pipes. As the result of thermal deformation cycle of welding, different structure is formed in the material of the welded joint. The structure, in turn, affects strength of the welded joint. More detailed study of the formed structure will help establish the dependence of the strength of welded joints on the size and characteristics of structural formations. To control the quality of welded joints, various non-destructive and destructive test methods are used. Analysis of methods for quality control of welded joints of polymer pipes showed that test procedures existing in current regulatory documents are not sufficiently informative and do not give quantitative assessment of strength [1, 2].

The purpose of this work is to study the structure and strength of the material of welded joints obtained under different welding conditions. Objects of study were butt-welded joints of polyethylene pipes for gas pipelines and hot element welded joints of polypropylene pipes for water pipes. Welding of polyethylene and polypropylene pipes was carried out at various ambient temperatures, including low ones (up to minus 50 °C). There is permissible temperature range to produce welding polymer pipes. Thus, according to the current regulatory documents, polyethylene pipes for gas pipelines can be welded at temperatures from minus 15 °C to plus 45 °C, and polypropylene pipes only at temperatures above zero. At lower values it is recommended to carry out welding work in heated shelters. In IOGP SB RAS, ways of solving this problem have been proposed, that is, technologies for hot element butt-welding of polyethylene pipes and hot element socket welding of polypropylene pipes have been developed, allowing welding to be performed at ambient temperatures below the standard in the open air without building heated shelters [3, 4].

2. Experimental part.

The structure of welded butt joints of polyethylene pipes was studied using universal optical microscope Altami POLAR 312 (Russia). The surface of samples for microscopic studies was prepared by etching in boiling toluene vapors for 3 min. Etching surfaces in the region of the fusion line (Figure 1) were studied at × 1000 magnification.

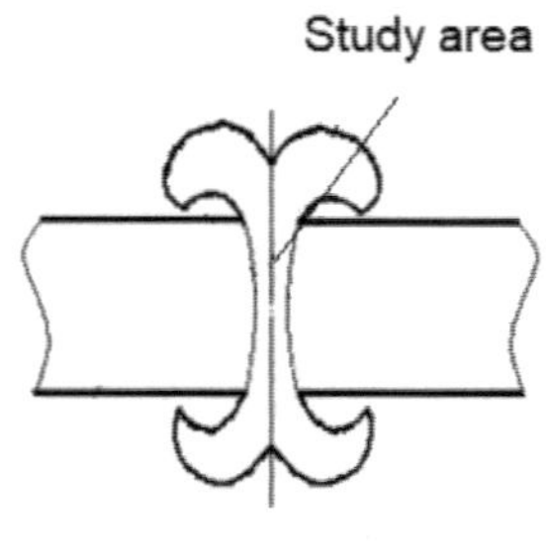

Figure 1. Schematic location of researched area of welded butt joints of PE pipes.

The supramolecular structure of material of samples of joints, welded via technology developed at IOGP SB RAS at low temperatures, is most identical to the structure in research weld area, performed using standard welding mode. Sizes of spherulites vary in the range of 5–15 μm (Figures 2a and 2c). It can also be noted that spherulites are slightly deformed. This is probably caused by shrinkage of weld material, which relates to the fact that welded pipe ends are rigidly fixed. The structure of material of welded joint obtained at minus 30 °C with violation of technology is characterized by presence of crystalline formation with dimensions of 3-5 μm. Fine-crystalline structure occurs with high cooling rate of welded joint.

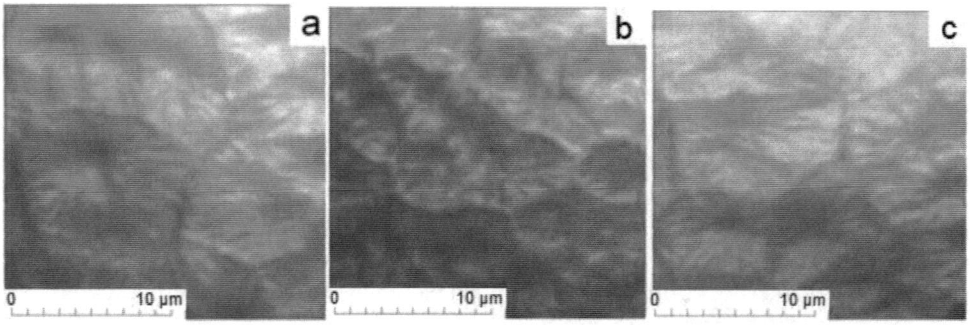

Figure 2. Material structure of welded joints manufactured under the following conditions: a) AAT = + 30 °C, standard welding; b) AAT = - 30 °C, with violation of technology; c) AAT = - 30 °C, developed technology.

Studies of strength characteristics of welded butt joints were carried out by tensile tests according to original method of quantitative assessment of strength of welding joint [5]. The essence of this technique is to create intended area of welding using pre-made template. The template is inserted between welded ends of pipes after heating operation during technological pause. After recommended time after welding, samples are made according to GOST 11262. In standard tests for axial tension, fracture took place exactly at the place of fusion. Figure 3 shows test results of welded joints of polyethylene pipes.

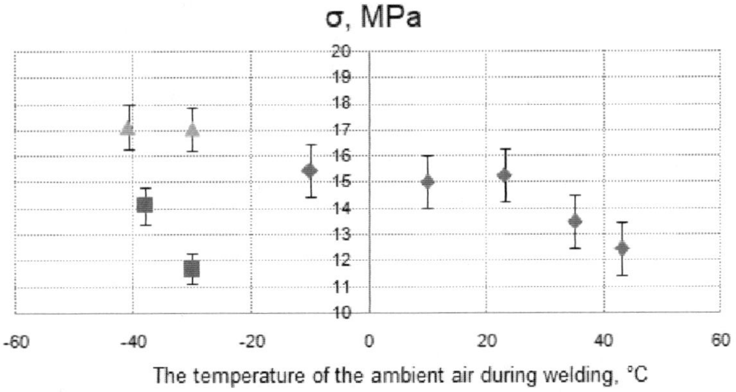

Figure 3. Values of destructive stress of the material of the seam of butt-welds produced at different ambient temperatures: ♦ - standard welding; ▲ - developed technology; ■ - welding with violation of technology.

As it can be seen from test results, strength of welded joints made at low temperatures by developed technology is higher than the same strength values of welded joints produced in the temperature range acceptable for welding. Strength indicators of welded joints obtained in violation of the technology are lower than the same indicators of welded joints made at low ambient air temperatures according to the developed technology and joints welded in the allowed temperature range. However, you should pay attention to the fact that at temperatures above 30 °C there is slight decrease in strength. This is probably because in this case cooling of welded joint is already too slow and crystalline formations in the structure of the material grow up to 50-100 microns. Thus, the formation of too small spherulites (up to 5 μm) in the material structure, as well as too large (up to 100 μm) negatively affects strength of welded joint of polymer pipes.

Hot element socket welding of polypropylene pipes was carried out at different modes: at temperatures above zero by standard technology, at temperatures below zero developed at IOGP SB RAS and at temperatures below zero by standard welding, i.e. with violation. The material structure of obtained welded joints was studied on electron scanning microscope JEOL JSM-7800F. The surface of samples for research was also prepared by etching in boiling toluene vapors. Figure 4 shows micrographs of sample material structure.

Figure 4. The structure of the material of the weld zone (fusion zone): a) welding at 23 °C; b) welding at minus 30 °C; c) welding at minus 30 °C, developed technology

Type and size of supramolecular formations in structure of the material of welded joints produced at minus 30 °C by the developed technology (Figure 4c) is most like the structure of welded joint obtained at room temperature by standard welding (Figure 4a). The structure is spherulitic, sizes of spherulites vary from 3 to 8 microns. In the process of welding at low ambient temperatures, violation of cooling rate is naturally high, and therefore the structure in the area under study is characterized with unformed small spherulites with sizes of 2-3 μm and the presence of frozen amorphous part (Figure 4b). That is, it can be concluded that the developed welding technology, which provides the same thermal process and cooling rate, contributes to the formation of structure in welded joint material the same as at the allowed temperature range of the welding production.

Studies of the strength of hot element welded joints were carried out by tensile shear tests using the original method, which allows to quantify the strength at the place of fusion of welded parts [6]. The essence of the technique lies in the stretching of standard sample, where the test hot element weld is in the middle of the sample. The sample was made according to Figure 5. On the coupling 1, an incision 2 was made to the point of fusion, parallel to the edge of the coupling. On the inner wall of the pipe 3 was cut notch 4, parallel to the edge of the pipe. The distance between slots 2 and 4 was calculated in such way that cross-sectional area of tested part A of hot element welded joint was equal or less than the minimum cross-sectional area of the sample outside the zone of the hot element joint. In shear tensile tests, the sample was fractured along the weld seam in area A.

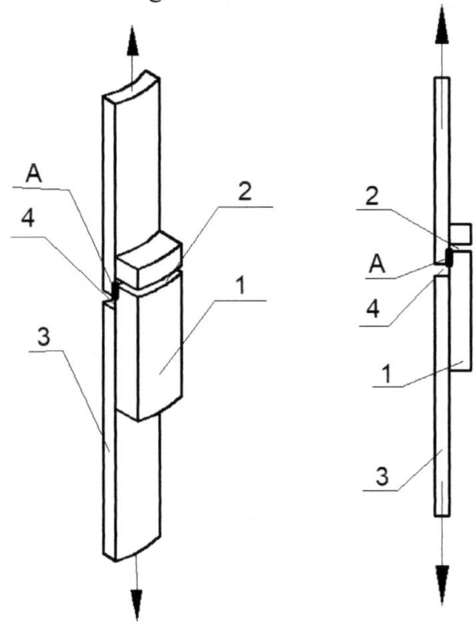

Figure 5. Sample for determination the strength of hot element welding under tension (the direction is indicated by arrows)

Welded joints obtained under different welding conditions of polypropylene pipes were tested. Test results are shown in Figure 6.

According to the results of mechanical tensile shear tests, breaking stress of hot element welded joints produced at temperature of minus 35 °C by the developed technology is not inferior to the strength of welded joints obtained by standard welding at room temperature.

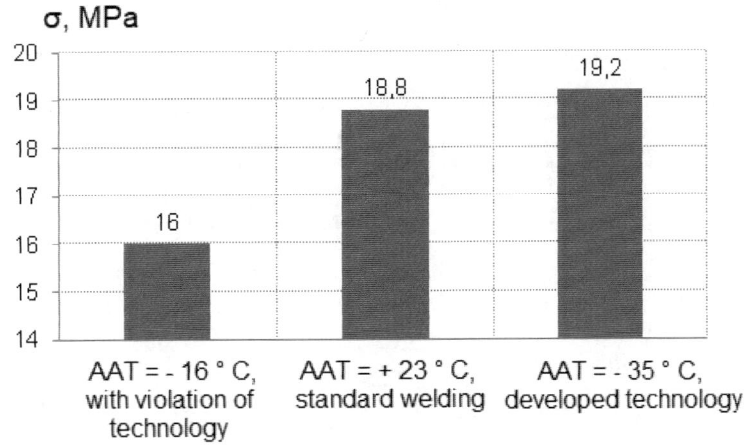

Figure 6. Destructive stress of welded hot plate joints of polypropylene pipes at various ambient air temperatures (AAT)

3. Conclusion

Studies of structure and strength properties of the material of welded joints have shown that the developed technologies for hot element butt welding of polyethylene pipes and hot element socket welding of polypropylene pipes ensure the formation of the same material structure of welded joints as during welding in regulated ambient temperature range.

Acknowledgment
The work has been accomplished within the framework of the State Order of the Federal Agency for Scientific Organizations of the Russian Federation (project No. AAAA-A17-117040710038-8 from 07.04.2017).

References:

[1] E.V. Danzanova, Issues of quality control of welded joints of polyethylene pipes for gas pipelines, Oil and Gas Business. 1 (2009) 1-9. Access mode: http://www.ogbus.ru/authors/Danzanova/Danzanova_1.pdf
[2] A.I. Gerasimov, E.V. Danzanova, G.V. Botvin, Testing of welded joints of polypropylene pipes, Vestnik ESSUTM. 66 (2017) 38-42.
[3] N.P. Starostin, M.A. Vasil'eva, E.V. Danzanova, O.A. Ammosova, Butt welding of polyethylene pipes at low temperatures, Welding International. 27 (2013) 318-320.
[4] N.P. Starostin, A.I. Gerasimov, G.V. Botvin, E.V. Danzanova, Welding of polypropylene pipes at negative temperatures, Construction of unique buildings and structures. 55 (2017) 7-18.
[5] A.I. Gerasimov, E.V. Danzanova, G.V. Botvin, RF Patent 2465560 (2012)
[6] A.I. Gerasimov, E.V. Danzanova, G.V. Botvin, Research Methods for Seal of Streams Polymer Pipes, Proceedings of the 12th International Conference on Mechanics and Materials (MRDMS-2018): AIP Conference Proceedings 2053, 030018 (2018); DOI: 10.1063 / 1.5084379 https://doi.org/10.1063/1.5084379

Investigation of the microstructure of the heat-affected zone of low-alloyed steel during pulsed arc aelding under conditions of low climatic ambient temperatures

N I Golikov[1, a], E M Maksimova[1, b] and Yu N Saraev[2, c]

[1] V. P. Larionov Institute of the Physical-Technical Problems of the North of the Siberian Branch of the RAS - Division of Federal Research Centre "The Yakut Scientific Centre of the Siberian Branch of the RAS", 1, Oktyabrskaya St., Yakutsk, 677980, Russia

[2] Institute of Strength Physics and Materials Science of the Siberian Branch of the RAS, 2/4, Academichesky ave., Tomsk, 634021, Russia

E-mail: [a] n.i.golikov@mail.ru, [b] nikiforova_em@mail.ru, [c] litsin@ispms.tsc.ru

Abstract. The development of a special technology of welding at low temperatures is an urgent scientific and technical challenge aimed at solving the problems of increasing the reliability and safety of products of equipment and structures operated under conditions of the North and the Arctic. One of the perspective technologies is pulsed arc welding. The structure of the heat-affected zones (HAZ) of welded joints produced by pulsed arc welding at an ambient temperature of 45 degrees Celsius below zero was studied. Two frequency modes were applied during modulated-current welding, and the results obtained were compared with the traditional method of manual arc welding with direct current. The features of the influence of the modulation frequency of the welding current and the temperature of the welding performance on the microstructure of the HAZ of 09G2S steel welded joints are revealed.

1. Introduction

In the regions of cold climate, the average annual ambient air temperatures are about 10 °C below zero and lower, and the period of negative temperatures lasts more than 210 days, so a significant amount of welding work during construction, installation and repair of structures is carried out at negative temperatures. Welding under conditions of low climate temperatures has its own characteristics associated with an increase in the cooling rate of welded joints [1]. As the cooling rate increases, the probability of formation of quenching structures increases, which can lead to brittle fracture of welded joints.

The metal of the near-weld zone is characterized by microstructural heterogeneity due to the gradient of temperature fields. The most problematic, in terms of providing a complex of properties, is the overheating area, which is formed as a result of thermal effects on the base metal [2]. Under the influence of high temperatures, an undesirable microstructural state is formed, which causes reduced cold resistance and embrittlement of the metal near the fusion line [3].

Improvement of the microstructural state of the HAZ of welded joints is possible using pulsed welding technologies [4, 5]. Pulsed welding technologies have significant advantages [4–9] and allow through programmable input of heat into the welded joint zone, control of the processes of melting and transfer of each drop of electrode metal, creation in the zones of welded joints of technical systems of a fine structure in the weld metal and HAZ, to significantly reduce the degree of residual deformations of welded joints.

The use of modulated current provides high efficiency control of the structure and properties of the weld metal when welding with current pulses of low frequency range from 5.0 Hz and below. However, the question of the influence of the current modulation frequency, as well as the effect of changes in the ratios of the parameters of current pulses, its amplitude, duration, modulation depth, on the formation of the structure, requires further study. According to the work of [10], the most effective frequency range, which ensures the reduction of structural components in the zones of the welded joint when welding is modulated by the current of low-alloyed steels, ranges from 1 to 2.5 Hz.

Research purpose is to study the structure of the HAZ of welded joints of low-alloyed steel, made in two frequency modes of pulsed arc welding at negative ambient temperatures.

2. Materials and equipment

Rolling sheet of 09G2S low-alloy structural steel, corresponding to State Standard 19281-2014, as the most widely used in structures operated in the conditions of the North and the Arctic, was chosen as the material for research. Steel of this grade is well welded in a wide range of welding modes, regardless of the thickness of the welded elements and the ambient temperature. The chemical composition and mechanical properties of 09G2S steel are presented in Table 1.

For research, manual arc welding of samples with a single-V butt weld was made. Welding was carried out in three passes using LB-52TRU and UONI 13/Moroz electrodes with diameters: 3.0 mm to form the root layer and 4.0 mm to form the filling and facing layers. The chemical composition and mechanical properties of the used welding electrodes are shown in Table 2.

Inverter welding power sources were used for welding the samples: FEB-315 "MAGMA" for pulsed arc welding and NEON VD-315 for direct current welding. The following welding modes were selected for research:

- pulse-arc welding with a modulation frequency of the welding current of 1.67 Hz (MCW–1.67) at a welding performance temperature of –45 °C;
- pulse-arc welding with a modulation frequency of the welding current of 5 Hz (MCW–5) at a welding execution temperature of –45 °C;
- direct current welding (DCW) at a welding temperature of –45 °C;
- direct current welding (DCW) at a welding temperature of +20 °C.

The main energy parameters (arc current and voltage) during the welding process were recorded with the AWR-224MD welding process recorder. During the technological experiment, the welding time was estimated and the average heat input was calculated in accordance with the generally accepted method. Heat inputs of welding the root layers were 660...980 kJ/m, and the filling and facing layers were 1320-2020 kJ/m.

Table 1. Chemical composition (wt%) and mechanical properties of 09G2S steel according to State Standard (GOST) 19281-2014

C	Si	Mn	Ni	S	P	Cr	V	N	Cu	As	Toughness (U-notch) [J/cm²]	TS [MPa]	Elongation [%]
<0.12	0.5 - 0.8	1.3 - 1.7	<0.3	<0.035	<0.03	<0.3	<0.12	<0.008	<0.3	<0.08	≥34 at 70°C	≥490	≥21

Table 2. Chemical composition (wt%) and mechanical properties of welding electrodes

Electrode	C	Si	Mn	Ni	Mo	S	P	Toughness (V-notch) [J/cm²]	TS [MPa]	Elongation [%]
UONI 13/Moroz	0.075	0.3	0.7	2.8	-	0.010	0.017	210	660	22.0
LB-52TRU	0.09	0.42	0.9	-	-	0.017	0.020	≥49 at -30°C	540	29

Metallographic studies of the microstructures of welded joints in the HAZ were performed on a high-resolution scanning electron microscope with a Schottky cathode JEOL JSM-7800F. To analyze the structure, a cross-section of the welded joints was made and metallographic sections were prepared. To reveal the structure of the samples, chemical etching of the surface with a 4% solution of nitric acid in ethanol was used.

3. Research results and discussion

The microstructure of the base metal is a ferritic-pearlite mixture with a ratio of the volume fraction of structural components in the amount of 90/10 (ferrite/perlite) (Fig. 1). The average grain diameter of ferrite is about 6 microns. The base metal has a homogeneous polygonal structure.

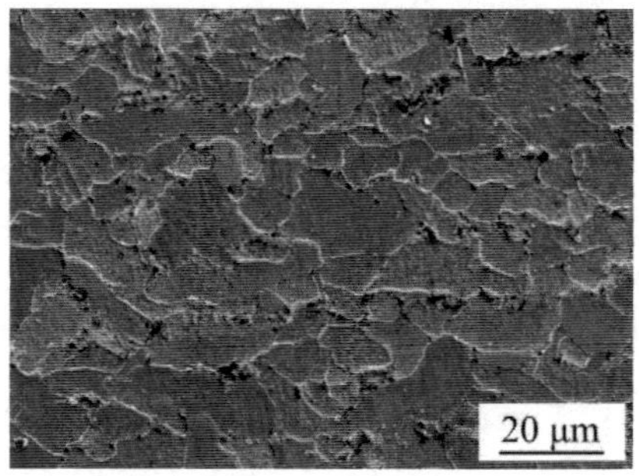

Figure1. Microstructure of the base metal.

The HAZ microstructures at the fusion line (FL) of welded joints at the level of the filling layer, obtained by weld with UONI 13/Moroz electrodes are presented in Fig. 2.

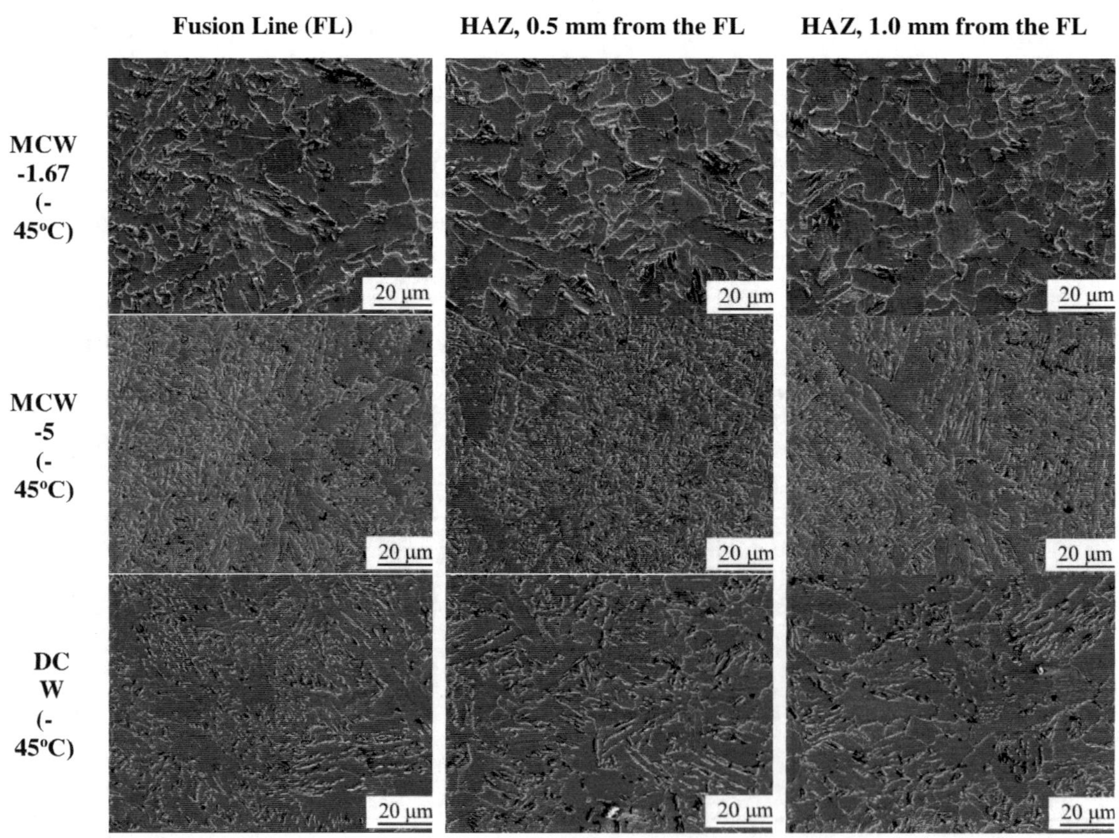

DC W (20°C)

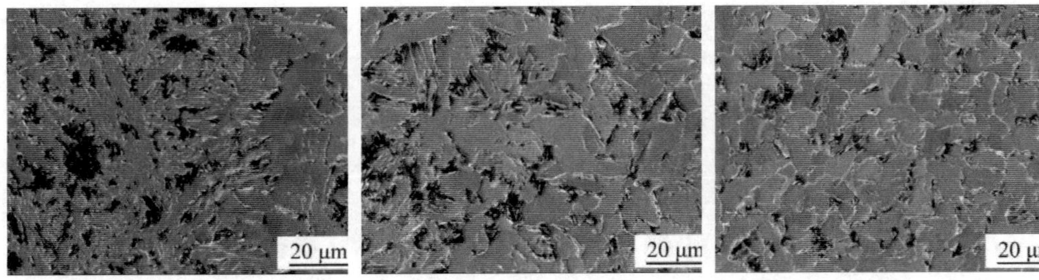

Figure 2. Microstructure of the HAZ of the welded joint at the level of the filling layer obtained using electrode UONI 13/Moroz.

The HAZ of the welded joints obtained at DCW (–45 °C) and MCW–5, on the fusion line and up to 1 mm from it, have a similar fine-needle microstructure consisting of ferrite, upper bainite and a small amount of lower bainite. In similar zones of the welded joint, obtained at the MCW–1.67 mode, a significantly different structure can be traced, consisting of a ferrite-carbide mixture, upper and lower bainite.

The structure of the HAZ of the specimen obtained at the DCW (+20 °C) differs significantly from the structure of the HAZ performed by the DCW (–45 °C). At the site of incomplete melting and at a distance of 0.5 mm from it, a ferrite-carbide mixture was found with a low content of bainite structure (Fig. 2).

Incomplete melting and overheating areas structure of the HAZ of welded joints obtained by welding with using electrodes LB-52TRU is a ferrite-carbide mixture with the presence of quenching structures of upper bainite (Fig. 3). The similarity of structural components of the HAZ of welded joints was revealed by MCW–5 and DCW (–45 °C), where the size of carbide components is larger than in the sample made by MCW–1.67.

The structure of the HAZ of the specimen performed at the DCW (+20 °C) significantly differs from the structure obtained at the DCW (–45 °C). At the zone of incomplete melting and at a distance of 0.5 mm from it, a ferrite-carbide mixture with a low content of bainite structure was found (Fig. 3).

According to the results of a calculated estimate, depending on the heat input of welding, it was found that under negative temperatures (at –45 °C), the cooling rate of the HAZ metal increases in the range of 30...70 % compared with a positive temperature (at +20 °C). The calculated estimate of the cooling rates during welding was performed according to the theory N. N. Rykalin. [14]. In this regard, in areas of incomplete melting and overheating of the HAZ of welded joints made under conditions of low climatic temperatures, the formation of quenching structures was detected.

Fusion Line (FL) **HAZ, 0.5 mm from the FL** **HAZ, 1.0 mm from the FL**

MCW -1.67 (-45°C)

MCW -5 (-45°C)

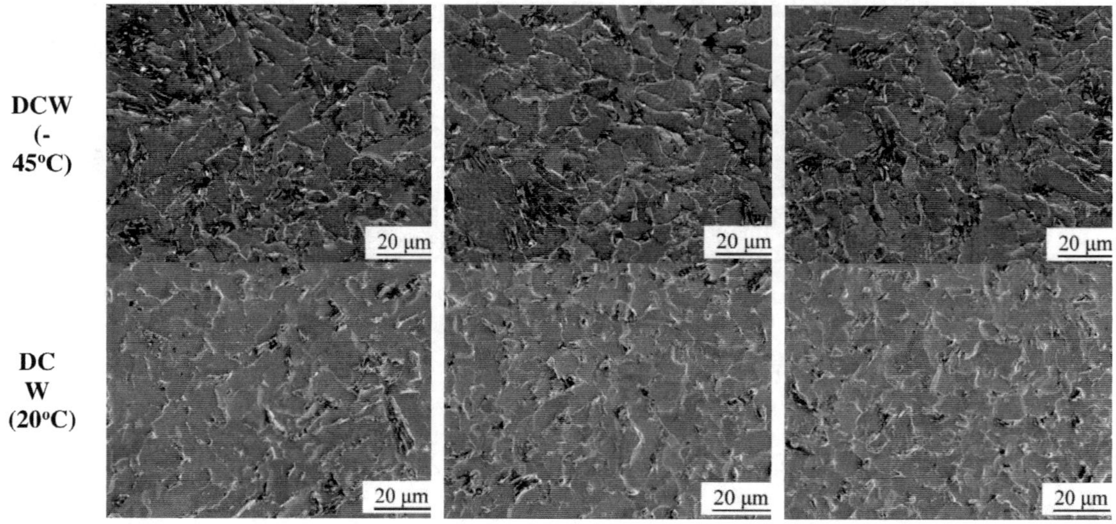

Figure 3. Microstructure of the HAZ of the welded joint at the level of the filling layer obtained using electrode LB-52TRU.

4. Conclusions

Analysis of the presented microstructures of welded joints zones, obtained during welding at direct current and in pulsed mode, shows that with an increase in the modulation frequency of the welding current, the heat input is equalized and, at a frequency of 5.0 Hz, the size of the HAZ becomes the same as during welding in DC mode. Apparently, this is due to the heat inertia of heat propagation from the heating source.

The study of the HAZ structures of 09G2S steel welded joints obtained by welding at different ambient temperatures showed that in zones of incomplete melting and overheating, welding at –45 °C forms ferritic-carbide mixtures with a certain content of quenching bainite structures that are more dispersed compared to structures at welding at room temperature.

Acknowledgement

The work was performed in the frame of an agreement on R&D cooperation between Tomsk Polytechnic University and ISPMS SB RAS and according to the research program of 2019 within RSF project No. 16-19-10010P.

References

[1] V.P. Larionov, Elektrodugovaya svarka konstruktsiy v severnom ispolnenii, Novosibirsk, 1986.

[2] L.A. Yefimenko, O.Yu. Yelagina, Ye.M. Vyshemirskiy, O.Ye. Kapustin, A.V. Muradov, A.K. Prygayev, Traditsionnyye i perspektivnyye stali dlya stroitelstva magistralnykh gazonefteprovodov, Moskva, 2011.

[3] X. Di, M. Tong, Ch. Li, Ch. Zhao, D. Wang, Microstructural evolution and its influence on toughness in simulated inter-critical heat affected zone of large thickness bainitic steel, Materials Science and Engineering: A. 743 (2019) 67-76.

[4] Yu.N. Sarayev, V.P. Bezborodov, A.A. Grigoryeva, V.A. Lebedev, S.Yu. Maksimov, N.I. Golikov, Svarochnyye soyedineniya tekhnicheskikh sistem otvetstvennogo naznacheniya metodov adaptivnoy impul'sno-dugovoy svarki, Voprosy materialovedeniya. 81 (2015) 127-132.

[5] Y.N. Saraev, V.P. Bezborodov, S.V. Gladkovskiy, N.I. Golikov, Improving the reliability of metallic structures in service in the conditions with low climatic temperatures by efficient application of advanced methods of modification of the zone of the welded joint, Welding international. 31 (2017) 631-636.

[6] A.M. Zhernosekov, V.V. Andreev, Pulsed metal arc welding (Review), The Paton welding journal. 10 (2007) 40-43.

[7] A.N. Smirnov, E.V. Kozlov, V.L. Knyaz'kov, Vliyaniye rezhimov svarki na elektricheskuyu napryazhennost i strukturno-fazovoye sostoyaniye v zone termicheskogo vliyaniya, Deformatsiya i razrusheniye materialov. 6 (2017) 28-33.

[8] E.M. Fedoseeva, T.V. Olshanskaya, Structure Formation and Nonmetallic Inclusions in Welded Joints when Welding Steel X65 by STT+API Technology, Materials Science Forum. 893 (2017) 229-233.

[9] T. Kurşun, Cold metal transfer (CMT) welding technology, The Online Journal of Science and Technology. 8 (2018) 35-39.

[10] Y.N. Sarayev, V.P. Bezborodov, Effect of the energy parameters of the welding process on the structure and properties of welded joints in low-alloy steels, Welding International. 27 (2013) 678-680.

[11] N.N. Rykalin, Raschety teplovykh protsessov pri svarke, Moskva, 1951.

IOP Conf. Series: Materials Science and Engineering **681** (2019) 012016 doi:10.1088/1757-899X/681/1/012016

Study of structural non-homogeneity impact on mechanical properties of dissimilar weld joints of carbon steel 20 and corrosion-resistant austenitic 12Kh18N10T steel

A L Goncharov[1], A Yu Marchenkov[1], E V Terentyev[1], I E Zhmurko[1] and A P Sliva[1]

[1] National Research University "Moscow Power Engineering Institute", 14, Krasnokazarmennaya str., Moscow, 111250, Russia

E-mail: goncharoval@mpei.ru

Abstract. The paper is devoted to identifying features of structural non-homogeneities formation in dissimilar welded joints of carbon steel 20 to corrosion-resistant austenitic 12Kh18N10T steel, obtained by tungsten inert gas welding (TIG), as well as studying of their influence on the mechanical characteristics of welded joints under static and cyclic loading. Differences in the formation of a diffusion carbide interlayer in weld metal alloyed with nickel and manganese are established. When austenite is alloyed with nickel, bulk carbon diffusion occurs, while in manganese austenite boundary diffusion becomes the main mechanism for carbon diffusion, which occurs mainly along grain and sub-grain boundaries of austenite. As a result of mechanical tests, it was established that under static tension welded joint works under conditions of strengthening by a soft (ferritic) layer due to the restraint of transverse deformations by more solid (carbide) layer. The limits of soft interlayer dimensions, under which the softening of welded joint does not occur, are established that will allow to proceed to the extreme temperature-time parameters of heat treatment and operation of welded joint.

1. Introduction

Nowadays welded structures from dissimilar steels are used in many industry branches. The expediency of their use is caused by the operating conditions and the ability to reduce metal consumption, improve efficiency and durability. One of the most common options for combining materials in a welded joint is a combination of carbon or low-alloy steel, usually pearlite class, and austenitic corrosion-resistant chromium-nickel steel. For example, the task of welding specified classes materials arises when joining pipelines and steam pipelines with elements of power equipment at nuclear power plants. At the same time, taking into account inevitable non-homogeneity of chemical composition in the welding zone of dissimilar steels, under the influence of such operational factors as temperature and operation time prerequisites are created for carbon diffusion movement and formation of structural non-homogeneity in the weld. The general regularities of such processes have now been investigated [1-4]. The main reason for carbon diffusion is the difference in its thermodynamic activity in solid solutions of the materials being joined, which is due to the different alloying of these solid solutions. Thus, in the presence of alloying elements in the weld metal, forming stable carbides (Cr, Ti, Nb, W), cementite dissociates in the less alloyed steel and released carbon diffuses through the fusion boundary into the more alloyed weld metal with the formation of resistant carbides of alloying elements in it [1, 3].Currently, work is underway to study the process of carbon diffusion and properties of dissimilar welded joints for steels of various structural classes [5-15]. As the main way to suppress the carbon diffusion movement in welded joints, it is proposed to perform barrier deposition method with a high nickel content on less alloyed steel [6, 11, 15]. Welded joints are considered either in the state after welding or after a short-term post weld heat treatment.

The aim of this paper was to study the process of structural inhomogeneities formation in welded joints of carbon steel 20 and corrosion-resistant austenitic 12Kh18N10T steel under prolonged temperature exposure and to establish the influence of soft ferritic interlayer width on the results of tension tests and cyclic tests for the samples with a weld joint.

2. Research methods

Dissimilar welded joints of plates with a thickness of 10 mm from steel 20 and corrosion-resistant austenitic 12Kh18N10T steel, obtained by TIG, were investigated in this paper. Two types of filler wire were used – an “Autrod 308LSi” filler wire with low carbon content (0.01%) and nickel as the main austenizer and “G102” wire with a carbon content of 0.4% and manganese as the austenizer (without nickel). Welding was carried out in several passes in the gap in order to obtain a weld of constant width (Figure 1a). The chemical composition of filler wires, along with chemical composition of the materials to be welded, is given in Table 1.

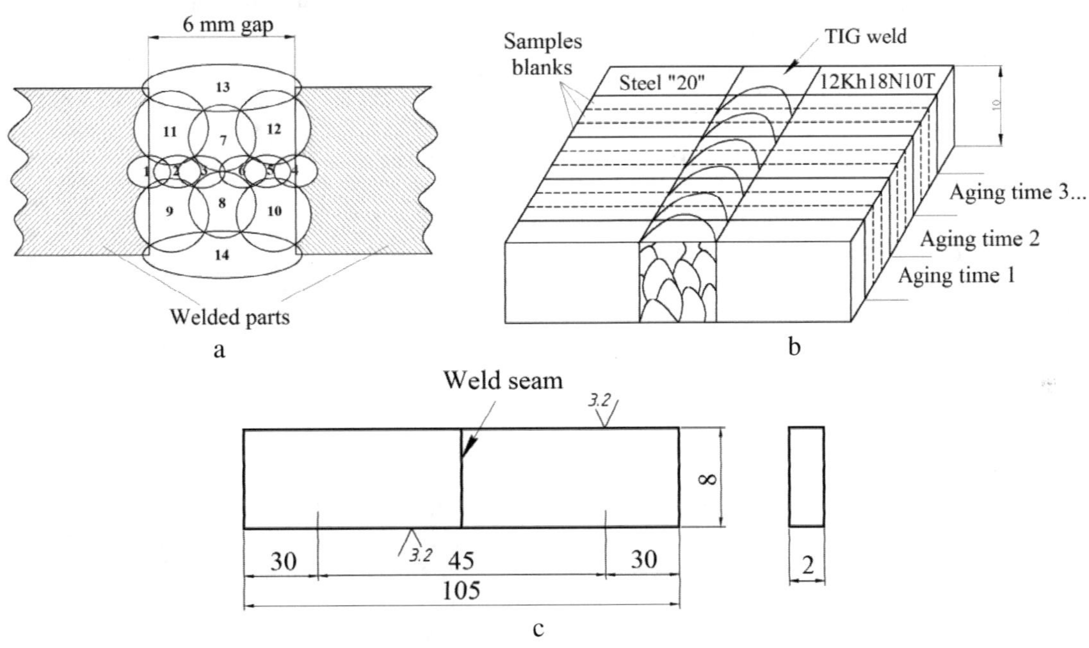

Figure 1. Gap filling scheme (a), welded plate cutting scheme (b) and sample for mechanical test scheme (c).

Table 1. The chemical composition of welded steels and filler materials

Material	Elements, wt%								
	C	Si	Mn	Ni	Cr	Cu	Ti	S	P
20	0.17-0.24	0.17-0.37	0.35-0.65	<0.25	<0.25	<0.25	–	<0.04	<0.04
12Kh18N10T	<0.12	<0.8	<2.0	10.0	18.0	<0.3	0.4-1.0	<0.02	<0.035
308LSi	<0.03	0.65-1.0	1.4-2.1	9.0-11.0	19.5-21.0	<0.3	–	<0.02	<0.03
G102	0.4	0.75	14	–	14.5	–	–	–	–

All the welded plates were subjected to heat treatment at 650°C with different aging times for the formation of structural interlayers of different width. After each exposure, a series of flat samples was cut off from the plate for metallographic studies and mechanical tests, after which the remaining part of the plate was heated again and held until the next aging time (Figure 1b). Before each heating, a protective carbon-free compound was applied to the surface of the plates in order to prevent intensive decarburization of samples surface due to its oxidation by the gas phase. Thus, the total maximum aging time was 300 hours. After cutting off a series of samples, a metallographic section was made, as well as 5 flat samples for mechanical testing (Figure 1c). The total number of test specimens was 55. It should be noted that spheroidization of cementite occurred in carbon steel during prolonged exposure at high temperatures, resulting in decrease in the ultimate stress of steel 20 base metal, as well as it was difficult to determine the boundary of ferritic interlayer. To restore the properties of carbon steel and the structure of its pearlite component, after the main heat treatment, a reducing heat treatment was carried out (heating up to 760 °C, holding for 15 minutes and air cooling). This made it possible to restore the pearlite component due to phase recrystallization and to increase the ultimate stress of carbon steel base metal. At the same time, the short-term exposure practically did not affect the picture of diffusion processes in the welded joint.

The microstructure of welded joints was studied using light microscopy methods. The length of the ferritic interlayer was measured by the bright field method on a Zeiss Axio Observer Z1m optical microscope. To identify the structure of carbon steel ferritic interlayer, a 4% solution of nitric acid in ethanol was used. In addition, this reagent made it possible to identify the overall structure of carbide interlayer without noticeably etching the grain boundaries of austenitic weld metal and base metal of 12Kh18N10T steel. Ultimate stress of welded joints was determined under uniaxial tension with the strain rate Vs = 2 mm/min. Cyclic tests were carried out by loading the samples with a pulsating tensile-unloading cycle (cycle asymmetry factor R = 0) up to the maximum cycle tension σ_{max} = 360 MPa with a frequency of f = 50 Hz and determining the number of cycles before destruction.

3. Results and discussion

In the process of samples aging after welding at temperature 650 °C in the weld zone of carbon steel with austenitic steel structural changes were observed due to the carbon diffusion movement from the more alloyed weld metal to carbon steel. At the same time, from the side of steel 20, near the fusion line there is a ferritic zone, and with distance from the fusion line, particles of structurally free cementite appear. The presence of structurally free cementite is due to the change of carbon solubility in ferrite during heating and cooling. At the holding temperature (650 °C) the solubility in ferrite is higher than at a temperature of 20 °C, at which observation was carried out. Therefore, carbon, which at the aging temperature is in solution, at the observation temperature is released from this solution in the form of tertiary cementite. The subsequent restorative heat treatment at 760°C promotes the formation of a pearlite in this zone, which is detected in the microstructure as small colonies in large ferrite grains.

Further from fusion line there is a zone with partially disintegrated pearlite, followed by the base metal structure, in which no diffusion changes take place. The typical structure and hardness distribution for the decarburized zone in steel 20 is shown in Figure 2. The structure of completely decarburized zone (ferritic layer) has a number of distinctive features. Firstly, ferrite grains in this zone reach very large sizes. Their height reaches 180-600 μm, and in horizontal direction ferrite grain has a size that is equal to or close to the width of ferritic interlayer. That is, in fact, the ferritic interlayer consists of ferrite columnar grains, herewith the initial size of these structural components (before aging treatment) is within 5-15 μm for ferrite and 3-8 μm for pearlite. The second feature of ferritic zone structure is that very close to the fusion line there is a thin zone of fine ferrite grains with dimensions of about 5-15 μm. The width of this zone is about 25 μm. Ferrite grain width and size in this zone do not depend on the exposure time. A similar structure of decarburized zone is observed in all studied samples, both when using filler material with Ni, and with Mn.

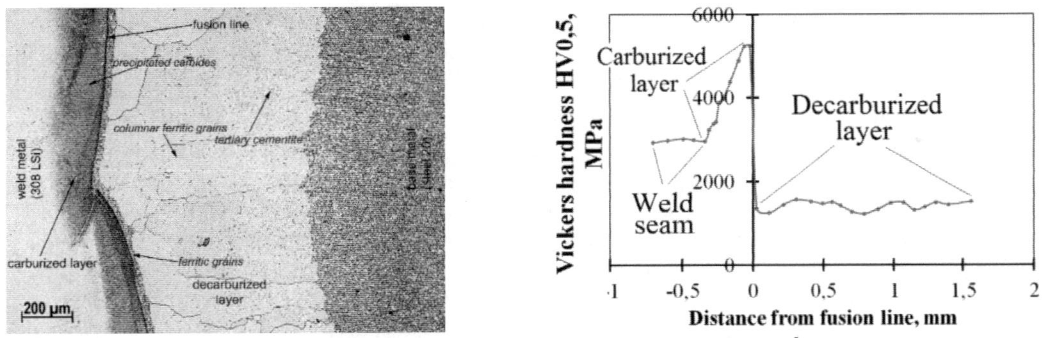

a b

Figure 2. Microstructure (a) and hardness distribution (b) of the decarburized zone in steel 20 after the aging at temperature 650 °C for 13 hours (filler wire - 308 LSi).

Near the fusion line in the weld metal there is a zone of increased etching, in which carbides of elements with a greater affinity for carbon than iron are formed. For the steels considered, such a basic element is chromium. Titanium, which is part of austenitic steel, is available in it in the amount, necessary for bounding carbon, which is also available in this steel.

Therefore, carbon coming from steel 20 through the fusion line boundary cannot be bounded into titanium carbides. This is confirmed by the absence of titanium carbides and carbonitrides well detectable by light microscopy in the structure of weld metal. The presence of a certain amount of these inclusions is explained by their incomplete dissolution during welding or by repeated release at the stage of crystallization and cooling of weld metal. The main high-carbon phase released in the weld metal and detected during metallographic study is $M_{23}C_6$ or M_7C_3 type carbides [16, 17]. Taking into account that carbon diffusion rate through the fusion line is substantially lower than carbide formation reaction rate, it can be assumed that the only «consumer» of carbon coming from steel 20 is carbide phase. Therefore, the zone of increased etching, which contains carbides, repeats the zone of carbon diffusion penetration into the weld metal. Figure 3 shows the carburized zones for two types of austenitic weld metal. It was found that in case of alloying weld by Mn (G102), the carbon penetration distance from the fusion line turned out to be significantly longer, and the nature of the carbide phase indicates that carbon moved mainly along the grain boundaries (figure 3, a). With an aging time of 300 hours, carbon penetrated almost the entire width of the weld. In the weld alloyed by Ni, carbide phase release in the carburized zone occurs equally both at the boundaries and inside the grains (Figure 3b).

Given the fact that decarburized ferritic zone has a lower hardness and strength, it is this zone in the welded joint that will determine its strength characteristics. It was found that in the samples, the welds of which were obtained with using a filler metal containing Mn, the decarburized zone length is greater at the same aging time. The growth rate of the ferritic interlayer was also higher in these welds.

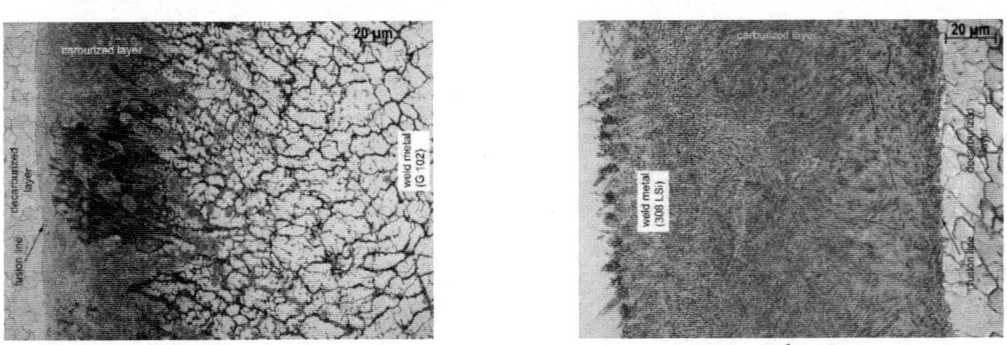

a b

Figure 3. Microstructure of the carburized zone after aging for 300 hours for two types of austenitic weld metal: alloying by Mn (a), x500; alloying by Ni (b), x1000.

Given the fact that decarburized ferritic zone has a lower hardness and strength, it is this zone in the welded joint that will determine its strength characteristics. It was found that in the samples, the welds of which were obtained with using a filler metal containing Mn, the decarburized zone length is greater at the same aging time. The growth rate of the ferritic interlayer was also higher in these welds.

From the results of static tests (Figure 4) it follows that at small values of diffusion interlayers width, the welded joint strength is determined by the ultimate stress of steel 20, which is equal to 434...476 MPa. With increasing interlayers width the welded joint strength is determined by the strength of ferritic interlayer's metal, and the values of ultimate stress decrease with increasing of interlayer width.

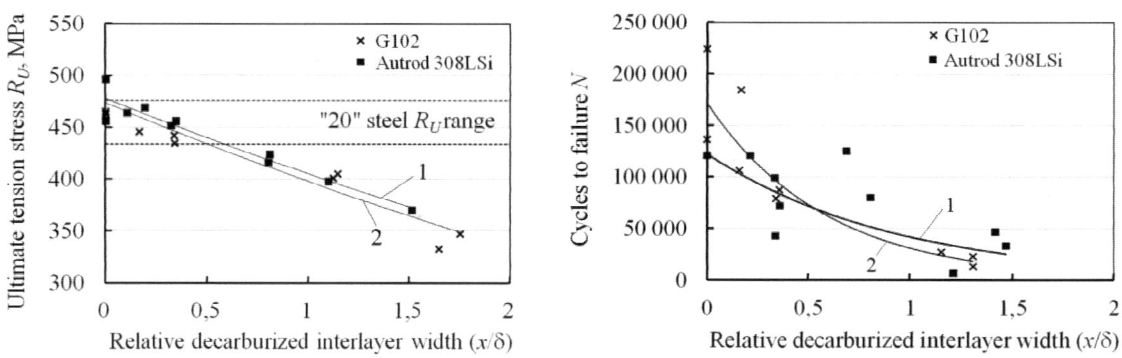

Figure 4. Results of mechanical tests of weld joints: impact of soft interlayer width (x/δ) on the ultimate stress R_u (a) and on the number of cycles to failure N (b).

Such nature of the change in the welded joints strength can be explained by the effect of soft interlayer contact hardening [18, 19]. In case of a sufficiently wide ferritic interlayer in the welded joint it goes through the plastic strain at stresses above 270 MPa, and in the central part of interlayer in the absence of durable adjacent sections, a transition to metal's concentrated plastic deformation occurs with its subsequent destruction. On the contrary, in case of very narrow ferritic interlayer the effect of contact hardening appears. At stresses of about 400 MPa the concentrated plastic deformation of the ferritic interlayer, which causes the appearance of transverse deformations of the sample, is hampered by the presence of adjacent metal sections with more durable ferrite-pearlite structure. The adjacent to a ferrite-pearlite structure sections, involved in concentrated deformation, like the metal of ferritic interlayer, is characterized by a much more intensive stress-strain state with large values of acting true stresses and strains. It also has a greater metal strain hardening and at the same time has a very significant plasticity reserve. A further increase in load during tensile tests leads to concentrated deformation in the base metal of steel 20, which, due to a smaller plasticity reserve, is destroyed earlier than metal in the area of the ferritic interlayer, which withstands greater plastic deformation without destruction.

It should be noted that the absolute values of the ferritic interlayer width are not given, since the character of the weld joint deformation is determined not by the absolute, but by the relative width of the soft interlayer. The relative width is the ratio of the interlayer width x to the smallest linear dimension δ of the sample cross-section (for the researched samples $\delta \approx 2$ mm). Thus, during short-term tests, it was established that welded joint metal softening occurs when the values of soft interlayer relative width are equal to 0.55 for both types of filler materials examined (Figure 4). However, given the higher growth rate of soft interlayer in a welded joint using filler material with Mn, metal's softening of such welded joint occurs at shorter aging time than welded joint metal using filler material with Ni.

Cyclic tests of welds under investigation showed that in samples welded with both types of filler material, resistance to cyclic loading also depends on the interlayer width. Thus, in welded joints with narrow interlayers, zone with such interlayers does not weaken the welded joint, and the destruction of all samples occurs along the base metal. With an increase in the interlayer width, the samples destroy along the interlayer with a monotonic decrease in the cycle number up to destruction (Figure 4).

4. Conclusions

The decarburized soft layer forms at the fusion line from the carbon steel side and consists of 3 structural zones (in the order of distance from the fusion line): large columnar ferrite grains (zone 1); large columnar ferrite grains and tertiary cementite inclusions (zone 2); ferrite-pearlite structure with partially dissociated cementite (zone 3). The greatest decrease in hardness occurs in zones 1 and 2, the decrease in hardness of zone 3 does not exceed 15% of the carbon steel base metal hardness. The growth rate of the ferritic interlayer is higher for welds alloyed with manganese instead of nickel, and the dependence of variation interlayer width on time differs from the parabolic one.

The diffusion movement of carbon in the weld metal alloyed with manganese occurs mainly along the grain boundaries. Under long exposures, carbon penetration is possible over the entire width of the weld. At the same time, the hardness of carburized zone in the welds with Mn is lower than in the welds with Ni due to the lower density of carbide inclusions.

The static short-term strength and cyclic strength of the welded joints under study is determined by the soft (ferritic) layer width. At small (less than 0.55) values of the soft interlayer relative width due to contact hardening, the strength of welded joints is determined by the strength of the least durable welded material, which is steel 20. With an increase in the relative width of the soft interlayer, its effect on the welded joint properties increases and destruction of welded joints occurs at lower stress (static tests), or with a smaller number of cycles (cyclic tests). The values of destructive stress under static short-term loading and the number of cycles up to destruction under cyclic loading monotonously decrease with increasing of interlayer width.

Acknowledgements

The research was carried out in the Moscow Power Engineering Institute (Moscow, Russia) with the financial support of the Ministry of Education and Science of the Russian Federation in accordance with base part of state assignment, project no. 11.8088.2017/8.9.

References

[1] Petrov G L 1963 *Weld joints inhomogeneities* (Leningrad: Sudpromgiz) p 206
[2] Manning J R 1968 Diffusion kinetics for atoms in crystals (Toronto: Van Nostrand) p 257
[3] Ryabov V R *et al.* *Welding of dissiminar metals and alloys* (Moscow: Mashinostroenie) p 239
[4] Gotalskiy Y N *Welding of dissiminar steels* (Kiev: Technika) p 183
[5] Rathod D W, Pandey S, Singh P K and Prasad R 2015 *Mat. Sci. Eng. A* **639** 259-68
[6] Rathod D W, Singh P K, Pandey S and Aravindan S 2016 *Mat. Sci. Eng. A* **666** 100-13
[7] Dinda S K, Basiruddin S, Roy G G and Srirangam P 2016 *Mat. Sci. Eng. A* **677** 182-92
[8] Vidyarthy R S, Kulkarni A and Dwivedi D K 2017 *Mat. Sci. Eng. A* **695** 249-57
[9] Li Y, Gong M, Wand K. Li P. Yang X and Tong W 2018 *Mat. Sci. Eng. A* **718** 260-6
[10] Liu G L *et al.* 2018 *Mat. Sci. Eng. A* **722** 182-96
[11] Wang W K, Liu Y, Zhang Q B, Zhang L J and Zhang J X *Mat. Sci. Eng. A* **747** 244-54
[12] Wang Q *et al.* 2019 *Mat. Characterization* **147** 434-42
[13] Wang Y, Cui H, Fan M, Chen Y and Lu F 2019 *Mat. Characterization* **151** 227-36
[14] Saini M, Arora N, Pandey C and Mehdi H 2014 *Int. J. Res. in Eng. and Techn.* **3(7)** 39-42
[15] Rathod D W *et al.* 2017 *Acta Metall. Sin. (Engl. Lett.)* **30(2)** 120-32
[16] Kwiatkowski A *et al.* 2018 *Acta Mater.* **147** 165-75
[17] Gudremon E 1966 Special steels (Moscow: Metallurgy) p 734
[18] Bakshi O A, Erofeev V V, Shakhmatov M V and Anisimov Y I 1983 *Welding production* **4** 1-4
[19] Shakhmatov D M and Shakhmatov M V 2018 *Welding and diagnostics* **1** 32-6

IOP Conf. Series: Materials Science and Engineering **681** (2019) 012017 doi:10.1088/1757-899X/681/1/012017

Investigation of thermo-EMF temperature dependences for construction materials of various structural classes

A L Goncharov[1], A V Chulkova[1], R V Rodyakina[1], V K Dragunov[1] and I S Chulkov[1]

[1] National Research University "Moscow Power Engineering Institute", 14, Krasnokazarmennaya str., Moscow, 111250, Russia

E-mail: reginarodyakina@mpei.ru

Abstract. The paper is devoted to the study of the thermo-EMF of various metals and alloys, which are widely used in welded structures. For the selected steels, the relationship between the absolute thermo-EMF and their structural class was established, and the dependences of the thermo-EMF on temperature were obtained. For some groups of steels, a method of predicting the thermo-EMF by chemical composition was developed.

1. Introduction

Thermoelectric phenomena, inseparably linked with the heat and electricity transfer in the welding zone [1–4], have a significant impact on the quality of welded joints in electron beam welding (EBW) of dissimilar alloys. Welding of such materials is remarkable by appearance of thermo-EMF and thermoelectric currents in the weld joint. They occur when the material is heated by heat flux arising during EBW. The resulting thermoelectric currents create a magnetic field on the surface and in the joint area of welded parts, leading to a change in the trajectory of the beam electrons through the joint thickness and as a consequence to the formation of incomplete welds in the welded joint.

The value of thermoelectric currents induction during EBW of dissimilar materials is determined by a large number of factors including temperature distribution in the welding zone, shape and thickness of the parts to be welded as well as thermo-EMF coefficient of the materials being joined [5, 6]. The diffusion thermo-EMF of a real metal is determined by a mechanism or several mechanisms of current carriers scattering on phonons, lattice defects, impurity atoms, magnons (in magnetically ordered metals), grain boundaries (in pure metals at low temperatures) and etc. Also, it is determined by electron spectrum, particularly of the Fermi surface geometry and temperature [7]. Diffusion thermo-EMF dominates in pure metals at temperatures above the Debye temperature. In accordance with the Boltzmann kinetic equation, the diffusion thermo-EMF for pure metals and a number of alloys is a linear function of temperature. However, in practice several mechanisms are involved in the formation of the total thermo-EMF in pure metals and especially alloys, and their quantitative influence is extremely difficult to take into account. In addition, for a number of metals current carriers may belong to several zones (for example, electrons and holes in transition metals) or several current-carrier groups belonging to the same zone but corresponding to different regions of the highly anisotropic Fermi surface. These circumstances lead to the fact that for most metals and alloys the dependence of thermo-EMF on temperature is nonlinear, and the direction of its change as a result of alloying is difficult to predict from priori considerations. Only some empirical regularities can be formulated being valid for individual groups of alloys. Moreover, alloying of transition metals at high

IOP Conf. Series: Materials Science and Engineering **681** (2019) 012017 doi:10.1088/1757-899X/681/1/012017

temperatures can lead to significantly more radical changes in thermo-EMF than alloying of alloys based on non-transition metals. The most noticeable effects of alloying are observed in alloys based on nickel, palladium and platinum with both transition and non-transition metals [8, 9]. Currently, there is no complete phenomenological or microscopic theory of thermoelectric phenomena in complex alloyed alloys. Therefore, fundamental equations in practice are not possible to be applied to determine thermo-EMF of alloyed steels and alloys.

The aim of this paper was an experimental study of thermo-EMF for steels and alloys of various structural classes, the determination of thermo-EMF magnitude and coefficient dependencies on temperature, on chemical and on structural-phase composition of the material and also working out the methods of predicting thermo-EMF value for a particular combination of materials.

2. Research methods

Using thermocouples, the temperature dependences of thermo-EMF for different materials of various structural classes were obtained. Temperature gradient was provided by placing thermocouple in a high-temperature Nabertherm furnace.

The temperature dependence of thermo-EMF was determined by the integral method (Figure 1) on thermocouples of the materials under study (Table 1), in which one of the branches was made of chromel or alumel.

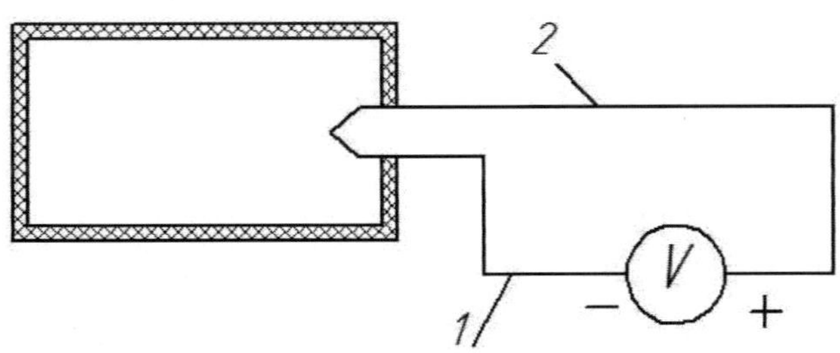

Figure 1. Scheme for integral thermo-EMF study: 1 – chromel, 2 – material under study.

The chemical composition of studied materials is presented in table 1 for alloys and in table 2 for steels.

Table 1. The chemical composition of alloys

No.	Material	Chemical composition [%]						
		C	Si	Mn	Cr	Ni	Ti	Other
1	36NKhTYu	<0.05	0.3–0.7	0.8–1.2	11.5–13	35–37	2.7–3.2	0.9–1.2 Al; 43.61–48.8 Fe
3	KhN60WT	<0.1	<0.8	<0.5	23.5–26.5	50.9–63.2	0.3–0.7	<4 Fe; 13–16 W; <0,5 Al
4	Alumel	0.1	0.85–1.5	1.8–2.2	–	93–96	–	<0.3 Fe; 0.25 Cu; 0.6–1.2 Co; 1.8–2.5 Al
7	SW117	–	–	0.8	–	4.5	–	Cu – base; 3.5 Fe
8	SW130	0,02	0,13	0,28	–	54	–	45 Fe
16	06Kh15N60M15	<0,08	<0,5	1,0–2	14–16	base	–	14–16 Mo
30	ChKh22	2.4–3.6	0.2–1.0	1.5–2.5	19–25	–	0.15–0.35	0.15–0.35 V

Table 2. The chemical composition of steels (M is martensite, F is ferrite, A is austenite, P is perlite)

No.	Material	Chemical composition [%]					Class	Thermo-EMF at 800°C
		C	Cr	Ni	Mo/V	Mn/Si		
2	EP517	0.13–0.18	11–12.5	1.7–2.1	1.35–1.75/–	<0.5/<0.5	M+F	7.07
5	308LSi	0.03	19.5–22	9–11	<0.75	1–2.5 / 0.65–1	A+F	–2.16
6	10Kh18N8G8S	0.1	18.5	8.5	–	8.5 / 0.8	A	–2.46
9	10Kh19N9G6	0.1	19	9	–	6 / 0.5	A	0.41
10	22Kh5	0.22	5.3	–	1.5/–	0.7 / 0.5	M	0.75
11	20Yu2	0.2	–	–	–	0.55 / 0.2	M	0.7
12	01Kh17N	0.01	17	0.25	–	0.4 / 0.3	F	5.79
13	12Kh18N10T	<0.12	17–19	9–11	–	<2 / <0.8	A	–2.18
14	08G2S	0.05–0.11	<0.2	<0.25	–	1.8–2.1 / 0.7–0.95	P	–0.67
15	12KhX13	0.09–0.15	12–14	<0.6	–	<0.8 / <0.8	M	6.37
17	10Kh19N11M4F	0.07–0.12	18–20	10–11.5	3.8–4.3/0.7–1.1	1–2 / 1–1.5	A	–1.67
18	09Kh16N4B	0.08–0.12	15–16.5	4–4.5	–	<0.5 / <0.6	A+F+M	4.41
19	08Kh21N10G6	<0.1	20–22	9–11	–	5–7 / 0.2–0.7	A+F	–2.3
20	03Kh12N9M2S	<0.03	11.6–12	8.5–8.9	–	0.6–0.9 / 1.4–1.7	A+M	2.81
21	06Kh19N9T	<0.08	18–20	8–10	–	1–2 / 0.4–1	A+F	–2.57
22	13Kh14N9S4F3G	0.15	13–15	8.5–9.5	4/2.7–3.1	0.9–1.3 / 0.2–0.4	A+F	–0.98
23	10Kh16N25AM6	0.08–0.12	15–17	24–27	5.5–7/–	1–2 / 0.6	A	–0.65
24	08Kh14GNT	<0.1	12.5–14.5	0.4–0.9	–	0.9–1.3 / 0.25–0.65	M+F	5.32
25	04Kh19N11M3	<0.06	18–20	10–12	2–3/–	1–2 / <0.6	A+F	–2.03
26	15Kh1M1F	0.1–0.16	1.1–1.4	<0.25	0.9–1.1/0.2–0.25	0.4–0.7 / 0.17–0.37	M	2.55
27	20Kh13	0.16–0.25	12–14	<0.6	–	<0.6 / <0.7	M+F	7.71
28	15Kh15NS	0.15	14.5	0.5	2.5	0.4 / 0.7	A+F+M	0.45
29	09G2S	<0.12	<0.3	<0.2	–/<0.12	1.3–1.7 / 0.5–0.8	P	–0.25

Thermo-EMF was measured with a V7-22 voltmeter when the hot junction was heated in the temperature range of 50–1100°C. Further, the relative Seebeck coefficient for pairs of conductors was calculated by the expression

$$S_r = \frac{dE_r}{dT}, \tag{1}$$

where E_r is measured relative thermo-EMF.

Then by the expression (2) was calculated the absolute Seebeck coefficient at each point:

$$S_{abs} = S_r - S_{chr}, \tag{2}$$

where S_{chr} is Seebeck coefficient for chromel.

According to expression 3, the absolute thermo-EMF was calculated for each material from Tables 1 and 2:

$$E_{abs} = \int_{50}^{T} S_x - S_{abs} \, . \qquad (3)$$

3. Results of study

Based on the analysis of the alloys presented in table 1, it is possible to identify alloys with a tendency to a thermo-EMF increase (alloy 16), to a thermo-EMF decrease (alloys 1 and 4) and alloys with an extremely weak thermo-EMF variation (alloys 3, 7, 8) (Figure 2a).

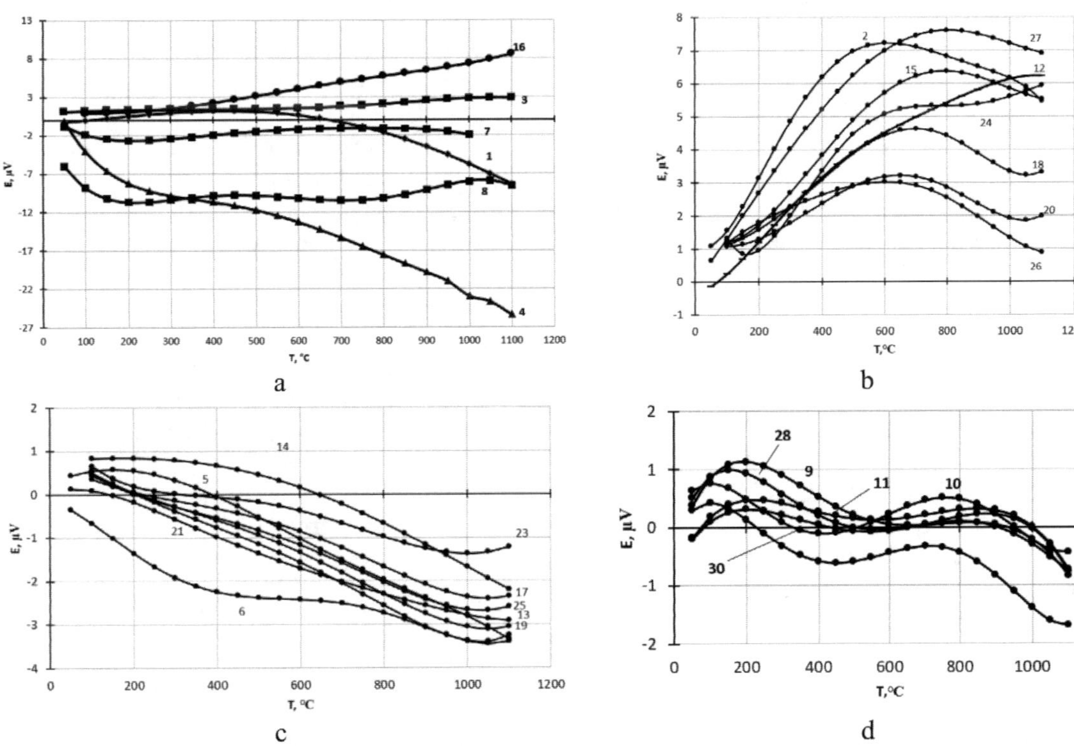

Figure 2. Experimental dependences of absolute integral thermo-EMF on temperature: a) for the alloys presented in table 1; b), c), d) for the steels presented in table 2. Number from 1 to 30 is serial number of the material (specified in table 1 or table 2).

Three groups with positive, negative, and sign-changing thermo-EMF values were selected from the studied steels.

In the first group all steels have positive dependences with an extremum, the maximum of which is reached at a temperature of 600-800 °C and after that the values of the thermo-EMF begin to decrease. Most steels have a high chromium content (11–17%). Steels of this group include martensitic class, comprising nickel, austenitic-martensitic, martensitic-ferritic, austenitic-ferritic-martensitic and ferritic classes (Figure 2b).

In the second group of steels thermo-EMF has nearly linear dependence on temperature, the values of which decrease approximately to 1000 °C and then there is a slight increase of them. The steels which have a tendency to a thermo-EMF decrease are the steels of austenitic and austenitic-ferritic classes (Figure 2c). Most steels have a high content of chromium (14-21%) and nickel (9-25%).

IOP Conf. Series: Materials Science and Engineering **681** (2019) 012017 doi:10.1088/1757-899X/681/1/012017

Thermo-EMF values of the third group steels begin to decrease with the temperature increasing up to almost 600 °C, then thermo-EMF values slightly increase reaching maximum at the temperature range of 900 °C, after which they start to decrease again (Figure 2d).

The obtained temperature dependences of thermo-EMF for construction materials of various structural classes indicate that the high-temperature region the thermo-EMF values are similar in each of steel classes. Thus, the results obtained confirm the thermoelectric properties of materials being connected not only with their chemical, but also with structural and phase state.

It was found that for nickel-based alloys absolute thermo-EMF temperature dependence is close to linear. Low-alloyed and nickel-chromium steel have a negative absolute thermo-EMF. Thermo-EMF of martensitic, austenitic-martensitic and martensitic-ferritic classes steels have extremum dependencies the maximum of which is reached at the temperature range of 600–800 °C, after that the values of thermo-EMF begin to decrease. The maximum values of thermo-EMF do not exceed 8 mV. The thermo-EMF values reach maximum in steels of martensitic-ferritic and ferritic classes with a similar chemical composition: Cr is about 19%, Ni is about 9% in austenitic-ferritic class steels and 0.25% in ferritic class steels.

Nickel-based alloys with doped chromium have a positive thermo-EMF value with a tendency to increase, while thermo-EMF absolute value of nickel alloys with a small amount of chromium tends to decrease and has mostly negative values. Alumel has the lowest value of thermoelectric power.

The main alloying elements in the considered steels are chromium and nickel. They also contain silicon, titanium, molybdenum, manganese and in some cases vanadium and cobalt. A stable statistical relationship between the thermo-EMF values of these steels and the main alloying elements in them was not found.

The experimental dependences obtained show that the low of thermo-EMF properties variation is different for different groups of structural materials and it is very difficult to use a single mathematical model for its calculating. Nevertheless, it was possible to obtain approximating dependences suitable for describing five grades of structural materials with positive thermo-EMF values and five grades with negative thermo-EMF values. On the basis of experimental data and statistical processing the parameter H was introduced to obtain the dependence of absolute thermo-EMF of steels on their chemical composition. This parameter is determined by expression (4) for steels with a positive thermo-EMF value and by expression (5) for steels with a negative thermo-EMF value.

$$H = Cr + 15C + 1.3Ni + 0.7Mo + 5\left(Mn + Si\right) + 15V \,, \tag{4}$$

$$H = Cr + 40C - 1.5Ni + 0.3Mo - 0.5Mn + 2Si \,. \tag{5}$$

The coefficients in these expressions were selected taking into account the dependence $E(H)$ calculated by the expression (6):

$$E\left(H\right) = k_1 H + k_2 \,, \tag{6}$$

where k_1 and k_2 are statistical coefficients of temperature dependence (Figure 3).

By comparing the thermo-EMF values E_c calculated using expression (6) with the experimental values of E_{exp}, it is possible to determine the thermo-EMF values at temperatures of 800–1100°C with sufficient accuracy. The average relative error does not exceed 5.5%.

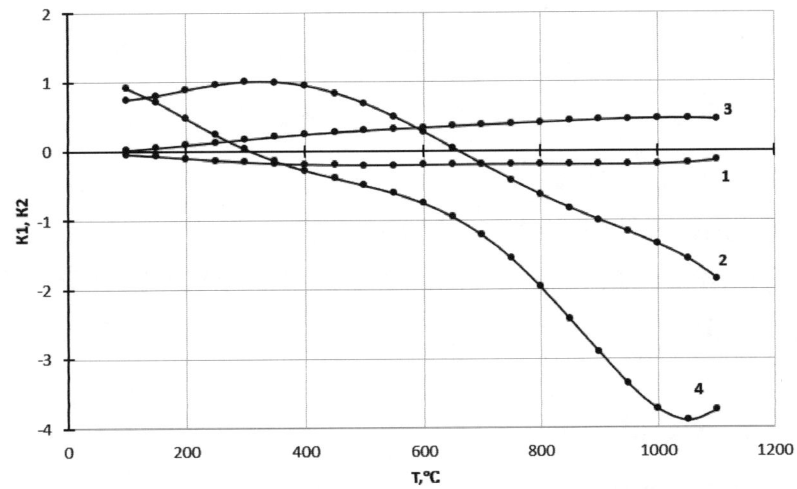

Figure 3. Temperature dependences of the coefficients k_1 and k_2 in the equation (6): 1 is temperature dependence of the coefficient k_1 for thermo-EMF negative values; 2 is temperature dependence of the coefficient k_2 for thermo-EMF negative values; 3 is temperature dependence of the coefficient k_1 for thermo-EMF positive values; 4 is temperature dependence of the coefficient k_2 for thermo-EMF positive values.

4. Conclusions

The obtained experimental values and methods for determining thermo-EMF of alloys can be used in practice to estimate the beam deflection angle in electron beam welding of dissimilar materials. The larger thermo-EMF value for a given pair of materials to be welded, the higher electron trajectory deviation is.

Acknowledgments

The research was carried out in National Research University «Moscow Power Engineering Institute» with the financial support of the Russian Science Foundation (project No. 18-19-00652).

References

[1] Dragunov V K, Muravyova T P and Rodionov Yu A 1990 *Welding production* **4** 2–4
[2] Nazarenko O K, Kaydalov A A, Kovbasenko S N *et al.* 1987 *Electron-beam welding* (Kiev: Nauk Dumka) p 256
[3] Nazarenko O K 1982 *Automatic welding* **1** 33–9
[4] Akopyants K S, Levitsky S M, Nazarenko O K, Neporozhny V Yu and Shilov G A 1989 *Let. to the J. of Tech. Phys.* **15(22)** 20–3
[5] Dragunov V K 1987 *Coll. of scient. papers of MPEI* **137** 33–7
[6] Dragunov V K and Chepurin M V 2001 *Welding production* **12** 8–16
[7] Scherbakov A V, Rodyakina R V, Kozhechenko A S, Gaponova D A, Goncharov A L and Dragunov V K 2017 *Let. to the J. of Tech. Phys.* **43(21)** pp 10–5
[8] Rogelberg I L and Beilin V M 1983 *Alloys for Thermocouples: Handbook* (Moscow: Metallurgy) p 36

Impact of amplitude modulation of heteropolar square current pulses on formation of permanent joints in metal structures under residual magnetization

A S Kiselev[1,2], S V Baklanov[2], A S Gordynets[1,2], Y N Saraev[1]

Institute of Strength Physics and Materials Science SB RAS [1], Academicheskiy ave., 2/4, Tomsk, 634055, e-mail: litsin@ispms.tsc.ru.
National Research Tomsk Polytechnic University [2], 30, Lenin ave., Tomsk, 634050

E-mail: kas@tpu.ru

Abstract. The work studies the impact of amplitude modulation of heteropolar current pulses on the formation of permanent joints of main pipelines done by manual shielded metal-arc welding. The results show that square pulsed current together with additional low-frequency current modulation facilitates formation of all weld layers in pipelines with residual magnetization. The work establishes that varying residual magnetization degree has no adverse effect on the quality of all layers of the permanent joints realized by modulated alternating square current, including their macrostructural and microstructural characteristics. This increases the technological stability of welding through elimination of electrode freezing during low-energy periods.

1. Introduction

In Russia, to transport oil and natural gas, large-diameter (1200–1420 mm) pipelines are used. Their condition during operation is monitored, as a rule, by pipeline magnetic pigs [1] that promote residual magnetization in the pipe. After assessing the condition of a pipe, damaged parts are replaced using manual shielded metal-arc welding. However, the residual magnetization in the pipeline after such inspection causes instable arc due to magnetic blowing during root layer welding. To eliminate such adverse effect, special equipment is used that lowers or cancels the strength of lateral magnetic field in the welding zone below critical 2 mT [2–6]. We should note that this method is fairly labor-intensive and has low efficacy due to application of special equipment which requires long preparation.

Therefore, an innovative engineering solution was suggested based on application of shielded metal arc welding by heteropolar square current pulses [7]. A special current switching principle in the welding circuit during polarity switch facilitates stable reinitiation of the arc and provides its spatial stability, including the cases of disturbing magnetic field with a strength of up to 100 mT [8, 9]. To implement the suggested technology, a special device IST-201 was developed that is connected to the welding circuit of a standard power supply. This appreciably increases the performance of repair works in main pipelines due to the absence of demagnetization operation. Nevertheless, despite the positive effect of heteropolar square current pulses on welding of magnetized parts by shielded electrodes, the issues of amplitude current modulation effect on the formation of permanent joints remain understudied. This is especially relevant to one-side welding of a root weld with formation of a back bead without back root pass or application of special backing.

At the same time, the amplitude modulation during shielded-electrode welding by unipolar current without external magnetic field has additional positive effect of arc reinforcing action on the molten pool. This guarantees formation of the root backing layer bead, simplifies welding process in all spatial positions, reduces requirements to the gap between the edges and their displacement, and eliminates freezing of the electrode [10, 11].

Aim of the work: to study the effect of amplitude modulation of heteropolar square current pulses on the weld seam formation during manual shielded metal-arc welding of magnetized pipelines.

2. Experimental.

To extend the technological capabilities of manual shielded metal-arc welding under magnetic field action, experiments were carried out on the impact of amplitude of heteropolar square current pulses on the formation of permanent joints. The functional scheme of the testbed is presented in Fig. 1.

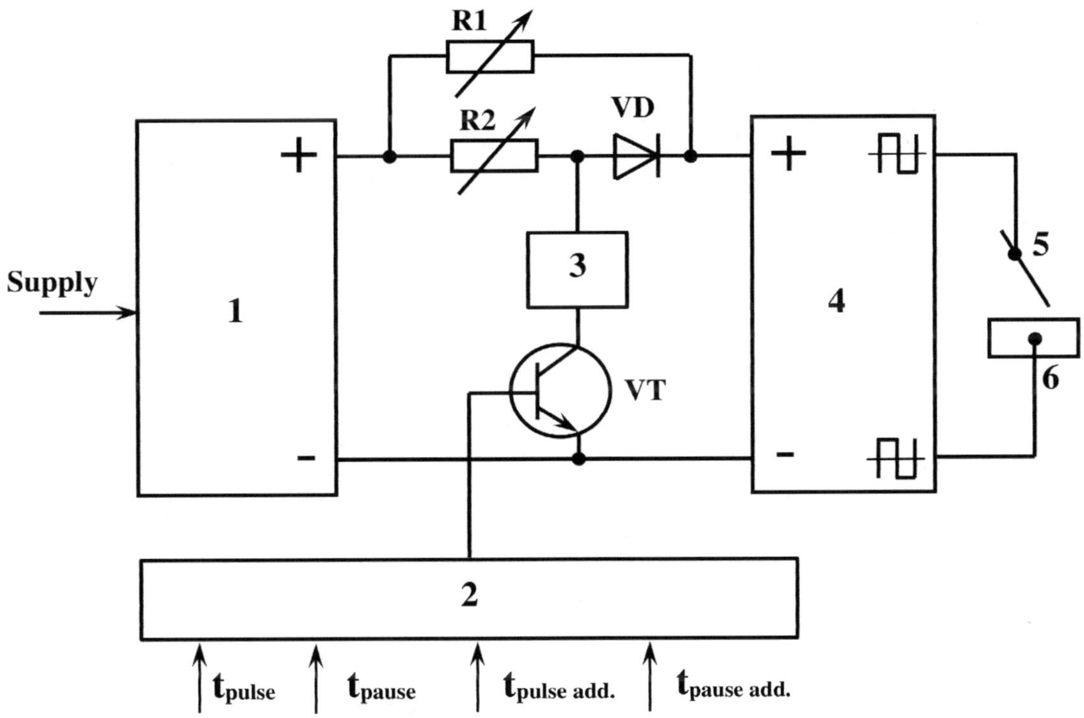

Figure 1. Functional scheme of testbed:
1) welding rectifier VD-306; 2) control unit; 3) voltage limiter; 4) welding current inverter; 5) electrode; 6) part; R1) pause current regulator; R1+R2) pulse current regulators; VD) bypass diode; VT) shunting transistor.

The root welding seam was done by 2.6-mm LB-52U electrodes. The initial magnetic field strength between welded edges (B = 100 mT) was set by an external inductor. The power supply was WD-306 device with a preset maximum current, which corresponding energy level (high or low) was limited by ballast resistors. The amplitude current modulation in the welding circuit was performed by a shunting transistor VT. The polarity was switched with the frequency of 500 Hz by IST-201 welding current inverter. A typical oscillogram of modulated alternating current is presented in Fig. 2. The welding regimes are presented in Table 1.

The testbed was used to weld 530-mm pipes from 12G2SB steel with the length of 1 meter and wall thickness of 10 mm. The mechanical properties are presented in Table 2.

Figure 2. Oscillogram of current in welding circuit ($\mu_i = 50$ A/unit, $\mu_t = 0.1$ s/unit).

Table 1. Parameters of root layer welding regime for permanent joint

No.	Current type and shape	Parameters of welding current	d_{el} [mm]	Electrode type
1	Unipolar, reverse	I_{weld} =116 A	3.2	LB-52U
2	Unipolar modulated	I_{weld} = 138 A; I_{pulse} = 192 A; I_{pause} = 30 A		
3	Heteropolar, square	I_{weld} = 137 A; frequency is 500 Hz		
4	Heteropolar, square, modulated	I_{pulse} = 192 A; I_{pause} = 30 A; polarity switch frequency is 500 Hz; t_{pulse} = 0.3 s; t_{pause} = 0.3 s.		

The pipe edges were prepared and assembled as per requirements of STO Gazprom 2-2.2-136-2007 standard. The initial magnetic field strength between welded edges (B = 100 mT) was set by an external inductor. The pipe was welded in non-revolution position in three passes (Figure 3). The electrodes and welding parameters were selected according to item 10.4 of STO Gazprom 2-2.2-136-2007 requirements.

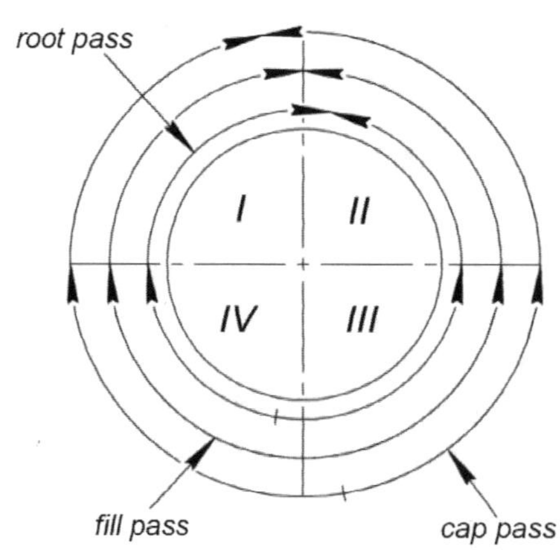

Figure 3. Scheme of manual arc welding of pipe joints

Table 2. Mechanical properties of 12G2SB steel (as per GOST 10705-80)

Ultimate tensile strength σ_u [MPa]	Yield stress σ_{02} [MPa]	Relative elongation δ [%]
490	343	20

Under the absence of magnetic field in the welding zone, direct current of inverse polarity was used. At initial magnetic field strength of 100 mT, amplitude modulation of heteropolar square current pulses, formed by IST-201 together with VD-306, were used. Chemical composition of the basic metal is presented in Table 3, while that for the seam metal is given in Table 4.

Table 3. Chemical composition of main 12G2SB steel

Elemental composition [%] (GOST 19281–89)									
C	Si	Mn	Cr	Ni	Co	Cu	Nb	Ti	V
0.12	0.42	1.46	0.12	0.08	0.01	0.15	0.05	0.01	0.06

Table 4. Chemical composition of seam metal

Welding conditions	**Chemical composition of seam metal [%]**							
	C	**Si**	**Mn**	**Cr**	**Ni**	**Co**	**Cu**	**V**
Magnetized pipe, heteropolar amplitude-modulated square current pulses	0.12	0.3	1.07	0.02	0.01	-	0.07	0.01
Demagnetized pipe, unipolar reverse-polarity current	0.12	0.3	1.20	0.04	0.02	0.01	0.04	0.02

The weld joints were initially subjected to visual measurement and x-ray inspection that revealed no defects. To analyze the impact of current type on the metal seam properties, its macro- and microstructure were investigated. The specimens for mechanical testing of the welded seams were cut from the pipe butt joints.

To determine the mechanical properties of the welded joints, the specimens were prepared as per STO Gazprom 2-2.2-136-2007 standard. The number of specimens and types of mechanical tests were chosen for 530-mm pipe with a wall thickness of 10.0 mm (Fig. 4).

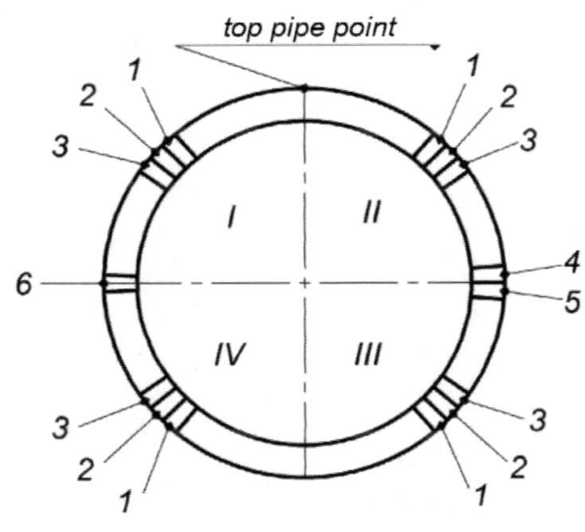

Figure 4. Scheme of specimen cutting for mechanical testing of round joint welds with a diameter of 530 mm: 1) specimen for static tension testing; 2) specimen for static face bend or fin bend testing; 3) specimen for static root bend or fin bend testing; 4) specimens for impact bending testing (along weld axis); 5) specimens for impact bending testing (along welding line); 6) specimen for hardness testing.

3. Results and Discussion.

The results demonstrate insignificant effect of welding conditions on the content of alloying elements in the seam metal. Indeed, static tension tests (Table 5) testify that regardless of the welding conditions, the basic metal of the specimens failed. Static bending tests showed no failure at a bending angle of 140° (Table 6). The impact bending tests of the welded joint demonstrated that at a temperature of minus 40 °C, the impact ductility of the seam metal in the case of amplitude modulation of heteropolar square current pulses is 1.12–1.8 times larger than that in the case of unipolar reverse-polarity current (Table 7). The application of amplitude-modulated heteropolar square current pulses increases the impact ductility of the metal from the heat affected zone.

Table 5. Welded joint static tension testing results

Welding conditions	**Butt/ quarter/ specimen**	**Mechanical properties of welded joints**		
		Ultimate tensile strength σ_u [N/mm^2]	**Relative elongation δ [%]**	**Relative reduction Ψ [%]**
Magnetized pipe, amplitude modulation of heteropolar square current pulses	b. 1/I/1	600	23	61
	b. 1/II/1	600	25	58
	b. 1/III/1	605	23	58
	b. 1/IV/1	572	30	66
Demagnetized pipe, unipolar reverse-polarity current	b. 2/I/1	583	23	61
	b. 2/II/1	600	25	66
	b. 2/III/1	588	25	58
	b. 2/IV/1	588	23	61

Table 6. Welded joint static bending testing results

Welding conditions	Butt/ quarter	**Bending angle [deg]**	
		Specimen 2 face bend	**Specimen 3 root bend**
Magnetized pipe, amplitude modulation of heteropolar square current pulses	b. 1/I	124	133
	b. 1/II	126	125
	b. 1/III	127	125
	b. 1/IV	126	124
Demagnetized pipe, unipolar reverse-polarity current	b. 2/I	128	126
	b. 2/II	128	124
	b. 2/III	134	126
	b. 2/IV	125	134

Table 7. Welded joint impact bending testing results

Welding conditions	**Location of concentrator**	**Butt/**	**Quarter/ specimen**	**KCV$^{-40°C}$ [J/cm^2]**
Magnetized pipe, amplitude modulation of heteropolar square current pulses	along weld axis	b. 1	II/4	142
			II/4	161
			II/4	132
	near-weld zone		III/5	213
			III/5	128
			III/5	219

Demagnetized pipe, unipolar reverse-polarity current	along weld axis	b. 2	II/4	97
			II/4	133
			II/4	118
	near-weld zone		III/5	118
			III/5	102
			III/5	168

Fig. 5 depicts macrostructure of root weld of permanent joints received at different levels of residual magnetization. Their analysis showed that the amplitude modulation of heteropolar square current pulses has no impact on the dimensions of the heat affected zone and does not deteriorate root welds of permanent joints in the case of parts with different residual magnetization level.

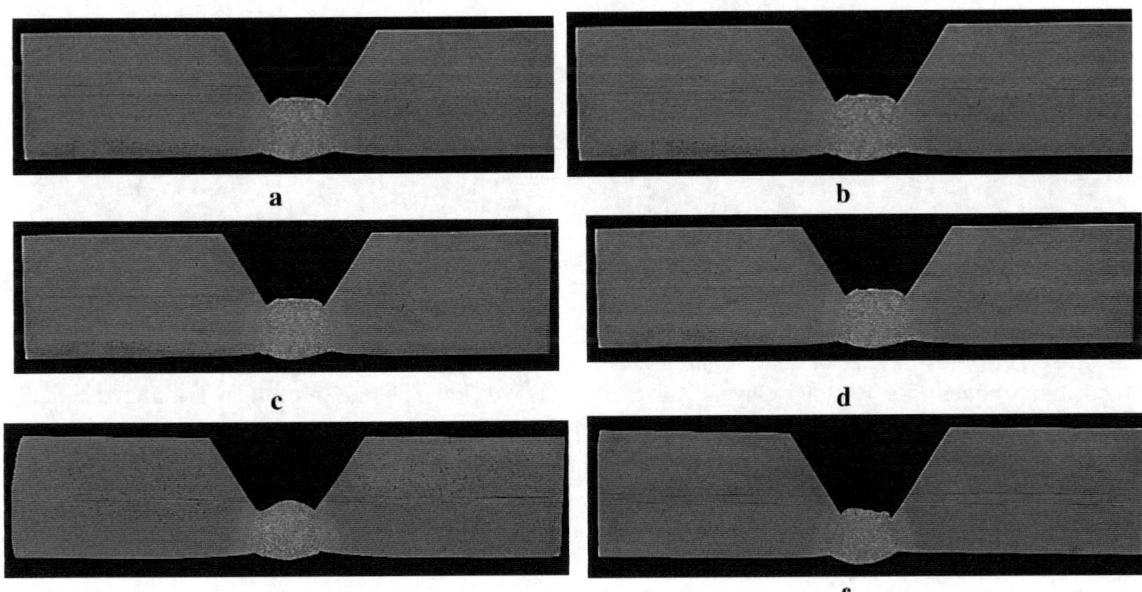

Figure 5. Macrostructure of root weld layers after vertical welding: a) demagnetized parts, unipolar current; b) demagnetized parts, unipolar modulated current; c) demagnetized parts, heteropolar square current pulses; d) demagnetized parts, amplitude modulation of heteropolar square current pulses; e) magnetized parts (initial magnetic field strength is 100 mT), amplitude modulation of heteropolar square current pulses; f) magnetized parts (initial magnetic field strength is 100 mT), heteropolar square current pulses.

Fig. 6 shows the macrostructure of the weld after consequent root, filler and cap passes under demagnetized and magnetized conditions.

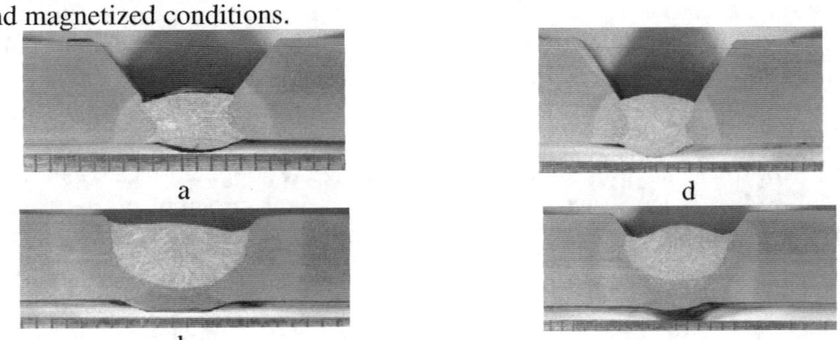

c f

Figure 6. Macrostructure of consequent formation of permanent joint layers: a, d) weld pass; b, e) root and fill passes; c, f) root, fill and cap passes; a, b, c) demagnetized parts; d, e, f) magnetized parts.

Fig. 7 depicts microstructure of root weld of permanent joints received at different levels of residual magnetization.

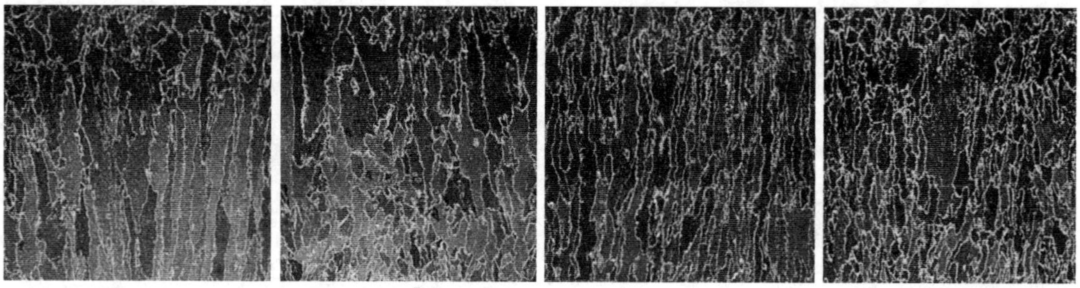

a b c d

Figure 7. Microstructure of weld metal: a) unipolar reverse-polarity current; b) unipolar modulated current; c) heteropolar square current pulses (500 Hz); d) heteropolar amplitude-modulated square current pulses (500 Hz).

The analysis of microstructure of weld metal demonstrates that consequent transition from unipolar reverse-polarity current, unipolar modulated current, heteropolar square current pulses to heteropolar amplitude-modulated square current pulses diminishes structural components at least 3–4 times. Interestingly, additional amplitude modulation of heteropolar square current pulses appreciably alters the structure of near-weld zone and provides smooth transition from large-grain structure of the weld metal to fine structure of the base metal [12, 13]. This evidences general decrease of structural nonuniformity, which justifies higher mechanical characteristics of permanent joints obtained by square current welding with additional low-frequency modulation.

4. Conclusion

Welding by heteropolar amplitude-modulated square current pulses is an effective technique that allows increasing the quality of root welds of main pipelines with residual magnetization. Variation of the residual magnetization degree has no adverse effect on the quality of all layers of the permanent joints, including their macrostructural and microstructural characteristics. In addition, the amplitude modulation of heteropolar square current pulses according to a preset algorithm improves stability of welding by eliminating electrode freezing during low-energy period.

Acknowledgement

The work was performed in the frame of an agreement on R&D cooperation between Tomsk Polytechnic University and ISPMS SB RAS and according to the research program of 2019 within RSF project No. 16-19-10010P.

References

[1] Gvozdetsky V.S., Mechev V.S. To the question of the movement of the electric arc in a magneticfield (in Russian) // Automatic welding, 1966, No. 6.

[2] Yushchenko K. A., Pestov V. A., Starushchenko T. M., Erashov A. S. Magnetic phenomena during welding of 0H9 steel and ways to eliminate their influence on the quality of welded joints (in Russian) // Automatic Welding, 1991, No. 8, P. 47-55.

[3] Korolkov P.M. The origin and methods for eliminating magnetic blast during welding (in Russian) // Welding industry, 1998, No. 5, P. 6-8.

[4] Leskov G.I., Protosey N.E., Novikov D.Yu. The conditions of demagnetization of products before welding and methods for their provision (in Russian) // Automatic welding, 1987, No. 7, P. 40-45.

[5] Kozlov V.V. Demagnetization of pipe joints during repair welding of pipelines (in Russian) // Welding production, 1999, No. 3, P. 42-44.

[6] RD-75.180.00-KTN-150-10 Regulation on cutting and inserting "coils", connecting parts, plugs, shut-off and control valves and connecting sections of main oil pipelines (in Russian).

[7] The method of arc welding (in Russian) / Patent for the invention No. 2245231, RU, C1, B23K9/09. Issued on January 27, 2005, Vol. no. 3.

[8] Gordynets A.S. Control of the arc welding process under the disturbing effect of a magnetic field (in Russian). Candidate's thesis (Engineering). — Tomsk, 2012. — 158 p.

[9] New Developments to Reduce Arc Blow during SMAW of Pipelines / Sergey V. Baklanov, Anton S. Gordynets, Alexey S. Kiselev, and Mikhail S. Slobodyan/ *Materials Science Forum Submitted: 2018-07-24/* ISSN: 1662-9752, Vol. 938, pp 96-103.

[10] Dedyukh R.I. Improving the efficiency of arc welding with coated electrodes in various spatial positions by modulating current (in Russian). Candidate's thesis (Engineering). — Moscow, 1982. — 203 p.

[11] Azarov N.A. Development of a high-speed process of welding with a modulated current of the root layers of a pipe joint with basic coated stick electrodes (in Russian). Candidate's thesis (Engineering). — Moscow, 1988. — 212 p.

[12] Exploratory studies to improve the reliability of critical metal structures operating under extreme loads and low climatic temperatures (in Russian) / Saraev Yu.N., Gladkovsky S.V., Golikov N.I., Kiselev A.S., Bezborodov V.P., Gordynets A.S., Lunev A.G., Chinakhov D.A., Khaidarova A.A., Trigub M.V. // High technology in projects of the Russian Science Foundation. Siberia. 2017. – 430 p., Tomsk, NTL. Managing editor Sharkeyev Yu. P. – 40 pages. Project of the Russian Science Foundation No. 16-19-10010.

[13] Saraev Yu.N., Lunev A.G., Kiselev A.S., Gordynets A.S., Semenchuk V.M. Investigation of the influence of the current type and modulation algorithm on the performance of welded joints (in Russian) // Welding industry. – 2019. - No. 3. – p. 3 - 8.

IOP Conf. Series: Materials Science and Engineering **681** (2019) 012019 doi:10.1088/1757-899X/681/1/012019

Digital visualization of thermal processes accompanying welding of small-diameter pipes

R A Krektuleva[1], M V Perovskaya[1], Y N Saraev[1], V M Semenchuk[1], R O Cherepanov[2], R O Bezginov[3]

[1]Institute of Strength Physics and Materials Science RAS, Tomsk, 634050, Russia
[2]Tomsk Scientific Center of SB RAS, Tomsk, 634050, Russia
[3]Gazprom transgaz Tomsk LLC, Tomsk, 634050, Russia

E-mail: rakrekt@mail.ru

Abstract. The work investigates the specificity of the thermal processes occurring during formation of a molten pool in curvilinear specimens. The process is exemplified by welding of small-diameter pipes. A numerical 3D-model of the interaction of a moving heating source and thick-walled pipes of various curvature was built. The material, diameter and thickness of pipes were selected as per active standards. The numerical calculations were performed. The practical experiment was performed using argon tungsten-arc welding. The results establish the effect of the welded part geometry on the dynamics of thermal processes in the molten pool.

1. Introduction

At present, the welded structures from small-diameter pipes are widely used in the industry (oil and gas, chemical, machine building, etc.), to create various articles for household purposes (metal furniture, advertisements, fences and other) and in various equipment (direct incuts into pipelines for mounting pressure gauges, piping of valve stations, pulse tubes for valve control). The household articles, in addition to operational requirements, should meet pleasant exterior requirements.

Interestingly, literature sources pay insufficient attention to welding of curvilinear surfaces of small diameter. Though, it is well known that welding seam formation in such articles is accompanied by metal over-heating, as compared to welding of flat specimens of the same thickness. It is believed, that overheating is imminent due to low heat dissipation inside welded pipes. In this connection, the manufacturing of relatively small critical structures often involves more labor-intensive mechanical processing (turning, machining, drilling, etc.), for instance in bearing production.

The aim of the work is to build a computational model for visualizing the formation of curvilinear welded joints and distinguishing the effect of the surface curvature on selection of welding regimes.

Materials and Methods.

The study compares welding of 09G2S steel pipes (as per GOST 19281–89) with welding of plates. The calculations were performed by argon-tungsten-arc welding as per GOST 2246–70. Curvilinear welding models were created by special software developed and tested by the authors for this work.

The calculations were validated by real experiments.

2. Problem solution method

Small-diameter pipes are usually welded by manual argon tungsten-arc welding. The standards allow welding flat parts without a filler wire; conversely, small-diameter pipes are welded only with a filler wire. Fig. 1 depicts steel specimens welded with and without a filler wire. Evidently, in the latter case, there is no weld bead, the seam dropped inside the pipe due to lacking heat sink and metal overheating. The overheat is known to be diminished by increased welding speed. However, in manual welding it is hard to measure and increase the speed.

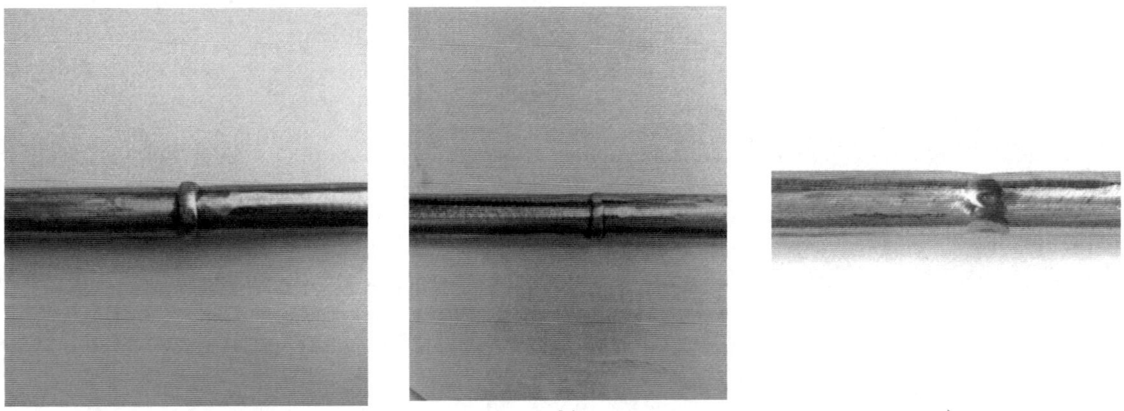

a) b) c)

Figure 1. Appearance of steel pipe welding seam: a) two-pass welding with filler wire; b) one-pass welding with filler wire; c) welding without filler wire.

At the same time, finding the arcing regimes that provide high-quality connection without a filler wire is of large interest. This is possible when the melt viscosity is sufficiently high and the melt has enough time to crystallize before it falls into the pipe under the gravity. This problem was solved numerically using finite difference method [1]. The contact zone of the welding parts was assumed to be absolutely smooth. In the calculations, it was distinguished by a line or plane of symmetry. Taking into account that the heat processes in the contact zone of arc with pipe surface are independent of the electrode movement direction, only linear trajectories of movement along the cylinder generatrix were studied. The calculation used a mathematical model of spatial heat conductivity problem with due regard to the moving heating source, presence of different states of aggregation (solid, liquid, gaseous, plasma) and structure-and-phase transitions [2]. The problem was solved in Cartesian rectangular coordinates by dividing the volume into cubic cells of different dimensions that were used to compose cylindrical panels of a given diameter (Fig. 2).

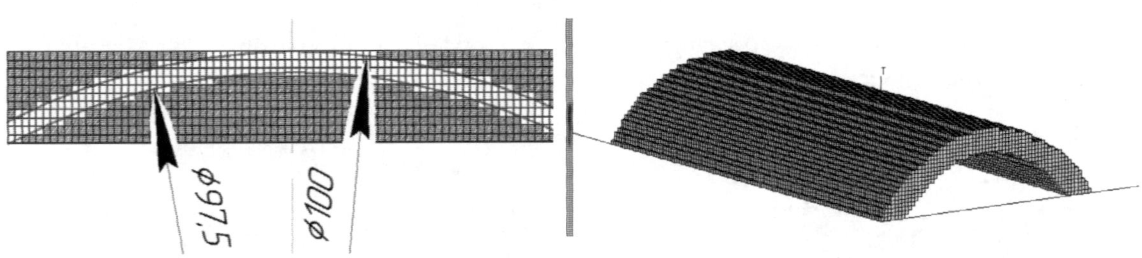

a) b)

Figure 2. Example of formation of calculated cylindrical panel: a) plane section for 100-mm pipe with wall thickness of 2.5 mm; b) 21.6-mm pipe fragment with wall thickness of 1.8 mm.

The calculations were made for pipes of different diameter: from 14 to 100 mm. The thicknesses were set as per GOST. Arcing regimes were selected to achieve a burn-through with the width of no less than 1–1.5 mm from the outside. The weld width was not specified; it was a consequence of preset conditions.

3. Digital Visualization Results and Discussion

When solving a problem of numerical modeling of heat processes in a molten pool, a particular difficulty is to output the results and represent the dynamics of spatial phenomena on a computer monitor. The main complexity is connected with a large volume of spatial-temporal information to be processed which appreciably inflates with increased number of input and output model parameters. The input parameters are arcing regimes, geometry of welded parts and physical material properties. The output parameters are the geometry of the welded joint, number and dimensions of the zones of structure-phase transitions. The compactness and readability of the whole information volume affects the understanding of phenomena hidden from an observer.

We will omit the methodology of information processing and give the study results. Some differences in welding of flat specimens and small-diameter pipes were distinguished. First of all, the mechanism of initial molten pool formation is different. For flat specimens, the molten pool shape in the cross-section is known to be semi-spherical or semi-elliptical. In pipe-shaped specimens with large curvature, it is pear-shaped. This is explained by different contact zones of the arc with the surface of the cylindrical panel and flat plate. Hence, the heat fluxes inside the metal will be different as well.

Fig. 3 depicts the molten zone profiles in horizontal sections of 33.5-mm pipe with the wall thickness of 2.8 mm at different distance below the surface.

For the purpose of comparison, we present the molten zone profile in a plate with the same thickness. Arcing regimes for the pipe were as follows: current I = 90 A, voltage U = 12 V, speed V = 155 mm/min. However, they were not suitable for plate welding, because the weld width on the back side turned out to exceed the standards, so the parameters were adjusted as follows: current I = 125 A, voltage U = 12 V, speed V = 200 mm/min.

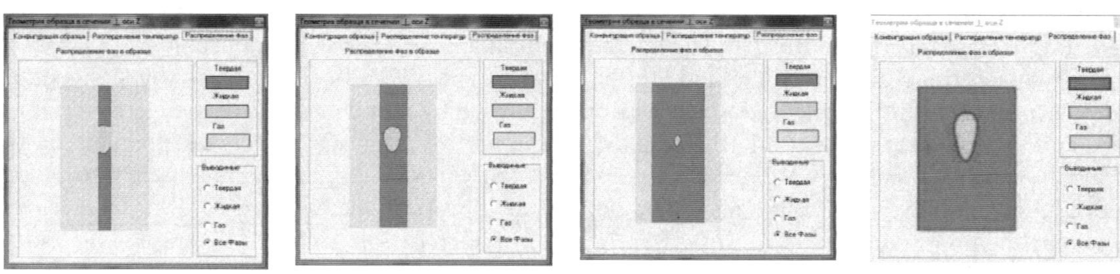

a) b) c) d)

Figure 3. Molten zone in horizontal sections of weld at different distance below the pipe surface: a) 0.2 mm; b) 1.0 mm; c) 2.8 mm; and in plate d) 0.2 mm

The calculations also revealed that the curvature of welded specimens affects the shape of the temperature front. It flattens with decreasing pipe diameter (Fig. 4).

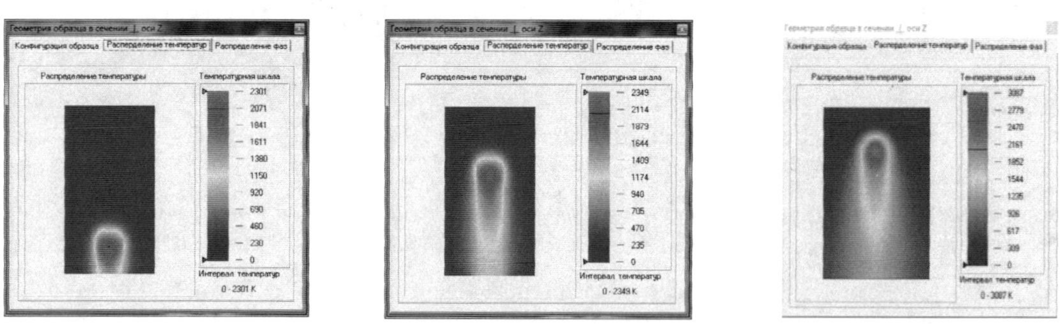

a) b) c)

Figure 4. Temperature fields in the subsurface zone at different moments of time from welding beginning: a) and b) cylindrical panel; c) plate.

To assess the validity of the calculations, line welding of a 33.5-mm steel pipe was performed at regimes described above and without a filler wire (Fig. 5). The pipe was made of 09G2S steel.

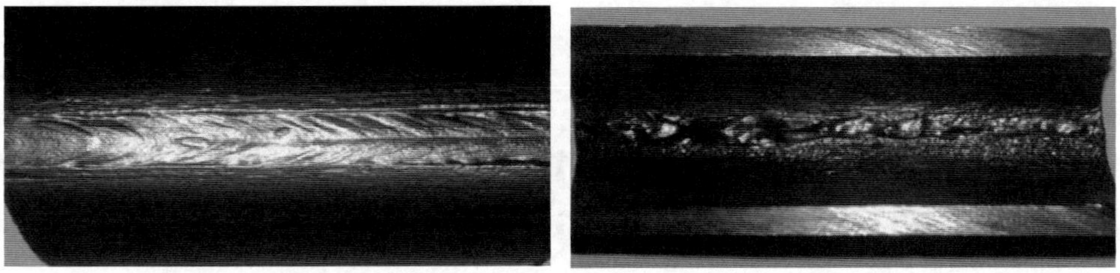

Figure 5. Photo of the weld: top view (left), bottom view (right).

Thus, the experiment has shown, that the arcing regimes were calculated correctly in general. The filler wire was not required; the weld seam formed without the melt dropping inside the pipe.

4. Conclusions

The work has suggested a numerical method for calculating heat processes accompanying the welding of nonlinear specimens in Cartesian rectangular coordinates. Such approach allowed visualizing the dynamics of the molten pool formation in small-diameter pipes and, most importantly, provide adequate comparison with similar processes when welding plates. The work has established the effect of welded specimen geometry on the thermal processes in the molten pool. The real experiment has testified the possibility to weld small-diameter pipes without a filler wire in the case of correct calculation of the arcing regimes.

Acknowledgement

The work was funded by RSF project No. 16-19-10010P (stage of 2019).

References

[1] Krektuleva R.A., Bezginov R.O., Cherepanov R.O., Cherepanov O.I. Thermophysical processes in contacting steel St3-aluminum pair in argon-arc welding by non-consumable electrode//Physical mesomechanics. 2015. 18 (3). 92–100/

[2] Saraev Yu.N., Krektuleva R.A., Kosyakov V.A. Mathematical modeling technological processes of pulsed argon TIG welding//Welding International.1997. 11 (10). 807–809.

Determination of mechanical properties of coatings obtained by non-vacuum electron beam cladding

T A Krylova[1] **and Yu A Chumakov**[1]

[1] Institute of Strength Physics and Materials Science of SB RAS, Tomsk, Russia

E-mail: kta @ispms.ru

Abstract. A study on the mechanical properties of composite materials obtained by the method of non-vacuum electron-beam cladding of refractory powders on a steel substrate was carried out. The mechanical properties of composite with coatings were investigated by a three-point bending test at room temperature. The strength of the obtained composite was 4.5-6 times higher compared to that of low carbon steel depending on the deposition modes.

1. Introduction

One of the advanced methods for applying hardening coatings is the method of electron beam cladding in air. This technology is based on the use of an accelerator of relativistic electrons [1, 2]. The method allows obtaining cladded layer with a uniform structure, possessing higher hardness, wear resistance, corrosion resistance, heat resistance and almost complete absence of cracks, cavities and micropores [3].

There are numerous investigations devoted to determining the elastic and strength characteristics of coatings obtained by the method of plasma and thermal spray, laser and electron beam cladding, and other methods [4-6]. However, deep knowledge is required in the field of structural strength under various loading conditions for widespread adoption of new coatings. The experience of their use is insufficient, and the influence of various factors has been little studied [7, 8].

The hardening coatings are basically used to protect metal elements of constructions and parts from various external influences (corrosion, wear, destruction, etc.). In order to establish the strength limits of the surface layers of most structural elements and parts, special studies must be carried out. One of the most common tests of materials and constructions is the quasi-static bending test. This method is used to determine the ability of a material to withstand a given plastic deformation and identify the ultimate bending strength. These properties complement the basic information about the coating (structure, phase composition, hardness) which predetermines its purpose. This type of test applies to cast iron, tool steel, steels after surface hardening and ceramics.

The aim of the work is to determine the mechanical properties of the composites obtained by the method of non-vacuum electron-beam cladding of the powder mixture (Cr_3C_2 + TiC) on low carbon steel. In this study, the ultimate strength and deflection of composite with coating were determined depending on the deposition modes.

2. Experimental

The powder mixture for cladding consisted of chromium carbide (Cr_3C_2, 54 wt%), titanium carbide (TiC, 6 wt%) and fluoride magnesium (MgF_2, 40 wt%, it was used as a flux, protecting the melt from oxygen). The particles of all powders were 2 – 5 μm in size. The electron-beam energy was 1.4 MeV. The beam current was 25 mA. The distance from the beam outlet to the workpiece was 90 mm. The

Gaussian diameter of the electron beam on the specimen surface was 12 mm. The width of the scanning was 50 mm. Five coatings were obtained corresponding to different velocities (v) of the samples under the beam and irradiation energies (W) (Table.1).

Table 1. Processing modes

Number	1	2	3	4	5
v (cm/s)	0.95	0.8	0.7	0.6	0.55
W (kJ/cm^2)	7.4	8.8	10.0	11.7	12.7
Layer thickness (mm)	1.7	1.3	1.5	3.7	4

The tests on quasi-static three-point bending (GOST 14019-03) were carried out on an Instron-1185 with automatic recording of the "load *P*- indenter displacement" diagrams. The indenter speed was 0.5 mm/min. The preload was 100 N. The distance between the supports was 35 mm. All tests were carried out at room temperature. The samples were fabricated using the electric spark machine. The final sample size of 5 × 5 × 50 mm was obtained manually by grinding and polishing. All samples were the same size, but the thickness of the cladding layer varied from 1.3 to 4 mm (see table 1). Bending tests were carried out before the first crack in the stretched zone of cladded layer.

The sample was installed on two supports. The load was applied exactly in the middle of the sample. Hardened cladding layer had tensile stresses during lowering of the upper indenter of the machine. The substrate subjected to compressive stresses. This method has become widespread due to the ease of implementation.

The strength of the samples was determined by the test results. The amount of deflection was also estimated. Simultaneously, the pattern of crack formation from the moment of their origin to the transition to the substrate metal with subsequent plastic deformation was investigated on the side polished surface of the samples.

The bending strength on the outer surfaces of the sample (on the cladding side) was calculated by the formula:

$$\sigma_b = \frac{3PL}{2bh^2}, \tag{1}$$

where *P* is the load (N), *L* is the distance between the supports (mm), *b* is the sample width (mm), *h* is the sample thickness (mm).

3. Results and Discussion

The dependence of the applied load *P* on the indenter displacement was obtained for different samples (Fig. 1). The maximum bending deflection for each sample was determined from the diagram (Table 2).

The deflection f can be divided into two regions (Fig. 1): f_{pl} determines a certain reserve of plasticity of the coating and f_d determines the nature of the fracture.

Thus, the analysis of the diagrams showed that samples 1, 3, 4, and 5 had increased ductility and strength. Composition with coating 5 was the strongest σ_b = 877.128 MPa. The maximum plasticity with the value f_{pl} = 1.256 mm was observed in sample 1. Sample 2, possessing some plasticity, was the most brittle. The maximum load on the curve of diagram was σ_b = 677.124 MPa.

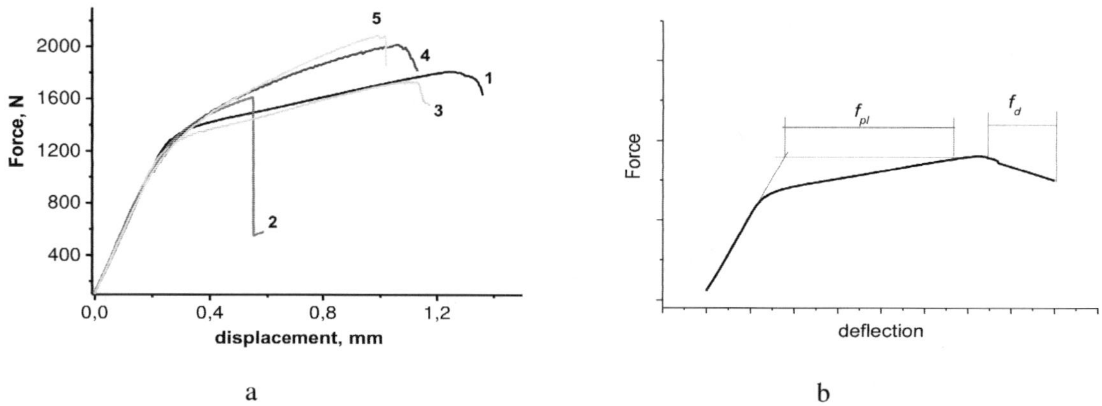

Figure 1. Force versus indenter displacement for samples with different thick coatings (a) and a schema the bending strain curve (b).

The character of coating destruction follows from loading curves (Fig.1). In our case of statistical bending, the destruction of two main types is possible: fragile and viscous. However, mixed destruction occurs most often in practice. Sometimes destruction can begin as viscous, and end as fragile.

Analysis of composite deflection indicated (Fig. 1) that a crack in sample 5 was formed due to brittle fracture. The length of f_d decreased to almost zero (see Table 2). In samples 1 and 4, segment f_d had the form of ductile fracture until the first crack appears. The fracture mode f_d before the appearance of the first crack corresponded to the case of ductile failure for the samples 1 and 4. The fracture mechanism of samples 2 and 3 was mixed (elastic-plastic).

Table 2. Bend test results

Number	P [N]	f [mm]	f_{pl} [mm]	f_d [mm]	σ_b [MPa]
1	1807	1.359	1.256	0.103	758.94
2	1612.2	0.585	0.548	0.037	677.124
3	1735.1	1.173	1.131	0.042	728.742
4	2021.7	1.13	1.064	0.066	848.82
5	2088.4	1.019	1.016	0.003	877.128
low carbon steel					150

As visual inspection of the samples after bending tests showed, the composite with coating 2 was broken with the formation of one crack passing through the entire coating both in width and in thickness (Fig. 2a). The crack was developed under the action of relatively low stresses but it development was brittle or quasi-brittle. The main crack penetrated into the substrate (low carbon steel) and branched into several small. This phenomenon is due to plasticity of steel. As shown in curve 2, Fig. 1a, the development of cracks was inhibited and stopped by the steel substrate. In samples 1, 3, 4 and 5, only nucleation of cracks were observed. Coatings were deformed at elevated loads fig. 1a before the formation of a single sinuous crack, while possessing some plasticity. Figure 2b shows a metallographic image of the side surface of the coating 4 after bending. The nature of the sinuous crack and the deformation curve of the coating 4 in Fig. 1a indicate a quasi-brittle fracture. Also, the presence of a small zone of plastic deformation in front of the edge of the crack and small secondary cracks shows a quasi- viscous fracture.

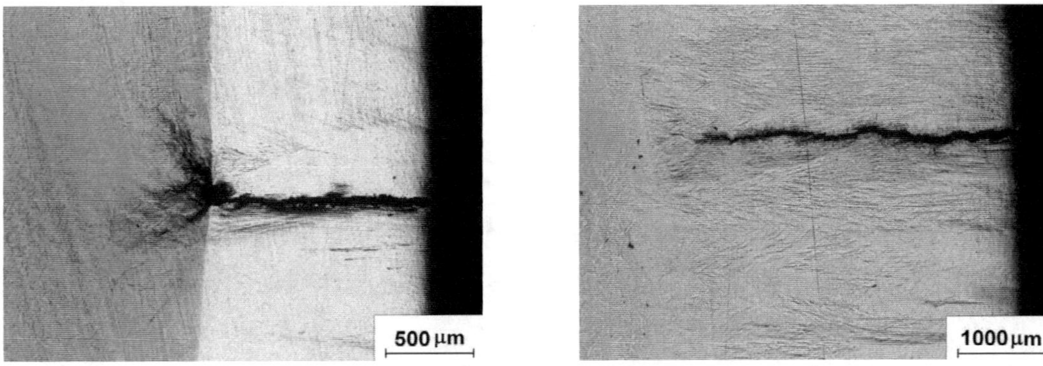

Figure 2. Metallographic image of a crack in coating after bending tests until the first crack appears for different samples: a) 2, b) 4.

Since this work investigated samples of the same size, but with different thicknesses of the deposited layer and obtained under different processing conditions. Further research is needed to understand the reasons for the observed effects. In particular, it is necessary to conduct a fractographic study of the specimen destruction - fracture, the establishment of the relationship between the mechanical properties, structure and microhardness of the deposited coatings. This is the subject of further research.

4. Conclusion

The method of non-vacuum electron-beam surfacing allows obtaining coatings with different thickness of the deposited layer. Depending on the processing mode, coatings have different properties. Mechanical tests of compositions with a three-point bending at room temperature showed that surfacing the powder mixture (Cr_3C_2 + TiC) on a steel substrate can increase the tensile strength by 4.5–6 times compared to the ultimate strength of low carbon steel.

Acknowledgments

The authors are thankful to Dr. M.G. Golkovsky (Budker Institute of Nuclear Physics of SB RAS, Novosibirsk) for electron-beam cladding.

The work was carried out in the framework of the Basic Research Program of the State Academies of Sciences for 2013–2020 (Project No. 23.2.1).

References

[1] Fominsky L P and Kazansky V V 1985 *Svarochnoe Proizvodstvo* **5** 13–15.
[2] Poletika I M, Borisov M D, Kraev G V, Vaisman A F, Golkovsky M G and Durakov V A 1996 *Russian physics journal* **39** 275-283.
[3] Bataev I A, Bataev A A, Golkovski M G, Krivizhenko D S, Losinskaya A A and Lenivtseva O G 2013*Applied Surface Science* **284** 472–481.
[4] Bashta A V 2012 *News of NTU "KhTU"* **48** 15-24.
[5] Kopylov V I 2016 *East European Journal of Advanced Technologies* **5** (5) 49–57.
[6] Panin S V, Belotserkovsky M A, Seifullina M P, Pochivalov Yu I and Ovechkin B B 2004 *Physical Mesomechanics* **7** 107-116
[7] Poletika I M, Golkovskii M G, Krylova T A and Perovskaya M V 2009 *Metal Science and Heat Treatment* **51**: 115-122.
[8] Krylova T A, Ivanov K V and Ovcharenko V E 2019 *Inorganic Materials: Applied Research* **10**

IOP Conf. Series: Materials Science and Engineering **681** (2019) 012021 doi:10.1088/1757-899X/681/1/012021

Analytical characteristics of the electron beam distribution density over the heated spot for optimizing the electron-beam welding process

S O Kurashkin[1], V D Laptenok[1], A V Murygin[1] and Yu N Seregin[1]

[1] Reshetnev Siberian State University of Science and Technology 31, Krasnoyarsky Rabochy Av., Krasnoyarsk, 660037, Russian Federation

E-mail: scorpion_ser@mail.ru

Abstract. The control of the electron beam energy distribution density over the heated spot, realized by using different scanning trajectories, can significantly improve the quality of welded joints. Experimental research by the authors showed that the best quality of welded joints is provided by scanning the electron beam in the form of a raster. Therefore, to optimize the process of electron-beam welding, trajectories of the classical raster, sinusoidal raster and truncated raster were proposed, and by increasing the scanning amplitude along the joint, the vapour-gas channel of penetration transforms into a stable cavity, on the front wall of which the metal melts, with it then being transferred along the side walls to the tail end of the weld pool. It is advisable to investigate the effect of the formation of the penetration cavity in the EBW of various materials and different thicknesses. For this, an apparatus must be created that implements scanning in the form of various rasters. For optimizing the process of electron-beam welding, the trajectories of the classic sawtooth raster, sinusoidal raster and truncated raster are proposed. For these scanning trajectories, analytical expressions and calculated characteristics of the electron beam energy distribution density are obtained over the heated spot. The trajectory in the form of a sinusoid and a truncated raster makes it possible to obtain double-humped energy distribution over the heated spot. The obtained characteristics facilitate the optimization process of electron-beam welding of different materials.

1. Introduction

The combination of energy and technological process parameters determine the quality of the weld during electron-beam welding (EBW). A deep penetration by a focused electron beam is created by forming a vapour-gas channel. Thus, the welding joint has a dagger shape with a characteristic mushroom-shaped formation in the upper part. The mushroom expansion of the weld in the upper part is associated with the periodic dispersion of the beam in the metal vapour. The pointed shape of the weld in the root part leads to the formation of root defects, or voids, arising due to the displacement of liquid metal by the penetration channel. Thus, to prevent such defects from occurring, it is necessary to form a wider root part of the welding joint. The simplest way to form the radius in the root of the seam is by defocusing the electron beam being used but it does not lead to the complete elimination of root defects.

To prevent root defects, it is necessary to form a paradynamic channel with a fairly wide lower part. The most effective way to influence the formation of the channel of penetration is by scanning the electron beam. The current trend in the development of welding technology is to use complex-shaped scanning for the formation of welding joints with a given cross-sectional shape. The use of digital signal generators allows us to create rasters of any shape with a sufficiently high resolution of up to 24 bits and also plays a major role in solving the problem of targeting the joint [1].

The method of choosing the scanning shape of the electron beam has not been developed yet. At the same time, a rather large amount of experimental data has been accumulated [2]. The most widely used forms of scanning beam trajectory are: longitudinal [3-4], x-shaped [5-7], circular and elliptical [8-10], and arc and clamp [11-12]. The scanning amplitudes typically vary between 1-3 mm.

Electron-beam welding by rotating an electron beam along a circular trajectory makes it possible to achieve a significant reduction in root defects and picoformation; however, since the power density in the central part of the heating zone is small, the use of a circular scan reduces the penetration depth compared to a stationary electron-beam welding beam.

In earlier published paper [13] present experimental research on the application of various scanning trajectories by electron beam with the focus on the high-quality formation of welded joints. The research showed that the best scanning trajectory for the quality of weld formation is a raster in which scanning of electron beam along the coordinate across the joint is according to the one-sided sawtooth law. For one scan along the junction, 16, 32 or 64 scans are carried across the junction along a two-sided sawtooth path. The scan period along the junction is equal to 1-2 ms.

More detailed studies with scanning in the form of a raster revealed that with an increase in the amplitude of scanning along the joint from 5 mm when welding small thicknesses to 15 mm when welding medium thicknesses, the vapour-gas channel is transformed into a stable cavity to the full depth of penetration [14]. The cavity shape was fixed by abrupt termination of the welding process by switching off the accelerating voltage.

Figure 1 shows a scheme of scanning by electron beam in the coordinate system associated with the weld pool. The shape of the resulting weld joint has almost parallel walls and a significant radius at the root, which allows root defects to be eliminated. The instability of the depth of penetration decreased by three to four times. The openness and stability of the penetration cavity leads to the degassing of the weld pool and a decrease in porosity. By reducing the proportion of energy spent on the evaporation of the material, the penetration capacity of the EBW process increases.

It is advisable to investigate the effect of the formation of the penetration cavity when welding various materials and different thicknesses. For this purpose, equipment should be created that implements scanning in the form of various rasters. In this paper, we consider three types of scanning: classic raster, sinusoidal raster and truncated raster. The issue of this research is to obtain the density of the energy distribution of the scanning by electron beam over the heated spot. It characterizes the distribution of energy in the weld pool.

2. Modelling method

The energy distribution in an electron beam is usually described by a normal distribution law characterized by the probability density of the species [15]:

$$f(x) = \frac{1}{\sigma * \sqrt{2 * \pi}} * \exp\left(-\frac{(x-m)^2}{2 * \sigma^2} \right), \quad (1)$$

where x is the abscissa, m is an expected value, and σ is a standard deviation characterizing the diameter of the electron beam and the concentration of energy in it. This value is determined by the electron-optical system of the gun and its focusing system. It depends on the welding current and is usually determined within $\sigma = 0.05$ — 0.5 mm. The smaller the diameter of the electron beam, the better the beam forms the gun.

In order to disregard the specific value of the standard deviation when constructing the characteristics of the energy distribution density over the heated spot, we introduce dimensionless coordinates and amplitudes of the beam scanning:

IOP Conf. Series: Materials Science and Engineering **681** (2019) 012021 doi:10.1088/1757-899X/681/1/012021

$$\bar{y} = \frac{y}{\sigma}, \tag{2}$$

where y – a coordinate along the junction, σ – a standard deviation, $\bar{y}$ – a dimensionless coordinate along the junction. In a system of dimensionless coordinates σ=1.

$$\bar{x} = \frac{x}{\sigma}, \tag{3}$$

where x – a coordinate across the junction, $\bar{x}$ – a dimensionless coordinate across the junction.

$$\bar{A} = \frac{A}{\sigma}, \tag{4}$$

where A – an amplitude, σ – a standard deviation, and $\bar{A}$ – a dimensionless amplitude.

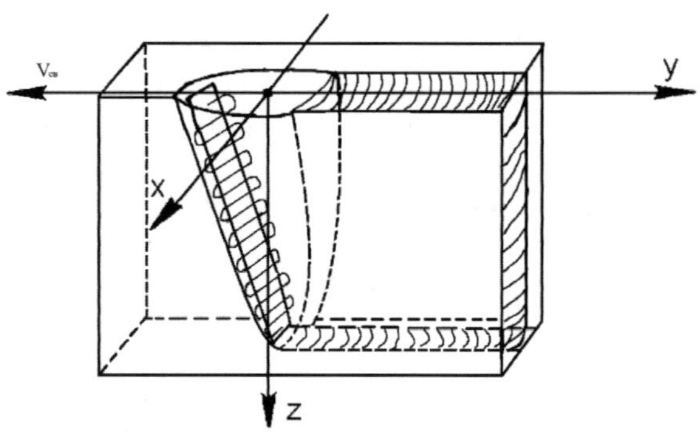

Figure 1. A scheme of electron-beam welding during scanning in the form of a raster.

The maximum ordinate of the normal distribution law (1), corresponds to the point $x=m$; the distance from point m of the distribution density decreases. The centre of symmetry of the distribution is the centre of dispersion m. The centre of diffusion characterizes the position of the distribution on the abscissa axis [15]. Therefore, by changing the centre of scattering according to the corresponding law, it is possible to obtain the density of energy distribution along the corresponding coordinate.

The electron beam energy is described by the following formula:

$$W(x, y) = U * I * \int_{-\infty}^{\infty}\int_{-\infty}^{\infty} W(x)W(y)dxdy, \tag{5}$$

where U – an accelerating voltage, I – a beam current, W(x) and W(y) – normalized energy distribution functions for the corresponding coordinates, $\int_{-\infty}^{\infty} W(x)dx = 1$, $\int_{-\infty}^{\infty} W(y)dy = 1$.

Scanning in the form of a classical raster while moving along the coordinate across the joint is carried out by a one-sided saw described by a piecewise linear function:

$$y = k * t + m, \tag{6}$$

where t – an independent variable, k and m – numbers, and provided that k = 1 for the time t = [-1;1] with period T, the following graph will be in the form of the saw in Figure 2.

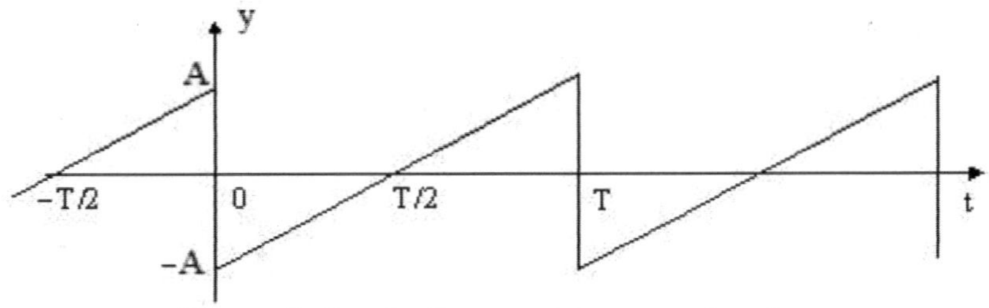

Figure 2. Graph of successive sawtooth pulses.

It is known that when we change the expectation in the formula (1), the graph shifts relative to its centre. Therefore, instead of the expected value, in the formula (1), we can enter the formula (6) and integrate it to get the energy distribution curve. We get the following mathematical formula:

$$W(\bar{y}) = \int_{-1}^{1} \frac{1}{\sigma * \sqrt{2*\pi}} * \exp\left(-\frac{\left(\bar{y} - \left(\overline{A_y} * k * t + m\right)\right)^2}{2*\sigma^2} \right) dt, \quad (7)$$

where $\overline{A_y} \in [0;25]$, $k = 1$, $m = 0$, $y \in [-30;30]$.

3. Experimental study

Figure 3 shows curves with a normal energy distribution for the case of a scanning electron beam trajectory along the junction in the form of a raster.

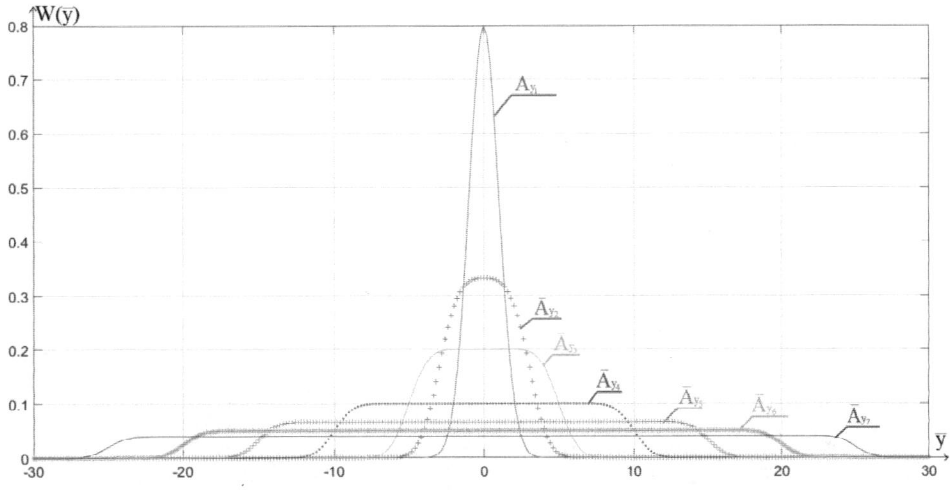

Figure 3. Energy distribution for the case of a scanning trajectory in the form of a raster: $\overline{A_{y_1}} = 0$; $\overline{A_{y_2}} = 3$; $\overline{A_{y_3}} = 5$; $\overline{A_{y4}} = 10$; $\overline{A_{y5}} = 15$; $\overline{A_{y6}} = 20$; $\overline{A_{y_7}} = 25$

Figure 4 shows a sinusoidal electron beam scanning trajectory. The sinusoidal trajectory is described by the following formula:

$$F(\varphi) = A * \sin(\varphi), \quad (8)$$

where $\varphi \in [0;2*\pi]$ – the phase.

Therefore, instead of the expected value, in the formula (1) we can enter the formula (8) and integrate it to get the energy distribution curve when applying the normal distribution law on the sinusoidal scanning trajectory (Figure 5). We get the following mathematical formula:

$$W(\bar{x}) = \int_{-\pi}^{\pi} \frac{1}{\sigma * \sqrt{2*\pi}} * \exp\left(-\frac{\left(\bar{x} - \left(\overline{A_x} * \sin(\varphi)\right)\right)^2}{2*\sigma^2} \right) d\varphi, \quad (9)$$

where $\overline{A}_x \in [1;10]$, $\overline{x} \in [-15;15]$, $\varphi \in [-\pi;\pi]$ – the phase.

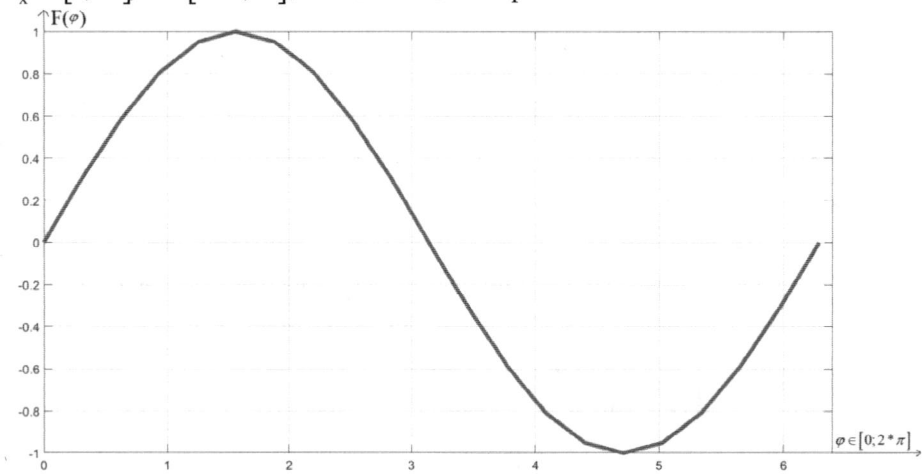

Figure 4. The shape of the controlling signal in the form of a sinusoid

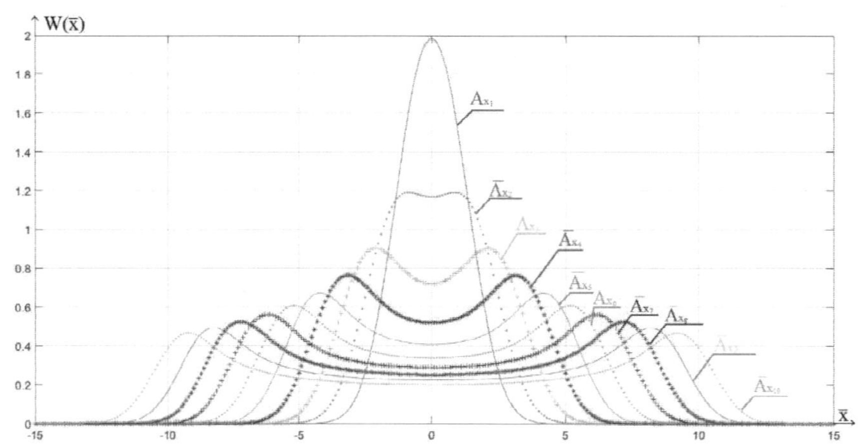

Figure 5. Energy distribution for the case of a scanning trajectory in the form of a sinusoidal: $\overline{A_{x_1}}=1$; $\overline{A_{x_2}}=2$; $\overline{A_{x_3}}=3$; $\overline{A_{x4}}=4$; $\overline{A_{x5}}=5$; $\overline{A_{x6}}=6$; $\overline{A_{x_7}}=7$; $\overline{A_{x_8}}=8$; $\overline{A_{x9}}=9$; $\overline{A_{x10}}=10$

When scanning with a truncated raster on certain areas for a certain period, the amplitude takes a fixed value. In formula (7) we add expressions with the help of which the peaks arising on values of equal amplitudes are described. We get the following formula:

$$W(\overline{x}) = \frac{T-2*M}{T} * \int_{-1}^{1} \frac{1}{\sigma*\sqrt{2*\pi}} * \exp\left(-\frac{\left(\overline{x}-\left(\overline{A}_x*k*t+m\right)\right)^2}{2*\sigma^2}\right)dt + \frac{M}{T} * \left(\exp\left(-\frac{\left(\overline{x}-\left(\overline{A}_x\right)\right)^2}{2*\sigma^2}\right) + \exp\left(-\frac{\left(\overline{x}+\left(\overline{A}_x\right)\right)^2}{2*\sigma^2}\right)\right), \quad (10)$$

where, $m = 0$, $k = 1$, M – a saturation time, T – a period of scanning.

Figure 6 shows curves with a normal energy distribution for the case of a scanning electron beam trajectory across the junction in the form of a truncated raster.

In order to use the obtained design characteristics, it is necessary to move from the dimensionless coordinates to dimensional coordinates, and for this, it is necessary to carry out the following steps:

$$y = \overline{y}*\sigma;\ x = \overline{x}*\sigma;\ A = \overline{A}*\sigma;\ W(x) = \frac{W(\overline{x})}{\sigma};\ W(y) = \frac{W(\overline{y})}{\sigma}$$

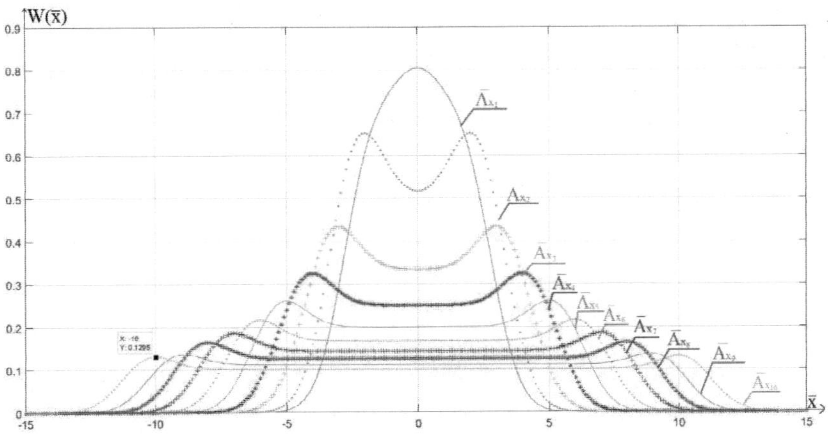

Figure 6. Energy distribution for the case of a scanning trajectory in the form of a truncated raster: $\overline{A_{x_1}} = 1$; $\overline{A_{x_2}} = 2$; $\overline{A_{x_3}} = 3$; $\overline{A_{x4}} = 4$; $\overline{A_{x5}} = 5$; $\overline{A_{x6}} = 6$; $\overline{A_{x_7}} = 7$; $\overline{A_{x_8}} = 8$; $\overline{A_{x9}} = 9$; $\overline{A_{x10}} = 10$

4. Conclusions

For research on the optimization of the process of electron-beam welding, it is advisable to use scanning by electron beam in the form of a raster, which allows a stable penetration cavity and high quality of welded joints to be obtained.

Sinusoid and truncated raster allow a double-humped distribution of energy along the coordinate across the joint to be obtained, which is necessary for the formation of a weld with parallel walls and a significant radius in the root, which excludes the appearance of root defects.

It is advisable to program the proposed scan forms in the control unit for electron beam scanning.

References

[1] Sobko S A, Kulikov V A 2013 *The method of control of electron-beam welding* (Russia: Moscow)

[2] Shcherbakov A B, Rodyakina R V, Novokreschenov V V and Lastovirya V N 2018 *Technology of material processing. Equipment of electron-beam complexes* (Moscow: Yurayt) chapter 4.3 pp 114-130

[3] Akopianc K S , Nesterenko V M 2002 Electron-beam welding of steels up to 600 m thick with longitudinal oscillations of the beam *Automatic welding* **9** 3-5

[4] Ohmine M, Hiramoto S, Jamane J 1983 Fundamental study on the pulsed electron beam welding // *Intern. Inst. of Welding* D **IV-348-83** p 13

[5] Belenkiy V Ya 1986 Scanning of an electron beam along an x-shaped trajectory as a means of reducing defects in the root of a weld in EBW *Automatic welding* **9** 35-37

[6] Barishev M S, Redchic A A 1990 Optimization of welding modes by oscillating electron beam *Int. Conf. on welding of non-ferrous metals* **4** p 34

[7] Zenker R 2002 Electron beam surface treatment and multipool welding – state of the art *Proceedings of the EBEAM. International Conference on High-Power Electron Beam Technology* **10** p 12-1–12-5

[8] Shilov G A, Akopianc K S and Kasatkin O O 1983 Influence of the frequency and diameter of the circular scan of the electron beam on the penetration in EBW *Automatic welding* **8** 25-28

[9] Akopianc K S, Emchenko-Ribko A V, Neporohin V Yu and Shilov G A 1989 Prevention of the formation of root defects in EBW with non-penetration of up to 60 mm depth *Automatic welding* **4** 30-34

[10] Komiro Y, Punshon C S, Gooch T G and Blakaley P S 1986 Effects of process parameters on centerline solidification in EB weld *Metal Constr* **2** 104–111

[11] Vasil'ev A M, Goncharov V A, Krivkov B G, Solyankin V V and Sharonov N I 1988 Electronic beam control device by welding *Shipbuilding industry: Series "Welding"* **6** 11-12
[12] Mara C l, Funk E R, McMaster R C and Pence P E 1974 Penetration mechanism of electron beam Welding and spiking phenomenon *Welding Journal* **53**(**6**) 246-251
[13] Seregin Yu N, Laptenok V D, Uspenskiy N V and Nikitin V P 2011 Experimental research on the optimization of the technology of electron-beam welding of aluminum alloys *Technologies and equipment of EBW-2011 Materials of the St.-Petersburg International Scientific and Technical Conference* p 71-80
[14] Seregin Yu N, Laptenok V D, Murygin A V and Bocharov A N 2019 Experimental research on electron-beam welding technology with a scanning electron beam *21st Int. Sci. Conf. Resetnev Readings-2017: IOP Conf. Series: Materials Science and Engineering* **467** p 1-8
[15] Venttsel' Ye S 1999 *Probability theory* (Moscow: Moscow State University Press) pp 116-119

IOP Conf. Series: Materials Science and Engineering **681** (2019) 012022 doi:10.1088/1757-899X/681/1/012022

Investigation of the stress state of flat truss rods. Numerical and physical modeling

T S Kiychenko[1], M V Tabanyukhova[1], N V Kharinova[1]

[1] Department of structural mechanics, Novosibirsk State University of Architecture and Civil Engineering (Sibstrin), Leningradskaya str., 113, Novosibirsk, 630008, Russia

E-mail: m.tabanyukhova@sibstrin.ru

Abstract. The paper presents the results of an experimental numerical study of the stress state of the rods of a flat truss and the results of the analytical calculation. The values of internal forces and stresses in the truss model were obtained using the SCAD software package, as well as experimentally using the polarization-optical method. The finite element method (FEM) is used to calculate two plane truss models. In the first case, the truss was modeled by rod elements, in the second – by plates. The fields of normal, tangential and principal stresses are obtained. With their help there were found the longitudinal internal forces in the rods of the farm. The interference bands in the model of a flat truss are obtained by the method of photoelasticity. Stresses are calculated with their help. The quantitative and qualitative analysis of the stress state, described in the numerical and physical experiment, is carried out.

1. Introduction

A truss is a structure consisting of perfectly rectilinear rods connected in nodes by ideal cylindrical hinges and working to perceive only the nodal load. The practice of calculations shows that replacing the hinge nodes in the farm with hard ones with this method of loading does not affect the stress-strain state of the structure. To determine the efforts in the rods of the trusses and movements there are many methods and specialized programs [1-15]. However, in all cases the truss is calculated as a rod structure with a constant longitudinal force within a single rod.

The point of interest is the stress state of a flat truss with rigid nodes under a nodal load transfer. For the study were used a numerical and physical experiment. As for the numerical determination of voltages, there is no high guarantee that they will be determined exactly. Solutions of such tasks, obtained with the help of software systems, require experimental checks. There are a number of experimental methods for studying stresses [16]. In this work, we used the polarization-optical method for studying stresses, namely the photoelasticity method [17–18].

2. Physical experiment

In work the facility PPU-7 (polarization-projection facility) was used. In order to study the stress state of a flat truss, a model was made of a polarization-optical material - epoxy resin (modulus of elasticity E = 3000 MPa, Poisson's ratio $\nu = 0.38$, price of a material band by stress $\sigma_0^{1,0} = 1.4$ MPa·cm) . The sample thickness is 5 mm, its dimensions in mm are presented at Figure 1, and the loading diagram is shown at Figure 2.

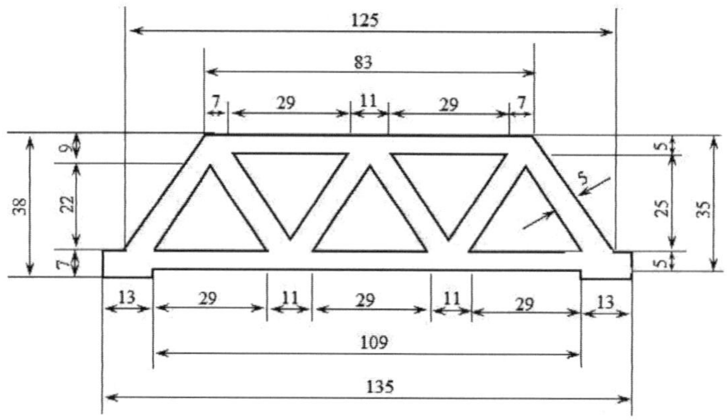

Figure 1. Truss model

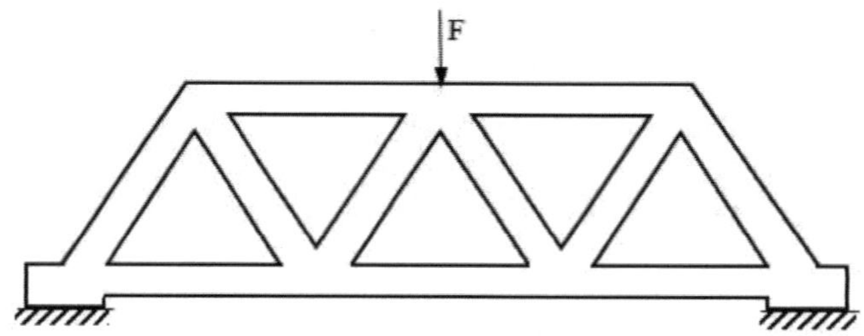

Figure 2. Loading diagram

The studies were performed using the photoelasticity method for direct radiography. During the experiment, there were obtained pictures of interference bands. A fragment of the interference band pattern for a model of epoxy resin is shown in Fig. 3, the pattern of the bands corresponds to a load of 0.4 kN. On the photo 1, 2, 3, ... denote the numbers of the stripes. Dark stripes correspond to the whole number of the strip, and light stripes – to the half.

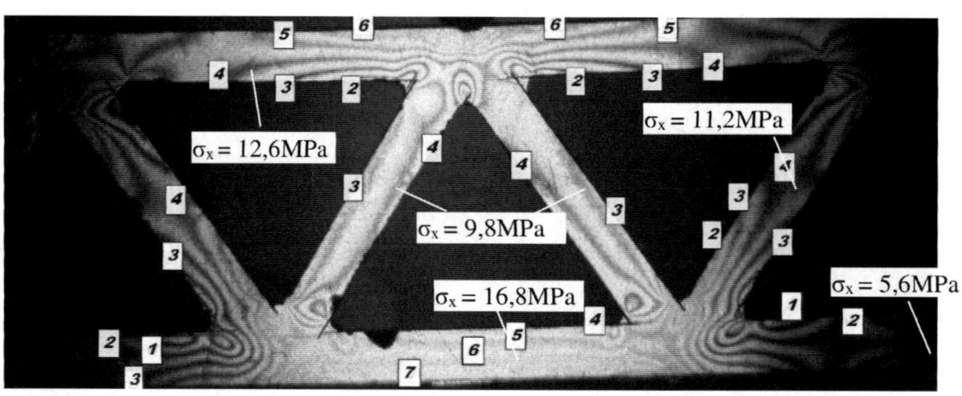

Figure 3. Fragment of the interference band pattern

On a fragment of the interference band pattern (Fig. 3) it can be seen that the rods are loaded unevenly. The stresses with the greatest magnitude appear under force (at the top on the vertical axis of symmetry). The most loaded is a horizontal rod located in the lower belt of the farm under power. Analyzing the pattern of interference bands, we can conclude that the normal stresses in the cross sections of the rods are not constant values. If they were constant, then a single-colored picture within each rod would be observed in the photo. In this case, we have bands of a different order. This suggests that the rods undergo eccentric tension (compression). With all this, assessing the order of the bands and their location within the rod, we can conclude that the eccentricity that occurs is insignificant. Using the data of the photoelasticity method, we calculated normal stresses in the rods of the truss model. The determination of stresses was calculated on the contour of the rod by multiplying the number of the strip by the price of the strip of material by the stresses. Such an approach is simple and convenient and is often used to determine stresses near geometric concentrators [19, 20]. The absolute values of the voltages found in this way are shown at Figure 3.

3. Numerical experiment

The finite element method (FEM) with the use of the SCAD software package was used to calculate two flat truss models. In the first case, the truss was modeled with rod elements (Fig. 4a), in the second - with plates (Fig. 4b). The triangulation of the plates is made of standard quadrangular elements, possibly rectangular in shape, one of the sides of which is oriented along the longitudinal axis of the truss rod, the cell size is 1x1 mm. Both models are loaded with a concentrated force applied to the node of the upper belt located on the axis of symmetry of the structure, node 6. The magnitude of the force was 0.4 kN. The material of the models, dimensions and their geometry are taken as in a physical experiment. The extreme left and right nodes of the lower chord of the trusses (nodes 1 and 4) are fixed from linear displacements. In the model specified by the core elements, no hinges were inserted in the nodes of their connections, i.e. farm knots are tough. In a truss modeled with plates, the joints of the elements are also rigid. As a result, two truss models with rigid nodes, working on the beam scheme, were obtained.

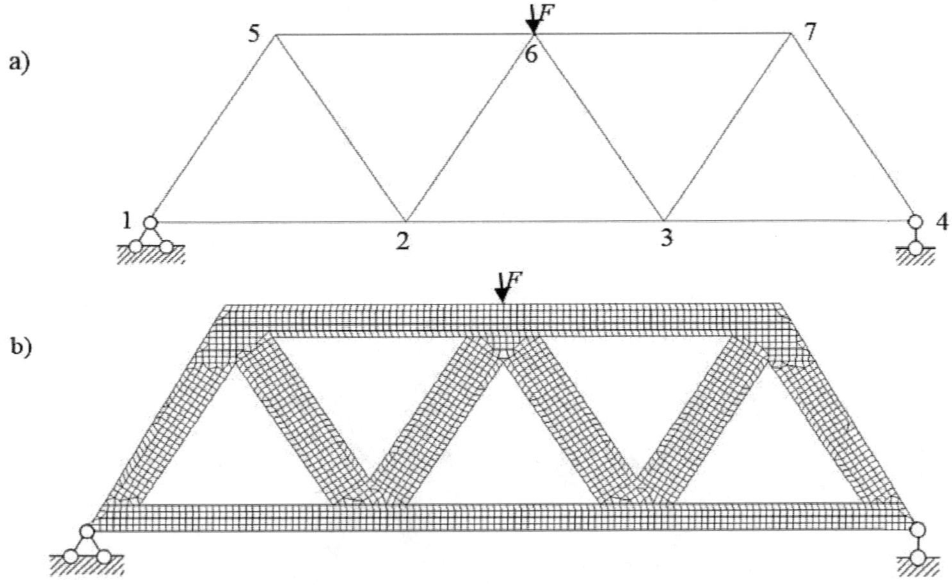

Figure 4. Load patterns of computational models

Fig. 5 shows the plot of the longitudinal forces in the truss rods. The efforts in the truss are obtained by a numerical method; they fully correspond to the similar efforts for a hinged truss under the same conditions of fastening and load, which can be obtained by methods of structural mechanics. As expected, with the beam operation, the truss structure is obtained: the upper belt is compressed, the lower one is stretched, the ascending braces are compressed, the descending ones are stretched.

IOP Conf. Series: Materials Science and Engineering **681** (2019) 012022 doi:10.1088/1757-899X/681/1/012022

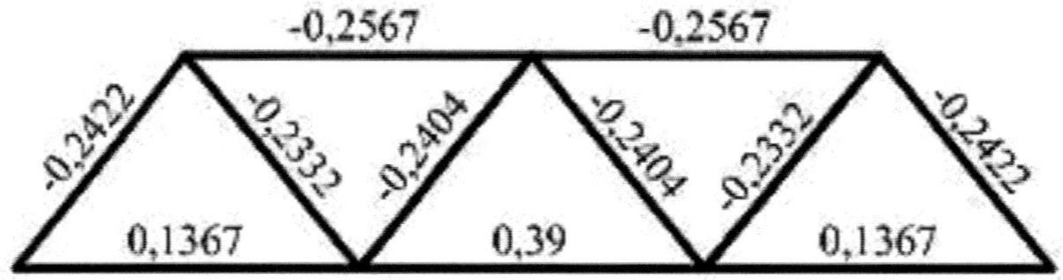

Figure 5. The forces in the truss rods (kN) obtained by the SCAD software package

According to the formula for determining stresses under axial tension-compression, using the received forces, the values of normal stresses in truss rods are determined. The values of the resulting stresses are shown in Fig. 6.

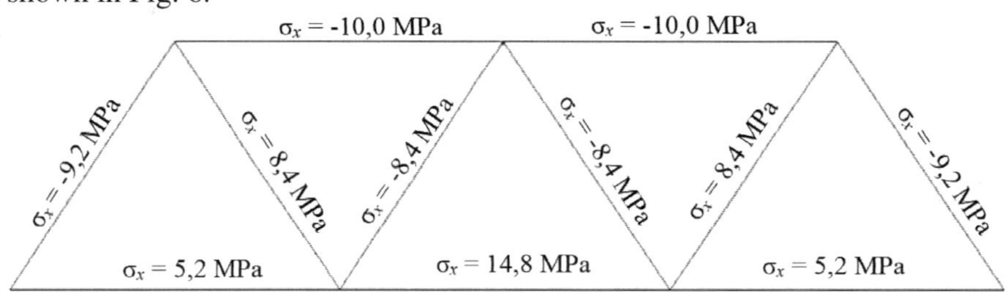

Figure 6. Normal stresses in truss rods (MPa) obtained by the SCAD software package

Fig. 7 shows the stress fields σ_1, σ_3, and τ_{xy} in the truss model, given with plates. In the upper zone of the truss, the stresses are negative, in the lower zone - positive, which coincides with the calculation of the truss, as a core structure, by methods of structural mechanics. The stresses in the bracings also coincide in sign with the calculation of the rod diagram.

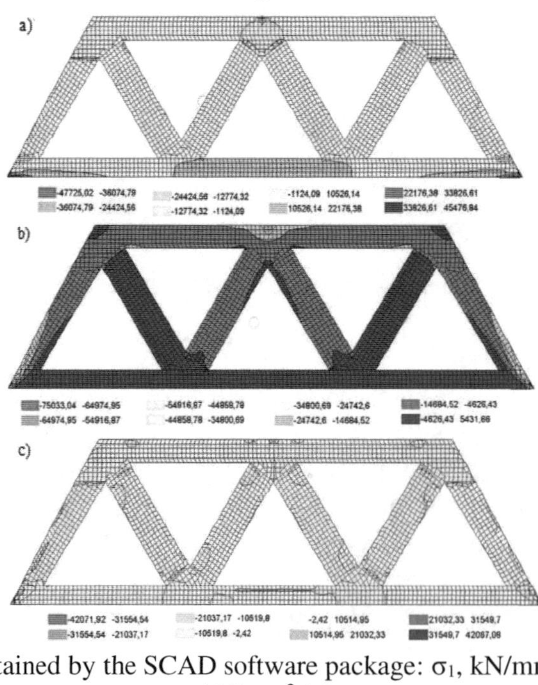

Figure 7. Stress fields obtained by the SCAD software package: σ_1, kN/mm^2 (a), σ_3, kN/mm^2 (b), τ_{xy}, kN/mm^2 (c).

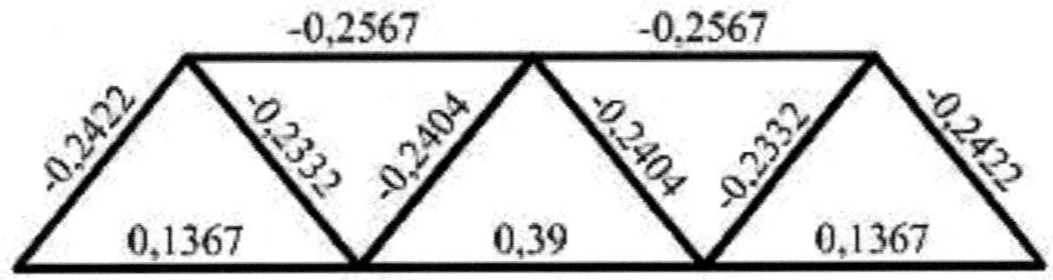

Figure 8. The forces in the truss rods (kN) obtained analytically.

Using stresses σ_x, oriented along the axis of the truss rod, longitudinal forces were found in the elements of the second model (Table 1). Longitudinal forces in the truss, given by the plates, vary slightly along the length of the element and differ from the values obtained by analytic methods of structural mechanics (Fig. 8) by 5-10%.

Table 1

Stress in the truss rods	$N_{1\text{-}2}$ [kN]	$N_{2\text{-}3}$ [kN]	$N_{5\text{-}6}$ [kN]	$N_{1\text{-}5}$ [kN]	$N_{2\text{-}5}$ [kN]	$N_{2\text{-}6}$ [kN]
Model in Fig. 5a	0.14	0.39	-0.26	-0.24	0.23	-0.24
Model in Fig. 5b	0.14	0.36	-0.26	-0.21	0.22	-0.22
% of divergence	4.97	7.62	0.59	12.38	4.12	9.51

4. Conclusions

Using the photoelasticity method, we obtained pictures of interference bands in a truss model. It should be noted that the picture of interference bands, in fact, is a field of tangential stresses in the model plane. It can be compared with the stress field obtained through the software package and shown at Fig. 7. Analyzing Fig. 3 and Fig. 7, it can be concluded that there is no qualitative correspondence between them. The finite element method found the efforts in the farm, that was made of the core elements and plates. The difference in the longitudinal forces in the structural elements obtained using these models is insignificant and is within the limits of the accuracy of engineering calculations. Comparison of the results of numerical and physical modeling showed good correspondence of the stresses in the rods of the trusses. The maximum divergency is within 12%, which is close to the experimental error. Despite the qualitative difference in the stress state of the truss rods, which, by and large, are plates in this work, this does not affect the quantitative assessment of the stresses.

References

[1] Belenya E.I. Kudishin, Yu.I., Ignatieva, V.S., et al. Metal structures: A textbook for students. higher studies. institutions / 10th ed., Sr. Publishing Center "Academy", 2007. 688 p.

[2] Heyman J. Design of a simple steel truss / J. Heyman // Proceedings of the Institution of Civil Engineers – Structures and Buildings. – 2010. –Vol. 163. – Is. 1. – Pp. 53–56. DOI: 10.1680/stbu.2009.163.1.53.

[3] Hutchinson R. G. The structural performance of the periodic truss / R. G. Hutchinson, N. A. Fleck // Journal of the Mechanics and Physics of Solids. – 2006. – Vol. 54. – Is. 4. – Pp. 756–782. DOI: 10.1016/ j.jmps.2005.10.008.

[4] V. Yu. Alpatov. Optimization of the geometric shape of spatially-rod structures / V. Yu. Alpatov, I. S. Kholopov // Metal structures. - 2009. - V. 15. - № 1. - P. 47–57.

[5] Camp C. V. Design of space trusses using modified teaching-learning based optimization/ C. V. Camp, M. Farshchin // Engineering Structures. – 2014. – Vol. 62–63. – Pp. 87–97. DOI: 10.1016/j.engstruct.2014.01.020.

[6] Kaveh A. Optimal structural design family by genetic search and ant colony approach / A. Kaveh, M. Shahrouzi// Engineering Computations. – 2008. – Vol. 25. – Is. 3. – Pp. 268–288. DOI: 10.1108/02644400810857092.

[7] Sonmez M. Artificial Bee Colony algorithm for optimization of truss structures / M. Sonmez // Applied Soft Computing. – 2011. – Vol. 11. – Is. 2. – Pp. 2406–2418. DOI: 10.1016 /j.asoc. 2010.09.003.

[8] D. Tinkov. Comparative analysis of analytical solutions to the problem of deflection of truss structures / D. V. Tinkov // Engineering and Construction Journal. - 2015. - № 5 (57). - pp. 66–73. DOI: 10.5862 / MCE.57.6.

[9] D. Tinkov. Analysis of exact solutions of the deflection of regular hinged-rod structures / D. V. Tinkov // Building mechanics of engineering structures and structures. - 2015. - № 6. - P. 21–28.

[10] Kiyko L.K. Analytical assessment of the arch of a truss under the influence of wind load / L. K. Kiyko // Scientific Herald. - 2016. - № 1 (7). - p. 247–254.

[11] Kirsanov M.N. Analysis of the deflection of a lattice girder truss type / MN Kirsanov // Construction Engineering Journal. - 2015. - № 5 (57). - pp. 58–65. DOI: 10.5862 / MCE.57.5.

[12] M. N. Kirsanov. Analytical calculation of a truss beam with a complex lattice / M. N. Kirsanov // Building mechanics and structural design. - 2015. - № 3 (260). - p. 7–12.

[13] M. N. Kirsanov. Static analysis and wiring diagram of a flat farm / M. N. Kirsanov // Bulletin of the State University of Maritime and River Fleet. Admiral S.O. Makarova. 2016. №5 (39).

[14] Rodionov I.K. Proshin I.V., Grak M.V. On the issue of reinforcement of stretched rods of steel trusses for industrial buildings / I.K. Rodionov, I.V. Proshin, M.V. Grac // Bulletin NGIEI. 2014. № 12 (43).

[15] Smolyago N.A. Improving the structure of a flat farm / N.A. Smolyago, O.A. Yakovlev // Bulletin of science and education of North-West Russia. 2015. №1.

[16] Agakhanov E.K. On the development of complex methods for solving problems of mechanics of a deformable solid. Bulletin of the Dagestan State Technical University. Technical science. - 2013. - № 2. - p. 39-45.

[17] Aleksandrov A.Ya. Polyarizacionno-opticheskie metody mekhaniki deformiruemogo tela / A.Ya. Aleksandrov, M.H. Ahmetzyanov. - M.: Nauka, 1973. - 576 s.

[18] Mindlin R. Stress studies by photoelasticity. -1940-UFN 23 p. 16–66. DOI: 10.3367 / UFNr.0023.1940001

[19] Albaut G. Models of plane elements with various singular features / G. Albaut, N. Kharinova, V. Pangaev, M. Tabanuykhova // Internatinal Journal of Fracture. – Kluwer Academic Publishers. Printed in the Netherland.- 2004. 128 (1). - P. 243 - 251.

[20] Tabanyukhova M.V. Photoelastic analysis of change mode I stress intensity factor in elements with angular notches / M.V. Tabanyukhova, V.V. Pangaev // In Proceedings of 16th European Conference of Fracture. Alexandroupolis, Greece, 2006. P. 447 - 448.

IOP Conf. Series: Materials Science and Engineering **681** (2019) 012023 doi:10.1088/1757-899X/681/1/012023

Reducing the distortion in thin-sheet structures made from high-strength steel based on CAE-welded joint analysis

R U Khaybrakhmanov[1,4], Yu S Korobov[2,3] and G A Bilenko[1]

[1] Postgraduate, Ural Federal University named after the first President of Russia B.N. Yeltsin, Yekaterinburg, Russian Federation
[2] Full Professor, Ural Federal University named after the first President of Russia B.N. Yeltsin, Yekaterinburg, Russian Federation
[3] Full Professor, M.N. Mikheev Institute of Metal Physics of the Ural Branch of the Russian Academy of Sciences, Yekaterinburg, Russian Federation

Email: [4]r.u.khaybrakhmanov@mail.ru

Abstract. High competition in the field of mechanical engineering requires the manufacture of lightweight thin-sheet structures made of high-strength steels ensuring the geometry of a certain accuracy. Supporting a given geometry is achievable by improving methods to reduce residual deformations. Therefore, the purpose of the investigation is to reduce the distortion based on the analysis of the influence of clamping rigidity on the stress-strain state of the assembly using computer simulation by the finite element method. The result of the study is a clamping model that allows to adjust the level of residual stresses and deformations, confirmed experimentally.

1. Introduction

Welding of thin-sheet parts from high-strength steel of martensitic-bainite class (panels, beams, shells) is accompanied by a change in the geometry of the assembly after welding. It is associated with the loss of structure stability due to the small thickness of the parts and the resulting stresses in the welded joints due to thermal loading of the parts by the welding arc and structural transformations accompanied by a change in the resulting structure volume. Common methods to reduce deformation are the use of welding devices which ensures maximum setup stiffness of structural elements [1].

This solution results in an increase in residual stresses in the welded joint. When reaching their critical value, cracks are formed. The aim of the work is to reduce the level of residual welding distortion without a significant increase in residual stresses. To achieve this goal, a scheme of adaptive clamps has been developed. The clamps change their rigidity depending on the stresses in the welded joint. To realize this scheme, a development estimate of the stress-strain state (SSS) over time was carried out by means of a finite-element analysis applied to the shell, which is a characteristic thin sheet part. Consider the effect of adaptive clamps on the shell SSS.

When modeling the welding process, an adaptive clamping force search was performed, in which after welding the level of part surface deformation in the considered section will be the smallest, while residual stress value will increase slightly so as not to increase the probability of cracking.

2. Methods and materials

In this study, a shell of high-strength steel 30HGSA (analog 14331), in which the longitudinal weld is made by tungsten gas welding (TIG) according to Fig. 1, is investigated.

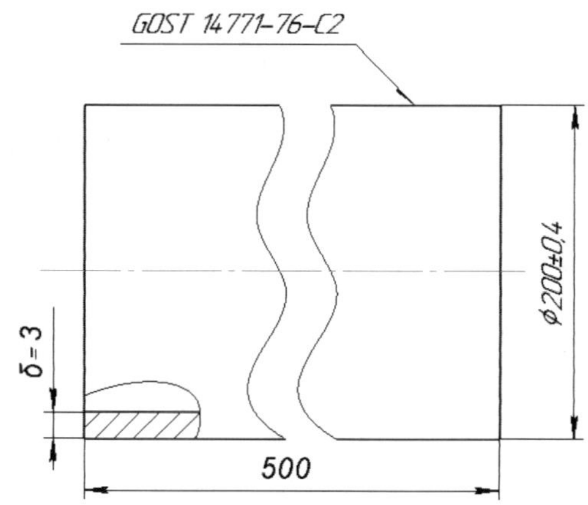

Figure 1. Shell scheme

The calculation was carried out in relation to the butt joint of plates with dimensions of 100x100x3 mm which gives results like a shell provided D / δ> 20 [2, 3], Fig. 2. The heating scheme is adopted to be a movable line heat source. The boundary conditions of the third kind are specified in the form of thermal interaction with the ambient air at a temperature of 20 ° C, in the absence of air movement.

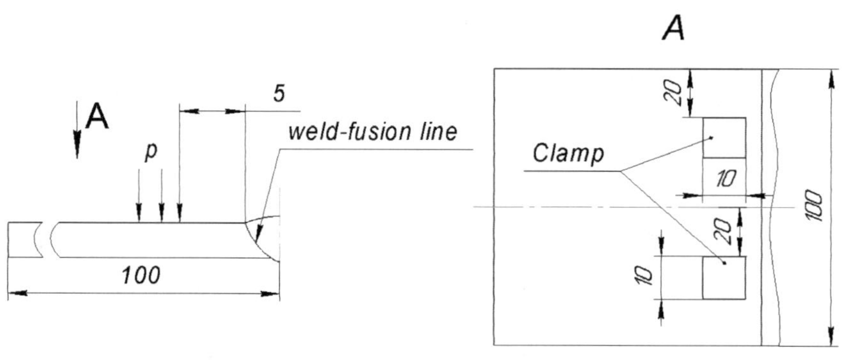

Figure 2. Scheme of the studied model

The calculations were performed using the SYSWELD software package. To solve the thermal and mechanical problem, volumetric elements of a prismatic shape were used, which together form the computational mesh according to Fig. 3. In the weld and heat affected zones (HAZ), the computational grid was thickened 10 times compared to the rest areas due to significant thermal and mechanical field gradients.

The software implementation of the clamp is shown in Fig. 4. Parts will tend to angular deformation during welding, this movement limits clamping. To simplify the model, the physical clamp is eliminated by increasing the rigidity of the parts in the fixing areas and introducing the conventional friction force along the surface of the parts. When expanding (heating part), it acts in the weld direction, while compressing (cooling part) does in the opposite direction.

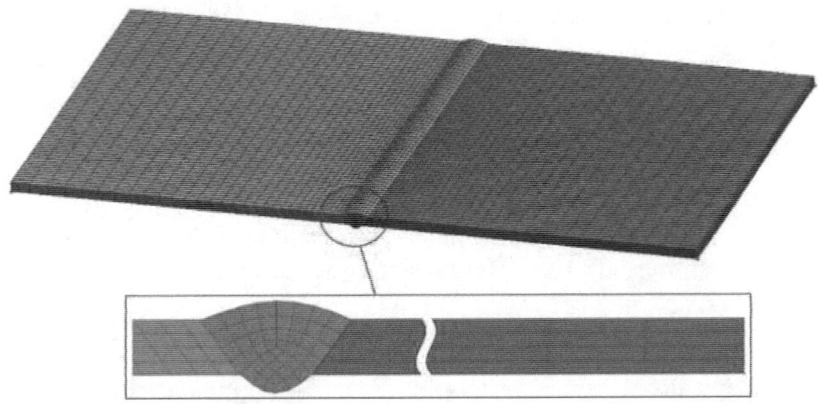

Figure 3. Mesh for solving thermal and mechanical problems

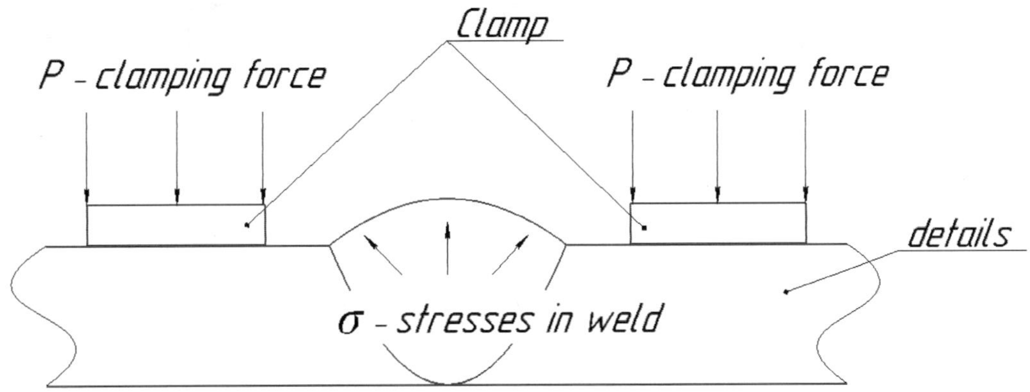

Figure 4. Adaptive clamp

The loading scheme includes the heat load forces arising due to uneven heating over the cross section of weld and the fixing point are replaced by four flat sections of the clamping action P = 0 - 5000 N, in which rigidity is increased according to the clamping force. A force of 0 N corresponds to free movement, and 5000 N corresponds to the absence of movement in the presence of force from thermal expansion. In the areas of the clamp action on the front and back sides of the plate, a friction force acts, which depends on the support reaction force (G) and the friction coefficient (μ). Friction force is determined by the formula [4]

$F_{fr} = \mu \cdot G$ (1)

It is customary to $\mu = 0,2$ [4].

Since the structure weight is negligible compared to the compression force, it can be disregarded, G=P. The input values for P and F_{fr} are given in Table 1.

30HGSA steel is a typical representative of the martensitic-bainitic class. A martensitic transformation occurs in it during welding, which is accompanied by an increase in the metal volume by about 3%, which significantly changes the deformation and stress kinetics [5, 6]. Structural changes were formalized in the model by changing the volume of the weld metal. Based on the thermokinetic diagram of 30HGSA steel, corresponding to the welding thermal cycle for different heat-input welding energies, the number of structural components (austenite, bainite, martensite) for different temperatures during cooling of the welded joint were taken. These data were entered into the calculation module in the SYSWELD software package in the analysis of SSS.

Based on the work done, an experimental check was made of the influence of the clamping rigidity on the stress-strain state of the plates, when the welding heat input is 500 J / mm, which corresponds to the welding parameters I = 120 A, U = 11.9 V, V = 2 mm / s. Welding was performed in automatic mode on a welding column of the MBL 2.0x2.0 series with a Lorch V40 AC / DC power source with an accuracy of 5%. Welding is produced for three samples. Three cases were tested: welding in a free state, with varying clamping rigidity, and in welding equipment, ensuring rigid fixing of the plates during welding. To ensure equal conditions for cooling all samples during the experiment, asbestos cloth was laid under the plates and clamps.

Loads of 20 kg mounted on each plate imitated the clamps providing constant pressure on the part. To estimate the residual stress level tests were selected for static bending and tension according to GOST 6996-66.

3. Result

Based on solving the thermomechanical problem, the maximum stresses and deformations were determined depending on the given effort, shown in Fig. 5. When analyzing the data, it was educed that there is such a critical value of the force Pcr, at which the maximum deformation region changes its localization. The graphs in Fig. 5 show that the smallest deformations are observed with a force of 600 N, residual deformations decreased by 45%, compared with the case of welding in a free state, and stresses increased only by 3%.

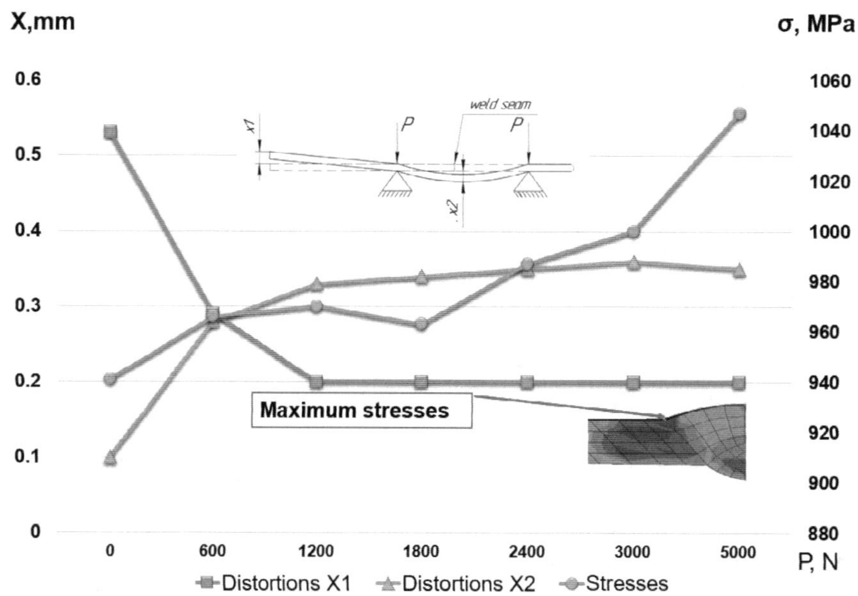

Figure 5. Change in maximum residual stresses and distortion depending on P

Mechanical tests have shown that when welding in a free state, the strength and ductility of the welded joint are increased in comparison with rigid fixing. Presumably, this is due to the favorable condition for the crystallization of the weld metal. The tensile strength of the welded joint with the use of a clamp has been increased by 67%, the bending angle by 20%, in comparison with rigid fixing.

X-ray inspection showed no cracks in the welded joints, i.e. the stress level did not achieve the ultimate strength in any of the cases. Deformations when welding in a free state amounted to 0.55 mm, using a clamp - 0.45 mm, with rigid fixing - 0.4 mm. The discrepancy between the calculated and experimental data did not exceed 11%.

4. Discussion

Based on the work done, it was proposed to equip the clamps of welding accessory with variable stiffness according to Fig. 4.

Clamps with variable stiffness in the welding accessory provide for ensure the feedback of the clamp elasticity and the level of stresses in the welded joint in the range of values close to the tensile strength. This allows to reduce the impact on the welded part during welding, which effectively reduces the deformation level and the likelihood of hot cracking. If the part due to thermal expansion begins to antagonize the clamp with a force greater than the determinate, the clamp provides some movement for the product deformation. This will allow to relax the stresses in the welded joints and heat-affected zone, as well as to limit the part deformations during welding. For example, feedbacks can be implemented in a pneumatic clamp by installing a check valve. This reduces the air pressure in the barrel diminishing the force on the part during welding. The part is deformed and those residual stresses are relaxed. And with the elastic deformation disappearance in the pneumatic cylinder barrel the pressure is increased the pressure leveling with the feed system. This scheme provides an impact on the part with a certain force that does not change over time.

The obtained results can be applied both for calculating the SSS of two plates when performing a butt weld and when welding a longitudinal joint of the shell.

5. Conclusion

The thin-sheet deformation construction model of high-strength steel during welding by computer calculations in SYSWELD has been developed.

Discrepancies between calculations and experimental data did not exceed 11%.

Experimental verification of the adaptive clamp effect on the SSS plates showed a similar strain value, and the absence of cracks in X-rays indicates that the residual stresses in the plates are below the tensile strength.

According to the calculation results, a decrease in the plate deformation by 45% was revealed when applying a clamp with variable stiffness, and the residual stresses did not significantly increase by 3%. It is assumed that the probability of cracking will remain at the same level.

According to the data obtained, it is possible to judge the residual stress and deformation level both during welding of the plate butt weld and during welding of the shell longitudinal weld. Adaptive clamping can be implemented in pneumatic, hydraulic, as well as in mechanical welding accessory.

Acknowledgments

The present study was supported by FASIE Grant №11995GU/2017. This work was done within the state order of IMP UB RAS on the subject “Laser”.

References

[1] Nikolaev G A, Kurkin S A and Vinokurov V A 1971 *Calculation, design and manufacture of welded structures* Samsonova M T (Moscow: «High school») p 760

[2] Konovalov A V, Kurkin A S, Makarov E L, Nerovnyy V M and Yakushin B F 2007 *Theory of welding processes* Nerovnyy V M (Moscow: Bauman Moscow State Technical University) p 752

[3] Feklistov S I 2003 *Simulation of residual welding deformations and stresses* (Moscow: State Research Centre CNIITMASH) p 122

[4] Popov V L 2013 *Mechanics of contact interaction and physics of friction. From nanotribology to earthquake dynamics* (Moscow: PHIZMATLIT) p 352

[5] Marsimets N A and Negoda E H 2007 *Welding technology for special steels* (Vladivostok: FEFU) p 149

[6] Gatovskikh K M and Karkhin V A 1980 *Theory of welding strains and stresses* (Leningrad Korablestroit. Inst) p 330

Development of electron-beam welding method with inserts

I A Kharitonov[1], V K Dragunov[1], A L Goncharov[1], E K Titarev[1] and M A Portnov[1]

[1] National Research University "Moscow Power Engineering Institute", 14, Krasnokazarmennaya str., Moscow, 111250, Russia

E-mail: ivan540@mail.ru

Abstract. The paper is devoted to the study of welded joints formation process both from homogeneous and dissimilar materials. The investigated welded joints were made by an electron beam using intermediate inserts of small thickness in a narrow gap. The influence of insert size as well as gap size on the stability of weld formation is shown. Electron beam oscillation configurations which eliminate incomplete penetration in weld root are determined. Values of thermal efficiency are experimentally found for different welding speeds. The results of structure study for dissimilar welded joints of plates from carbon steel 20 and austenitic corrosion-resistant AISI 316L steel are presented. The values of penetration rate for base and filler metal are calculated and experimentally confirmed.

1. Introduction

Currently, in spite of all existing difficulties of welding dissimilar materials, different combined constructions are widely used in modern industry. This tendency is specified not only by economic factors, which often play one of the most important roles in such construction creation and development, but also by operational requirements. Improvement of performance parameters of various power plants necessitates the use of modern expensive materials. However, in market economy conditions, cost of expensive materials can greatly restrain release of new products as well as increase cost of finished product. Therefore, it is necessary to rationally develop such constructions in which not the whole product is manufactured from more expensive and scarce materials, but only those components or parts that directly reside in specific operating conditions (high loads, temperatures or corrosive environments) [1]. In combined constructions it is advisable to join individual parts by welding because their mechanical connection is often difficult and unreliable in operation. For connection dissimilar metals various methods of welding in solid phase and fusion are used.

Use of an electron beam as a heat source during welding solves a number of very important problems of joining dissimilar materials. Since electron beam is independent source according to the nature of heat exposure, melting processes for base and filler metal can be regulated independently of each other. As a result, penetration rate of base metals can be adequately regulated, thereby ensuring formation of required chemical composition and structure of weld metal by changing electron beam position relative to the joint [2]. But this method has several disadvantages: in addition to the possibility of incomplete penetration in weld root, there is impossible to use this method for all pairs of materials to be welded. The most widely used method for regulating chemical composition in welding dissimilar metals is based on the use of intermediate inserts between the parts to be welded. This method allows welding such materials, direct connection of which is impossible. Inserts can be used as transition layer or as filler material, the main purpose of which is weld metal alloying. Also the weld can be alloyed by pre-applied of a thin coating by galvanic method [3] or by surfacing [4]. But the use of these techniques leads to a serious rise in cost of finished product as well as a decrease in productivity.

Nowadays, all existing techniques used in electron-beam welding cannot fully provide the required physical and chemical properties of the weld. In this regard, development of new techniques to ensure required composition and structure of weld metal due to its alloying has scientific and practical significance.

The aim of this work is to study the process of welded joints formation both from homogeneous and dissimilar materials. The investigated welded joints were made by an electron beam using intermediate inserts of small thickness in a narrow gap.

2. The research methods

For our research the ELA-60/15 power supply was used. As the base material for welding plates with a thickness of 20 mm from steel 20 and corrosion-resistant austenitic AISI 316L steel were used. Intermediate inserts were made of plates with a thickness of 1 mm and 1.5 mm from steel 20 and precision 36NKhTYu alloy (Table 1). Intermediate insert was installed in a slot weld groove with a gap. Weld groove was made of two types: square and V-shaped (Figure 1).

Table 1. The chemical composition of materials (wt%)

Material	C	Mn	Cr	Ni	Ti	Mo	Si	Al	Fe
20	0.20	0.5	–	–	–	–	–	–	base
AISI 316L	0.02	1.5	17	12	0.3	2.5	0.8	–	base
36NKhTYu	0.04	1.2	12.5	36	3	–	0.5	1.1	base

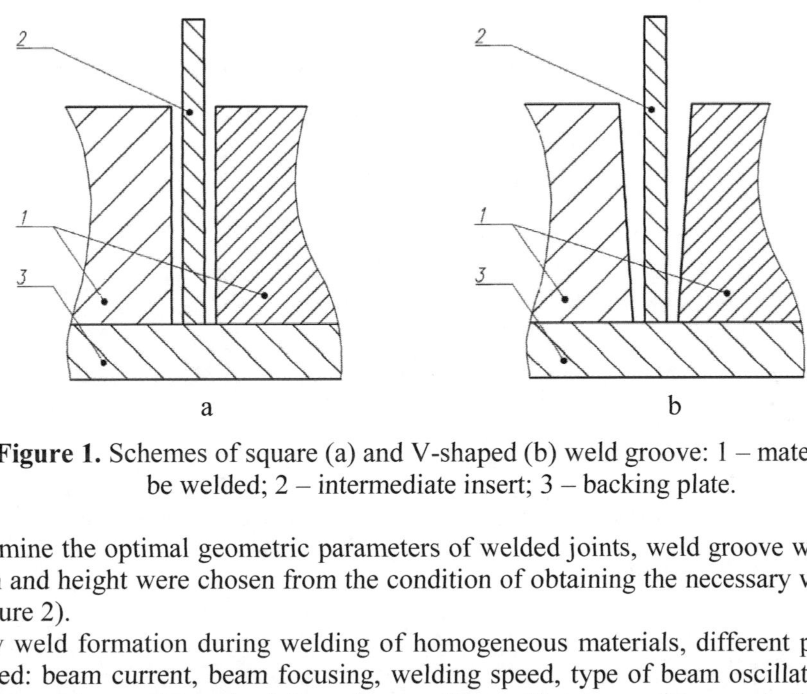

Figure 1. Schemes of square (a) and V-shaped (b) weld groove: 1 – materials to be welded; 2 – intermediate insert; 3 – backing plate.

To determine the optimal geometric parameters of welded joints, weld groove width was changed. Insert width and height were chosen from the condition of obtaining the necessary volume to fill weld groove (Figure 2).

To study weld formation during welding of homogeneous materials, different process parameters were changed: beam current, beam focusing, welding speed, type of beam oscillation. Beam current and welding speed were determined from the condition of joint complete penetration. Electron beam focusing was set in such a way that ensures maximum penetration depth at the selected welding modes. Electron gun was placed vertically. Welding was performed both with beam oscillation along various trajectories with different parameters and without it. Oscillation parameters were chosen from the condition of elimination poor fusion in weld root. Electron-beam welding (EBW) modes are presented in Table 2. After welding macro-sections were made from welded samples and macrostructure of welded joints was investigated. Investigation of homogeneous welded joints formation was carried out on the plates from carbon steel 20. An insert for regulation weld chemical composition and structure was made also from steel 20.

a b

Figure 2. Welding samples.

Table 2. Electron-beam welding modes

Number of mode	U (kV)	I (mA)	v (m/h)	Welded materials	Beam oscillation	Weld groove type	Section number
1		195	30	steel 20	Transverse line segment, Amplitude 1 mm, Frequency 400 Hz	Squared shape	3(a)
2		115	15	steel 20	Transverse line segment, Amplitude 0.8 mm, Frequency 400 Hz	Squared shape	3(b)
3	60	185	30	steel 20	Transverse line segment[a], Amplitude 0.8 mm, Frequency 400 Hz	V-shaped	3(c)
4		190	30	steel 20 and AISI 316L steel	Transverse line segment, Amplitude 1 mm, Frequency 400 Hz	Squared shape	4(a)
5		195	30	steel 20 and AISI 316L steel	Transverse line segment[a], Amplitude 1 mm, Frequency 400 Hz	Squared shape	4(b)

[a] Sinusoidal signal change in the deflection system

The most important characteristic of any process is its efficiency. In electron-beam welding, thermal efficiency characterizes penetration efficiency. This characteristic was based on the heat balance condition [5, 6]:

$$\eta_T = \frac{Fv\rho \cdot (c(T_m - T_0) + L)}{\eta_{ef} IU}, \tag{1}$$

where η_T is thermal efficiency; F is penetration area (mm^2); v is welding speed (mm/s); ρ is metal's density (g·mm^{-3}); c is specific heat capacity (J·kg^{-1}·K^{-1}); T_m is melting temperature (K); T_0 is initial temperature (K); L is specific heat of fusion (J/g); η_{ef} is EBW efficiency ratio; I is beam current (mA); U is accelerating voltage (kV).

Investigation of dissimilar welded joints formation was carried out on the plates from carbon steel 20 and austenitic corrosion-resistant AISI 316L steel. An insert made of high-nickel 36NKhTYu alloy was used for regulation weld chemical composition and structure.

One of the important characteristics of dissimilar weld is penetration rate, which shows filler metal contribution in the whole weld volume. The following relation was used for its determination:

$$\gamma_{ins} = \frac{F_{ins}}{F}, \quad (2)$$

where γ_{ins} is filler metal contribution in weld metal; F_{ins} is cross-sectional area of insert material (mm^2); F is total penetration area (mm^2).

After welding metallographic sections were made of weld samples and macrostructure and microstructure of welded joints were studied.

3. Results of studies

As a result of the analysis of welds obtained from homogeneous steels, the criteria of assembling for welding were determined. It has been found experimentally that the key condition for quality formation of any thickness welded joints is presence of a protruding insert's part above the surface of welded samples. Protrusion size should not exceed 12 mm. Under these conditions, metal completely fills the entire weld groove volume. Sizes of gap and insert are selected according to geometrical parameters of welded materials (Table 3). The optimum gap size between base material and insert is 0.3–0.5 mm. Reducing the gap size in the root part of the weld leads to cavities formation, not filled with liquid metal. To eliminate this problem, it is necessary to increase input power by 15-20%. Insert size is chosen on the basis of required volume to fill the entire weld groove. The optimal expansion angle of V-shaped weld groove is 1.5–2°.

Table 3. Recommended insert and gap parameters

Weld thickness (mm)	Insert size (mm)	Gap size (mm)
10 – 20	0.5 – 1.0	0.2 –0.3
20 – 25	1.0 – 1.5	0.3 –0.5
25 – 30	1.5 – 2.0	0.3 –0.5
>30	>2	0.5

Macrostructures of welded joints, obtained at the modes 1, 2, 3 (Table 2), are shown in Figure 3. In welded joints made without beam oscillations, incomplete penetrations in weld root were found. The reason is that in this case all input energy is spent on insert heating, and base metal heating and melting occur only due to heat transfer from molten insert.

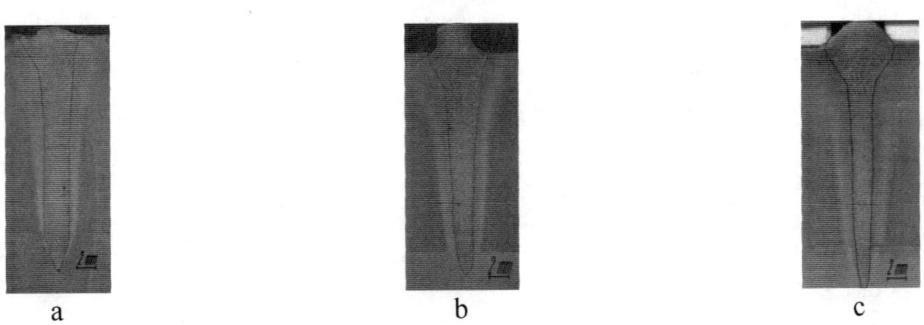

Figure 3. Macrostructure of welded joints, obtained at the modes 1, 2, 3.

The use of beam oscillations makes it possible to distribute input energy between base metal and insert in such a way that common weld pool is formed and such defects as incomplete penetration are eliminated. For example, use of a sinusoidal signal in deflecting system at transverse beam oscillation leads to a change in power density distribution: maximum value will be observed at maximum beam deflection, and minimum will be observed in the insert center. Thus, insert material isn't overheated and uniform melting of the edges being welded and filler material is achieved.

The values of thermal efficiency for welding process at the selected welding modes are presented in Table 4. Comparison of the results obtained in our case and presented in the paper [5] shows that thermal efficiency value for welding process with an intermediate insert is 4–5% higher than without it. Improving efficiency at the mode No. 3 compared to the mode No. 1 is due to the type of beam oscillation.

Table 4. Thermal efficiency values for homogeneous welded joints

Number of mode	Effective power (kW)	Penetration area (mm^2)	Thermal efficiency
1	10.6	78.3	0.46
2	6.8	81.2	0.41
3	10.1	77.9	0.48

When welding dissimilar materials, in addition to changing the conditions of welded joints formation, the use of insert makes possible to effectively alloy weld metal. Figure 4 shows macrostructure and microstructure of dissimilar welded joints obtained at the modes No. 4 and No. 5. Weldability analysis, realized using Schaeffler structural diagram (Figure 5), shows as follows: if penetration rate of welded materials is equal 0.5, then without use of an insert quenching structures inevitably form in weld metal.

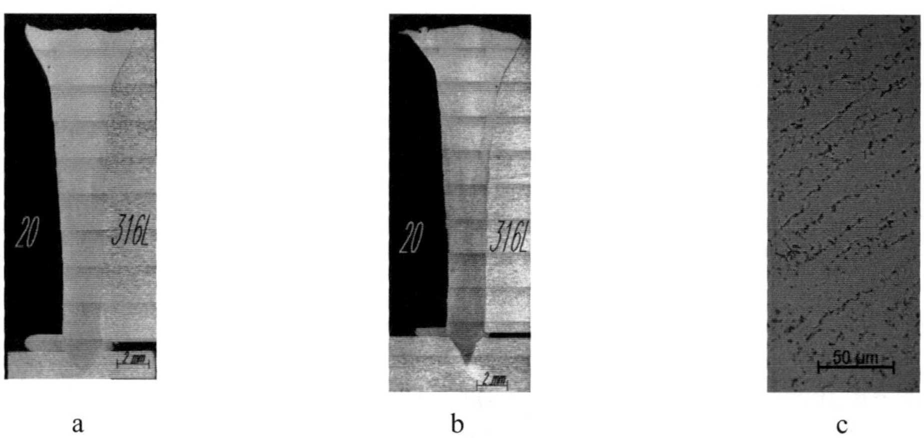

Figure 4. Macrostructure (a, b) and microstructure (c) of welded joints, obtained at the modes 4 and 5.

It is possible to eliminate the appearance of such structures in the weld by ensuring penetration rate of austenitic steel equal to 0.9. However, in practice, it is difficult to obtain such penetration rate due to displacement of electron beam because in such case the probability of root defects formation is high. Therefore, for obtaining the austenitic structure of weld metal, inserts from precision 36NKhTYu alloy were used. In this case, for the obtained penetration areas (weld cross-sections) (Table 5), austenitic structure of weld metal will be formed, if the part of filler material (precision 36NKhTYu alloy) in it is at least 23%.

Study of welded joints microstructure showed that all the welds have an austenitic structure. Penetration rate of insert material for studied welded joints is 35–40% (Table 5). In Figure 5, point 5 in the austenitic region on Schaeffler structural diagram corresponds to such weld composition.

Table 5. Part of filler material in weld metal

Number of mode	Original transverse area of insert (mm^2)	Fusion area (mm^2)	Part of filler metal (%)
4	28	75.5	37
5	29	76.2	39

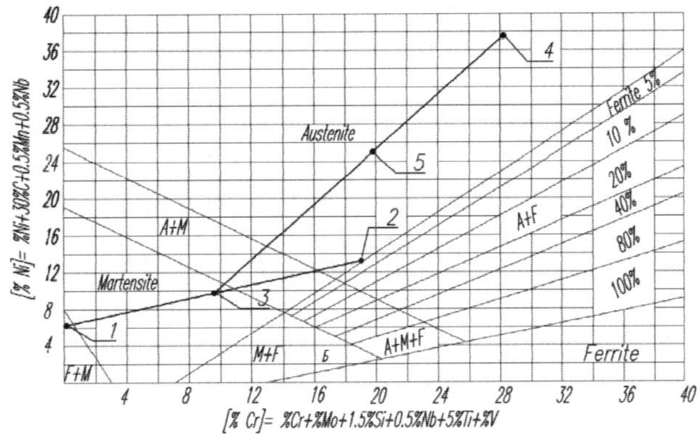

Figure 5. Schaeffler structural diagram: 1 – steel 20; 2 – austenitic AISI 316L steel; 3 – weld metal without alloying (penetration rate is 0.5); 4 – precision 36NKhTYu alloy; 5 – weld metal with additional alloying.

4. Conclusions

Study the process of welded joints formation both from homogeneous and dissimilar materials was realized. The investigated welded joints were obtained by an electron beam using intermediate inserts up to 1.5 mm thick in a gap, which size was approximately 2.5 mm. As a result of the study, maximum size of intermediate insert has been established. Its protruding part should not exceed 12 mm for all weld thicknesses, and optimum gap size between insert and welded edges of base metals should be 0.3–0.5 mm. Under these conditions, metal completely fills the entire weld groove volume. For elimination lack of fusion in the root part of welded joints beam oscillation are required.

Acknowledgments

The research was carried out in National Research University «Moscow Power Engineering Institute» with the financial support of the Russian Science Foundation (project No. 18-19-00652).

References

[1] Ryabov V R, Rabkin D M, Kurochko R S and Strizhevskaya L G 1984 *Welding of Dissimilar Metals and Alloys*(Moscow: Mechanical Engineering) p 239

[2] Dragunov V K, Muravyova T P and Rodionov Yu A 1990 *Welding production* **4** 2–4

[3] Beibert A, Borchers E, Borchers E, Hohn W and Klar S 1985 *Verfahren zum Verschweiben von hochleglerten mit un-oder niedrigietenStalen* (Pat. 217162 DDR, MKI B 23 K 31/00).

[4] Goncharov A L, Sliva A P, Portnov M A, Terentyev E V and Gribkov M S 2013 *Bull. of Perm National Research Polytechnic University. Eng., Mat. Sci.* **15(4)** 74-87

[5] Martynov V N, Gribkov M S, Kozhechenko A S and Shcherbakov A V 2015 *Electrometallurgy* **9** 30-5

[6] Shcherbakov A V, Rodyakina R V, Kozhechenko A S, Gaponova D A, Goncharov A L and Dragunov V K 2017 *Techn. Phys. Let.* **11** 958–60

IOP Conf. Series: Materials Science and Engineering **681** (2019) 012025 doi:10.1088/1757-899X/681/1/012025

Finite element modeling of multiple electrode submerged arc welding of large diameter pipes

A Khudyakov [1], Yu Korobov [1,4], P Danilkin [2,3] and V Kvashnin [2,3]

[1] Ural Federal University named after the first President of Russia B.N. Yeltsin, Yekaterinburg, Russian Federation
[2] The Russian Research Institute of the Tube & Pipe Industries, Chelyabinsk, Russian Federation
[3] South-Ural State University (National Research University), Chelyabinsk, Russian Federation
[4] M.N. Mikheev Institute of Metal Physics of the Ural Branch of the Russian Academy of Sciences, Yekaterinburg, Russian Federation

E-mail: Aohudyakov@gmail.com; Yukorobov@gmail.com; DanilkinPA@gmail.com.

Abstract. For the process of multiple-electrode submerged arc welding (MESAW) of large-diameter pipes (LDP) a mathematical model has been developed with a numerical solution in the software product "SYSWELD". The paper defines the parameters of double ellipsoid model of heat sources by J. Goldak. Using regression analysis of experimental data empirical formulas for calculating geometrical parameters of heat sources model by J. Goldak and deposition rate were developed for welding with direct current on positive polarity and alternating current. The model allows to predict the shape of weld and to estimate analytically the cooling rate in the heat affected zone (HAZ). The developed model and its numerical solutions with sufficient accuracy for practical calculations reflect the process of longitudinal welding of LDP.

1. Introduction

Modern production of large diameter pipes implies welding of a longitudinal weld in two passes by means of MESAW after assembling by gas metal arc welding. Welding is accompanied by a large number of physicochemical processes associated with heating, melting, cooling, crystallization of the weld metal and structural transformations of the base metal in the HAZ. Thermal impact has the decisive influence on the nature and kinetics of thermomechanical and physical processes of arc welding, and subsequently on the microstructure and mechanical properties of welded joints [1, 2, 3]. According this there is no doubt the relevance of the calculation of temperature fields during MESAW of high-strength LDP for their further use in solving problems related to structure formation, mechanical properties of welded joints and dimensions of the weld and HAZ.

2. Finite element modelling

During MESAW heat input into the product is simultaneously carried out by several volumetric sources, therefore the temperature fields in the studied plane are determined according to the principle of superposition by the sum of temperature fields from each heat source (figure 1):

$$T_a(q_\Sigma, r, t) = \sum_{n=1}^{i} T_i(q_i, r, t), \tag{1}$$

Where Ta – field of temperatures on the studied plane; q_{Σ} – total heat input of welding; q_i – heat input of i-th heat source; r – radius vector; t – time.

For modelling of volumetric density of heat generation in the vicinity of each welding arc, the model of volumetric heat generation proposed by J. Goldak [4, 5, 6] was used. The model of the heat source by J. Goldak is described by the normal distribution of the specific thermal power over all coordinate axes in the volume of the body having the form of a double ellipsoid (Figure 2).

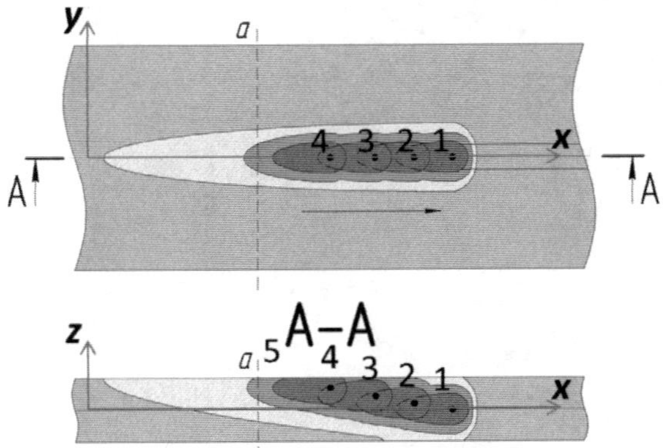

Figure 1. Scheme of thermal fields during MESAW

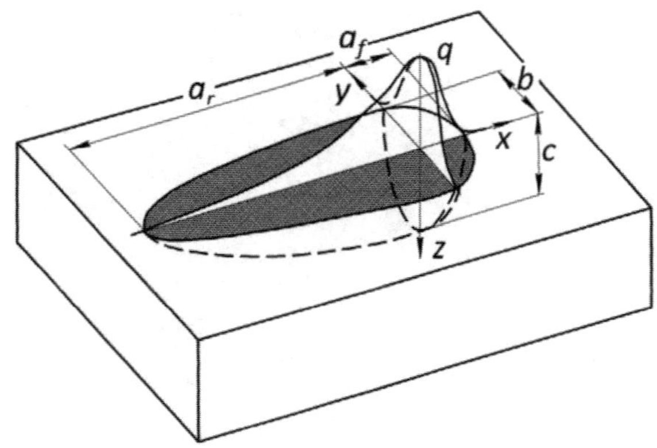

Figure 2. Model of double ellipsoidal heat source by J. Goldak

The independent distribution of the specific heat power qv in the frontal (index f) and tail (index r) parts of the ellipsoid is described by the equations [7, 8]:

$$q_{v,f} = f_f \cdot \frac{6 \cdot \sqrt{3} \cdot q}{a_f \cdot b \cdot c \cdot \pi^{3/2}} \cdot \exp\left(-3 \cdot \left(\left[\frac{x + v \cdot (t - \tau)}{a_f}\right]^2 + \left[\frac{y}{b}\right]^2 + \left[\frac{z}{c}\right]^2\right)\right), \tag{2}$$

$$q_{v,r} = f_r \cdot \frac{6 \cdot \sqrt{3} \cdot q}{a_r \cdot b \cdot c \cdot \pi^{3/2}} \cdot \exp\left(-3 \cdot \left(\left[\frac{x + v \cdot (t - \tau)}{a_r}\right]^2 + \left[\frac{y}{b}\right]^2 + \left[\frac{z}{c}\right]^2\right)\right), \tag{3}$$

where q – the effective thermal power of the heating source; τ – the delay time, counted from the beginning of the source action; t – the current time; v – source movement speed (welding speed); x, y, z – semiaxes of the ellipsoid in the direction of the coordinate axes OX, OY, OZ; ff and fr – the coefficients that determine the ratio of heat introduced into the frontal and tail parts of the ellipsoid; af, ar, b, c - the corresponding radiuses of the normal distribution.

According to the theory of heat propagation in a solid [9] a scheme of a fast moving point heat source in a flat layer was taken. The heat power distribution was represented by Goldak's model of a double ellipsoid. Common dependencies of the specific heat, thermal conductivity and total surface heat transfer [10, 11] on body temperature for low carbon steel were used in the developed finite element model. The introduction of heat into the edges during MESAW was represented as separate volumetric heat sources from each arc. Figure 3 shows adopted scheme of the temperature boundary problem.

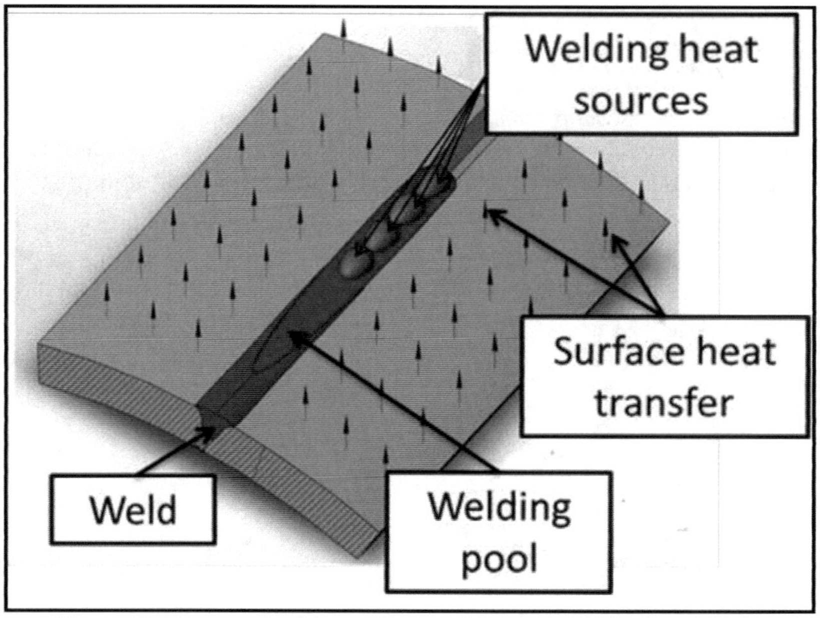

Figure 3. Temperature boundary problem for MESAW of LDP.

Welding of a longitudinal seam of a LDP is carried out in a special edge bevelling. Since each welding arc deposits a certain amount of molten metal, each welding heat source acts on a different depth. The first heat source acts in the root of the edge bevelling, the second source acts on the surface of the weld metal deposited by the first arc, the third - on the surface of the weld metal deposited by the second arc and etc. The scheme of heat sources distribution in depth is presented in Figure 4.

The Z coordinate of each subsequent heat source will be determined as the height of the triangle with an area equal to the cross-sectional area of the metal deposited by the previous welding arcs. The scheme for determining the Z coordinate of heat sources during MESAW is shown in Figure 5.

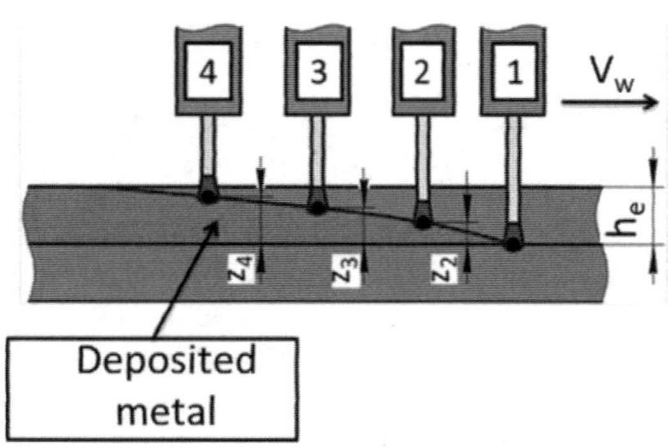

Figure 4. The scheme of heat sources distribution in height: h_e - the depth of edge bevelling; Z_i - coordinates of the i-th heat source.

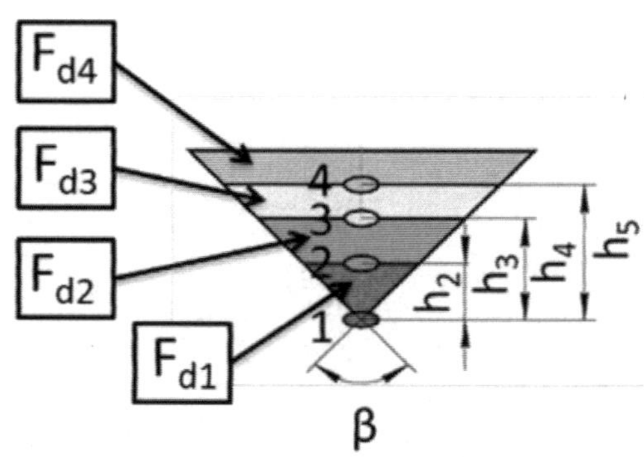

Figure 5. The scheme for determining the Z coordinate of heat sources: F_{di} – the cross-sectional area of the deposited metal by the i-th arc; β – angle of edge bevelling.

The Z coordinate of each heat source was determined from the expression of the height of a triangle by a known area using the equation:

$$Z_i = \sqrt{\frac{\sum_{i=1}^{n}(F_{d_{i-1}})}{\tan(^{\beta}/_{2})}}, \quad (4)$$

where F_{di} – the cross-sectional area of the deposited metal.

The cross-sectional area of the metal Fn deposited into the welding pool by each arc is determined by the equation:

$$F_d = \frac{\alpha_m \cdot I_w}{\gamma \cdot V_w}, \quad (5)$$

where α_d – the deposition coefficient, for submerged arc welding is equal to the melting coefficient α_m; I_w - welding current; γ - the density of the deposited metal; V_w - welding speed.

To calculate the cross-sectional area of the deposited metal empirical dependences of melting coefficients were obtained for welding with direct current of positive polarity ($\alpha_{d(DC+)}$) and alternating current ($\alpha_{d(AC)}$) for electrode ejection of 40 mm. Dependences were obtained using the method of regression analysis of experimental data.

$$\alpha_{d(DC+)} = 0{,}053 \cdot \frac{I_w}{d_e^{0{,}88}}; \quad (6) \qquad \alpha_{d(AC)} = 1{,}257 \cdot \frac{I_w^{0{,}539}}{d_e^{0{,}514}}, \quad (7)$$

where I_w – welding current; d_e – electrode diameter.

Basing the developed model, simulation of the cooling rates in CGHAZ was carried out using SYSWELD software package (license agreement No 2910-2014 ДУ). In the Sysweld software the power distribution of volumetric heat sources is carried out according to the algorithm in accordance with equations (2, 3). The heat of the electric arc enters the edges through the surface of forming crater, i.e. through the volumetric heat source. That is why in the developed finite element model it was assumed that the geometrical parameters of the heat sources by Goldak are equal to geometrical parameters of the crater of the weld pool that forms under a welding arc. Therefore, to determine the parameters of the crater depending on welding modes parameters additional studies were conducted. To determine the parameters of the crater a technique based on the use of inserted tungsten indicators [12, 13] was used. For this tungsten wires were inserted into the plate subjected to welding in the plane perpendicular to the welding direction (figure 6). Wires with a diameter of 0.5 mm and with a step of 1.5 - 2 mm were used.

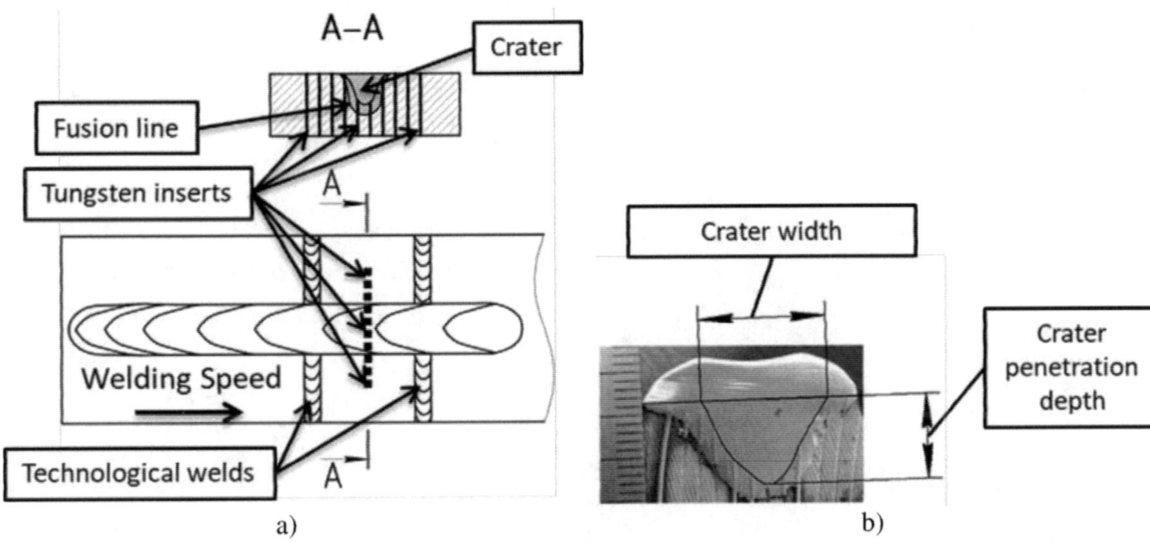

Figure 6. Determination of geometrical crater parameters: a) Scheme of plate with tungsten inserts; b) Crater geometrical parameters.

The results of the experimental data were processed using regression analysis, as a result of which empirical dependences of the geometric parameters of the crater from the parameters of welding modes were obtained:

$$h_{cr}^{DC+} = 2{,}123 \cdot \frac{I_w^{1{,}516}}{V_w^{0{,}528} \cdot d_e^{0{,}488}} \quad (8); \qquad b_{cr}^{DC+} = 54{,}205 \cdot \frac{U^{1{,}216} \cdot d_e^{0.904}}{I_w^{0{,}829} \cdot V_w^{0{,}571}} \quad (9);$$

$$h_{cr}^{AC} = 0{,}642 \cdot \frac{I_w^{1{,}168}}{U^{0{,}756} \cdot V_w^{0{,}756} \cdot d_e^{0{,}923}} \quad (10); \qquad b_{cr}^{AC} = 6{,}865 \cdot \frac{U^{0{,}851} \cdot d_e^{0.6}}{I_w^{0{,}851} \cdot V_w^{0{,}32}} \quad (11),$$

where h_{cr}^{DC+}, h_{cr}^{AC} – crater depth during welding with a direct current on positive polarity and alternating current, respectively; b_{cr}^{DC+}, b_{cr}^{AC} – crater width during welding with a direct current on positive polarity and alternating current, respectively; U - voltage.

The modeling of the welding process was performed in accordance with the industrial standard welding modes of LDP. Graphic images of the numerical solution of the finite element model are presented in Figure 7. The developed finite element model with sufficient accuracy for practical purposes provides prediction of the geometric parameters of the weld (Figure 8) and the cooling rate (Figure 9) in coarse grained heat affected zone. The error does not exceed 10%.

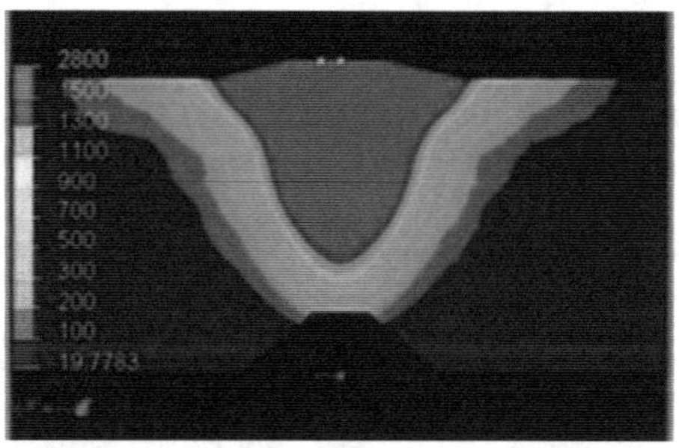

Figure 7. Model of multiple-electrode SAW weld shape applying to the 1153x30.9 mm pipe, inner weld.

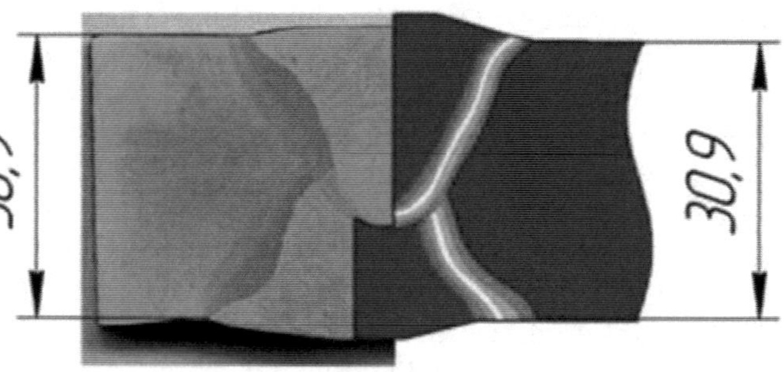

Figure 8. Comparison of the geometric parameters of the real weld and its model applying to the 1153x30.9 mm pipe.

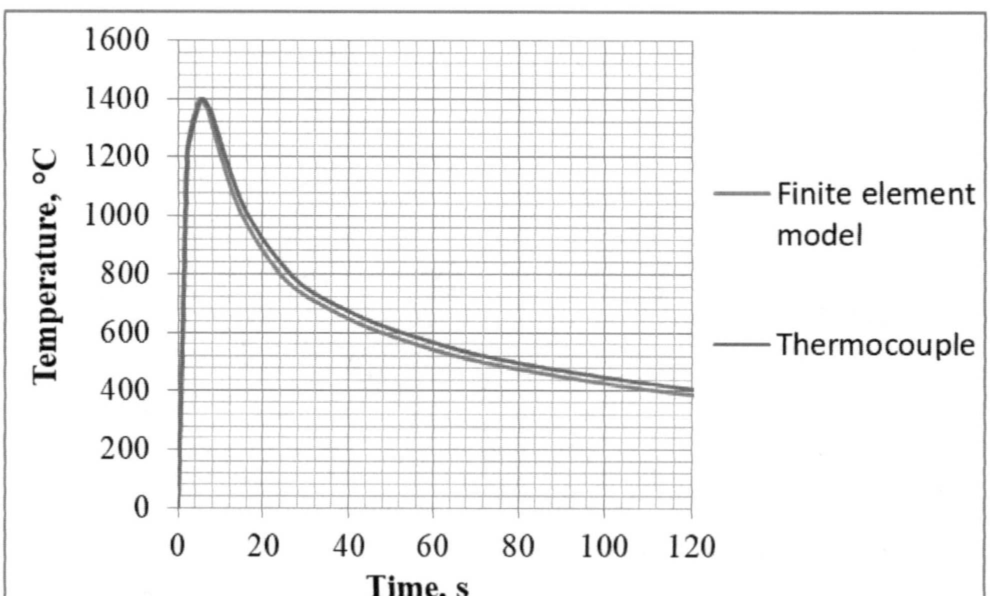

Figure 9. Thermal cycle of welding in coarse grained heat affected zone.

3. Conclusions

The parameters of heat sources by J. Goldak were determined for the finite element modeling of the heat distribution process during multiple electrode submerged arc welding of large-diameter pipes. Also using regression analysis of experimental data empirical equations have been developed for calculating melting coefficients for welding with direct current on positive polarity and alternating current.

A mathematical model of multiple electrode submerged arc welding with a numerical solution in the SYSWELD software was developed. The model allows to predict the shape of weld and to estimate analytically the cooling rate in the heat affected zone (HAZ). The developed model and its numerical solutions with sufficient accuracy for practical calculations reflect the process of longitudinal welding of LDP.

Acknowledgments
This work was done within the state order of IMP UB RAS on the subject "Laser".

References

[1] B.M. Berezovsky, 2006, Mathematical models of arc welding, Vol. No1, Chelyabinsk, 547 p. (in Russian).

[2] B.M. Berezovsky, 2006, Mathematical models of arc welding, Vol. No5, Chelyabinsk, 645 p. (in Russian).

[3] V.A. Karhin, 2013, Thermal processes during welding. St. Petersburg, 646 p. (in Russian). Goldak J., Chakravarti A., Bibby M. A new finite element model for welding heat source // Metallurgical Trans. B. — 1984. — 15B. — pp. 299-305.

[4] Goldak J. Akhlaghi M. Computation welding mechanics, Springer Science + Business Media Inc., Boston, 2005.

[5] Goldak J. Computer modelling of heat flow in welds / J. Goldak, M. Bibby, J.Moore, R. House, B. Patel // Metallurgical Transactions, 1986. V.17B. P. 587-600.

[6] Koch F. Simulation of the temperature field and the microstructure evaluation during multi-pass welding of L485MB pipeline steel / F. Koch, M. Enderlein, M. Pietrzyk // Computer methods in material Science. Vol. 13, No1, 2013, pp. 173-180

[7] O.A. Slyvinskyy, A. O. Prepiialo, V.L. Bondarenko, V. P. Slyuta, 2014, "Calculation and experimental analysis of thermal processes of thin-sheet stainless steel welding using TIG and CMT methods", Technological Systems, No 1, 2014, pp. 76-82.

[8] N.N. Rykalin, 1951, Calculations of thermal processes during welding. Moscow, 296 p. (in Russian).

[9] Konovalov, A. Kurkin, E. Makarov, V. Nerovnij and B. Yakushin, 2007, "Theory of welding processes" Moscow, 752 p. (in Russian).

[10] K. Bagryanskij, Z. Dobrotina, K. Hrenov, 1976, "Theory of welding processes", Kiev, 424 p. (in Russian).

[11] A.V. Savinov, 2013, "Improving the technological properties of the arc at TIG welding", Phd diss. Volgograd, Russia. 243 p. (in Russian).

[12] V.I. Atamanuk, 2008, "Development of ways and means to improve the stability of welds formation during welding with non-consumable electrode" Phd diss. Volgograd, Russia. 153 p. (in Russian).

Numerical modeling of a plate with an oval cut and cracks on the contour

A A Lazarev[1], N V Kharinova[1]

[1]Novosibirsk State University of Architecture and Civil Engineering (Sibstrin), Novosibirsk, Russia

E-mail: n.kharinova@sibstrin.ru

Abstract. A numerical study of a plane model with an oval cut and cracks-cuts along the contour, made at different angles, was performed. The change in the stress-strain state with the successive appearance of cracks-cuts was considered.

1. Purpose and objectives of the study

The purpose of the work was to study the stress-strain state of the model with an oval cut, depending on the appearance of cracks-cuts on its contour. Objectives of the study: to assess the effect of successive appearance of cracks in the contour cut on the stress state of the element; to calculate the stress concentration coefficients near the cracks top. The SCAD software package was used for the numerical study.

Previously, models with other concentrators were studied [1-3].

The studied model was made by the polarization-optical material – Plexiglas brands of E2 (modulus of elasticity E = 3500 MPa, Poisson's ratio ν = 0.4) with dimensions of 93mm x 84mm x 6mm. At the intersection of the axes of symmetry cut out the oval hole. At different angles to the contour of the oval cut cracks-cuts were made. A photo of the model and the sequence of cracks is shown in Figure 1.

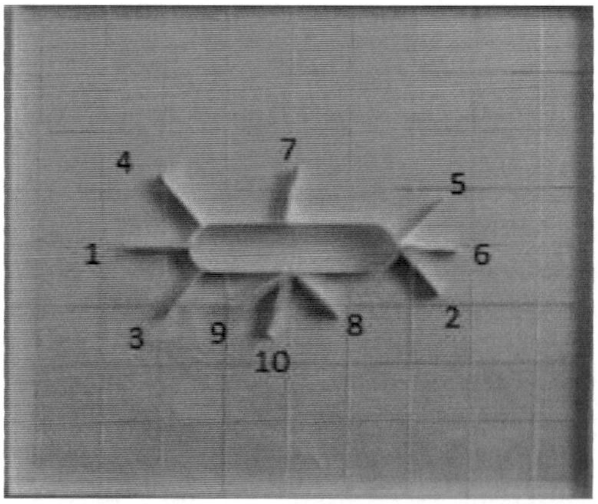

Figure 1. The model under the study.

2. Numerical experiment

The finite element method (FEM) using the software complex "SCAD" was used to calculate a group of models with different number of cracks-cuts on the contour of the oval cutout. The order of occurrence of cracks is shown in Figure 1. Cracks № 7-10 have deviations from the straight axis, other cracks - straight. The finite element scheme of the computational model and the type of loading are shown in Figure 2. The model was loaded with a compressive rated load q_0=10.75 kN/m in the vertical direction. The vertical axis of symmetry is fixed from the horizontal displacements, and the horizontal axis from the vertical ones. The models had a square base grid with a cell size of 1x1 mm.

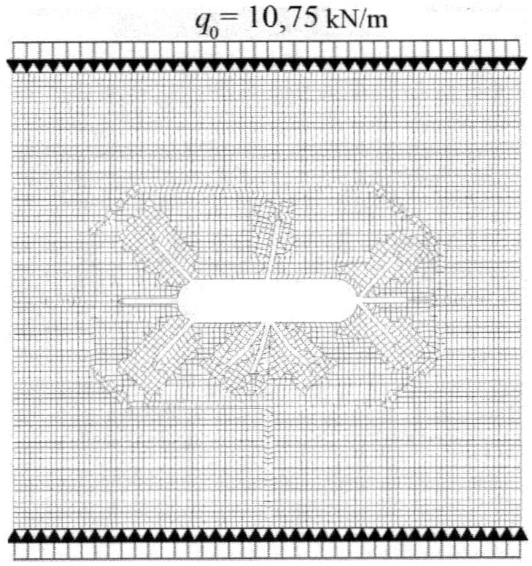

Figure 2. Loading scheme of the calculation model

3. The results of the numerical experiment

Figure 3 shows the stress fields σ_1 and σ_3 in the models with different number of cracks-cuts. The following it was noted analysing the results. The zone of influence of concentrators was not great, the stresses in the model were quickly aligned and become uniform. At the tops of horizontal cracks-cuts marked the highest stress values. Stresses at the tops of cracks № 5 and 2 were much higher than that of cracks № 3 and 4.

Cracks № 7-10 were in the zone of positive stresses σ_1. On tops of inclined cracks No. 2, 3, 4, 5 pass lines of change of a sign of tension σ_1.

With the advent of new cracks-cuts stress at the tops of existing cracks, as a rule, increased. The increase in stress was 5-20%. Most with the advent of new cuts increase stress at the top of the crack № 4 (2.2 times), at the tops of cracks №1 and 3 maximum stress increased 1.4 times. The appearance of crack № 6 unloads the stress state at the tops of cracks № 2 and 5, and crack № 10, respectively, cracks № 8 and 9.

4. Kinetics of stress concentration coefficients

According to the results of numerical experiment, the stress concentration coefficients at the crack-propyl tops were determined by the formula (1)

$$K_\sigma = \frac{\sigma_{max}}{\sigma_0}, \quad (1)$$

where σ_{max} and σ_0 - respectively the maximum and nominal stresses.

The calculated maximum stresses were taken from the results of numerical calculation of FE at the crack tops, nominal, respectively, at a distance from the concentrator.

The resulting concentration factors were listed in Table 1. Horizontal indicates the number of cracks in the model. Vertically - at the top of a crack the concentration coefficient was obtained.

The highest value of the concentration coefficient obtained from the top of the horizontal crack-cut: No. 1 $K_\sigma = 19{,}45$ and No. 6 - $K_\sigma = 15{,}23$. At the same time, with the appearance of new cracks in the model, the concentration coefficients increased most strongly at the top of crack No. 4 (2.2 times). The concentration coefficients at the tops of cracks № 5 and 9 decreased by 30 %.

The concentration coefficient in the model with an oval hole without cracks along the contour was $K_\sigma = 5{,}2$. Thus, the appearance of the first horizontal crack increased the stress concentration in the model by 2.7 times. At the end, when there were already 10 cracks on the circuit, this value reaches 3.8 times.

In reference books on stress concentration [4, 5] the type of tasks described in this section of the article was absent.

Tabel 1. Stress concentration coefficients

Cracks number	**1**	**2**	**3**	**4**	**5**	**6**	**7**	**8**	**9**	**10**
1	13.74	15.05	14.99	15.51	17.14	17.59	18.47	19.27	19.28	19.45
2		13.02	13.20	13.39	13.55	8.87	8.72	9.11	9.81	10.02
3			4.98	4.61	4.35	4.49	5.85	6.48	6.43	6.60
4				2.99	4.30	5.95	6.75	6.80	6.61	6.58
5					14.22	8.87	9.86	9.45	9.43	9.38
6						14.65	14.70	14.71	15.18	15.23
7							7.03	7.69	7.94	8.07
8								6.95	7.57	6.35
9									6.79	4.66
10										8.64

IOP Conf. Series: Materials Science and Engineering **681** (2019) 012026 doi:10.1088/1757-899X/681/1/012026

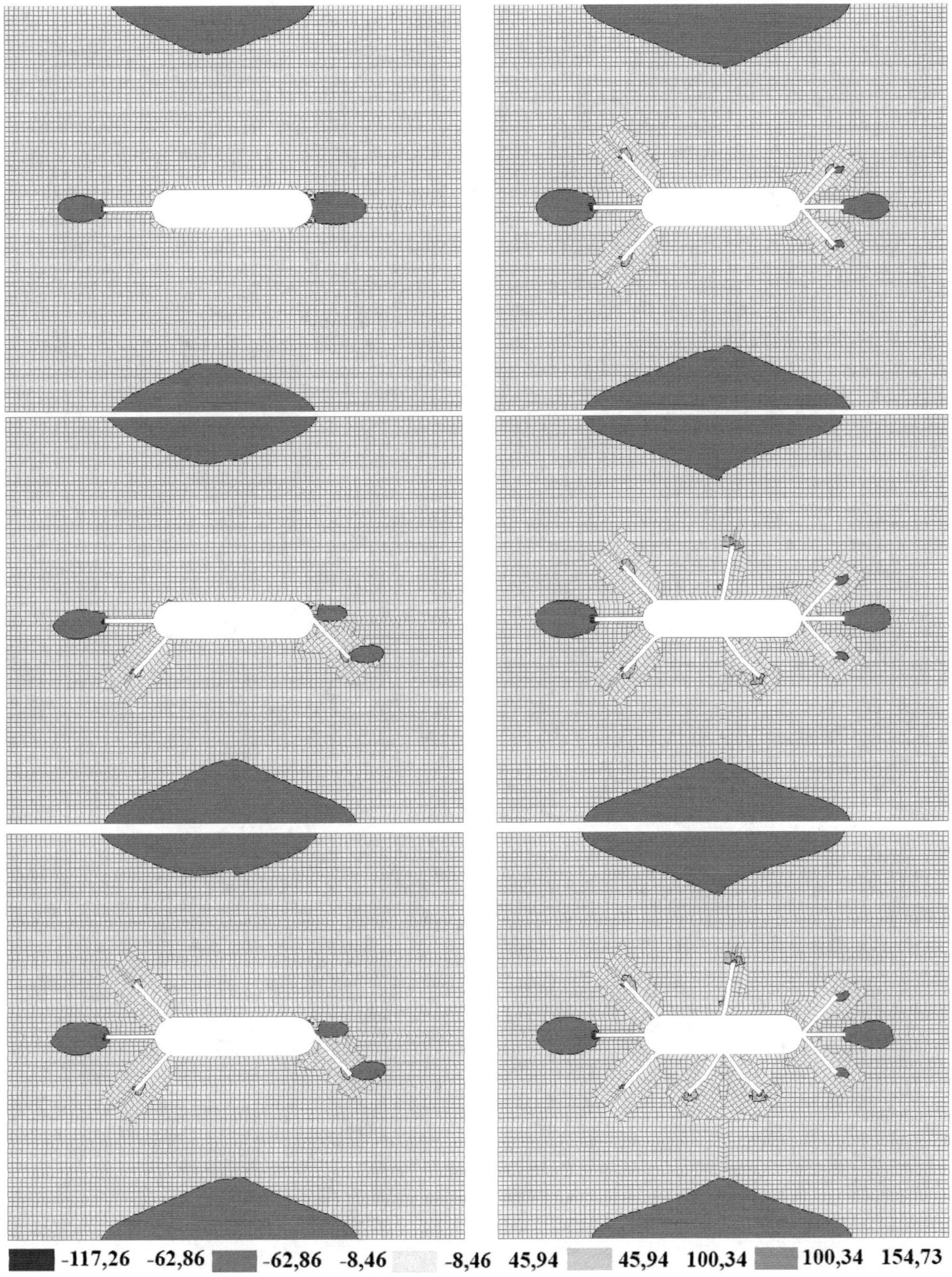

Figure 3. Stress fields σ_1

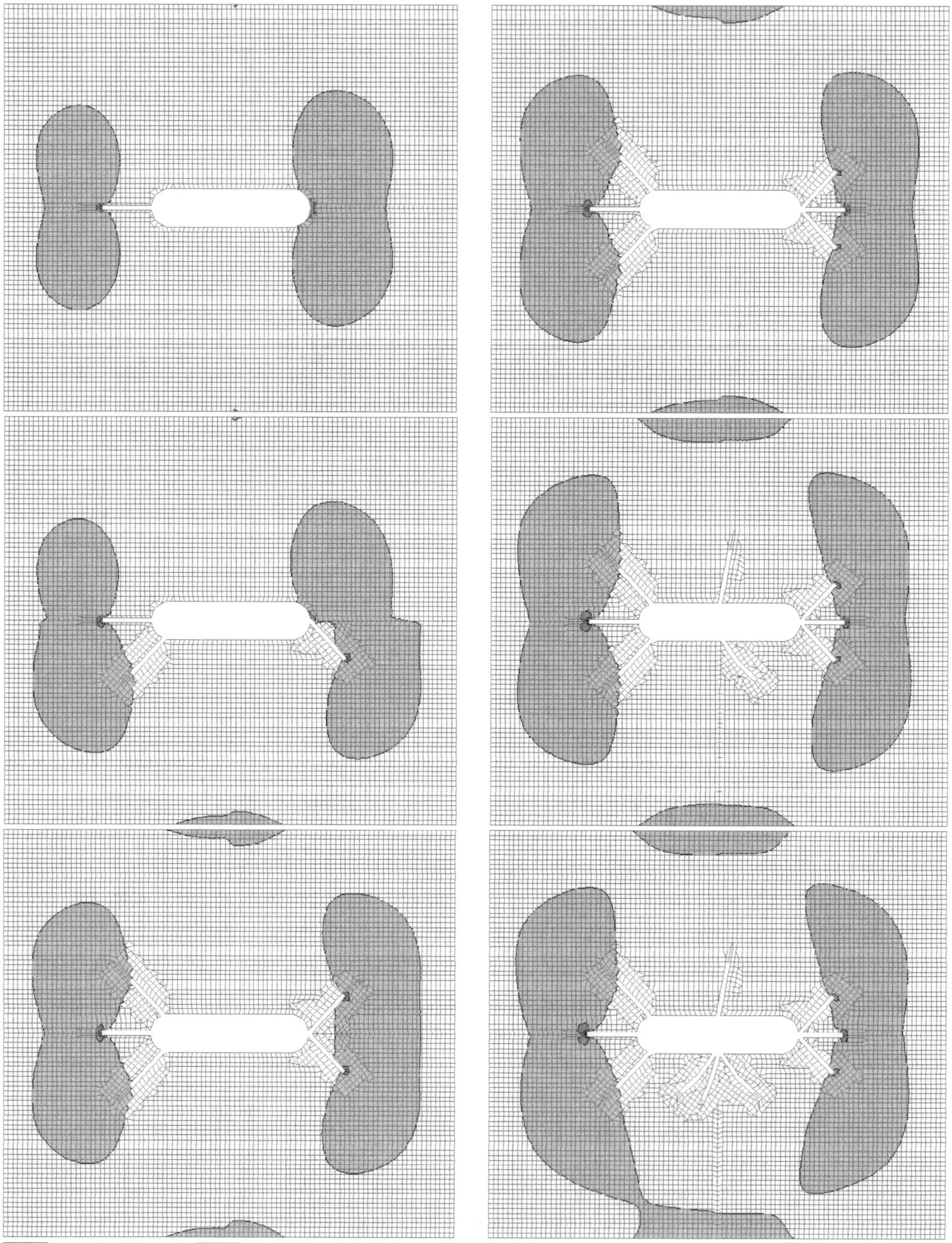

-348,54 -267,6 -267,6 -186,67 -186,67 -105,74 -105,74 -24,8 -24,8 56,13

Figure 4. Stress fields σ_3

IOP Conf. Series: Materials Science and Engineering **681** (2019) 012026 doi:10.1088/1757-899X/681/1/012026

5. Some of the findings of the research

In this paper, the influence of cracks arising along the contour of holes in structural elements on the stress state and concentration coefficients was investigated. Considered the change of the stress state in the model in the process of the emergence of new cracks.

It was noted that the vertical and close to them cracks have tensile stresses at the vertices, which was more dangerous for the structural strength. Although their values were significantly less than those of the horizontal ones.

Analysis of the data presented in Table 1 shows that with the appearance of new cracks, the stress concentration at the tops of existing cuts tends to increase. Cracks that appear between the slanted cuts, coming from the same point, lead to their discharge. The highest values of the concentration coefficients were at the tops of the horizontal cracks-cuts. At the tops of vertical cracks concentration coefficients were much smaller, but have a positive sign, which makes these cracks no less dangerous than horizontal.

References

[1] G. Albaut Model determination of stress levels in elements of building structures with angled notches / G.N. Albaut, M.V. Tabanjuhova // News of Higher Education Institutions. Civil Engineering. - 2006. - №10. - P. 107-112.

[2] G. Albaut To the problem of the stress intensity factors determination in the elements of building structures / G.N. Albaut, V.V. Pangaev, M.V. Tabanyukhova, N.V. Harinova // Izvestiya Vuzov. Civil Engineering. - 2005. №1. - P. 97 - 102.

[3] Tabanyukhova M.V. Photoelastic analysis of change mode I stress intensity factor in elements with angular notches / M.V. Tabanyukhova, V.V. Pangaev // In Proceedings of 16th European Conference of Fracture. Alexandroupolis, Greece, 2006. P. 447 - 448.

[4] Peterson R. Coefficients of stress concentration. Graphs and formulas for calculating structural elements for strength / R. Peterson - M.: Mir, 1977. - 450 p.

[5] Savin G.N. Reference book on stress concentration / G.N. Savin, V.I. Tulchy - Kiev: Vishcha school, 1976. - 412s

IOP Conf. Series: Materials Science and Engineering **681** (2019) 012027 doi:10.1088/1757-899X/681/1/012027

Regularizing method for the analysis of CT images of welded joints of pipelines

A V Likhachev[1,2], T S Kiychenko[1] and N A Likhachev [1]

[1]Novosibirsk State University of Architecture and Civil Engineering (Sibstrin) Leningradskaya Street 113, Novosibirsk, Russia, 630008
[2]Institute of Automation and Electrometry SB RAS Academician Koptyug ave. 1, Novosibirsk, Russia, 630090

Abstract. The aim of the paper is to improve the accuracy of the estimation of the shape and size of cracks in the welded seams of pipelines from their X-ray tomograms. Based on the results of previous studies, the procedure of deconvolution of projection data before reconstruction is considered. The influence of the value of the regularization parameter in the deconvolution algorithm based on the Fourier transform on the quality of the tomograms is investigated. In the numerical simulations, the estimates of the optimal values of the regularization parameter for different data recording conditions are derived.

1. Introduction

One of the possible tools of monitoring the condition of engineering construction being in operation is X-ray sensing [1 – 3]. With this diagnostics, cracks, cavities and other internal defects can be detected and localized. Their size and shape can be estimated with a high degree of accuracy by applying tomography techniques to the sensing data [4 – 6].

The mathematical basis of tomographic methods is the two-dimensional Radon transform [7, 8]. This transformation matches a function of two variables $g(x,y)$ with a set of its integrals along straight lines crossing its support. It is written as follows:

$$f(p,\varphi)=\int_{-\infty}^{\infty} g\left(\sqrt{p^2+l^2},\varphi+\arctan\left(\frac{l}{p}\right)\right)\mathrm{d}l\,, \tag{1.1}$$

where $f(p,\varphi)$ is the value of the integral from $g(x,y)$ along the line, which is specified by the parameters p and φ – which are the distance to the origin and the angle of inclination of the normal to the abscissa axis, see Fig. 1, on the left. The left part of the expression (1.1) in tomography applications is usually called projection data from $g(x,y)$, and the function $f_\varphi(p)$ obtained from $f(p,\varphi)$ if one fixes the angle φ, is a parallel projection.

Suppose that the function $g(x,y)$ describes the X-ray attenuation coefficient distribution in the weld cross section under study. Under certain assumptions about the propagation of X-rays in a substance, it can be shown that the intensity of the radiation measured by the detector at point D from the source at point S is shown in Fig. 1, left, given by the formula:

$$I = I_0 \exp\left(-\int_S^D g(l)\,\mathrm{d}l\right), \tag{1.2}$$

where I_0 is source intensity. Integration is carried out on the segment connecting the source and the detector. It is easy to see that the expression (1.2) is reduced to the integral from along the line. Thus, as a result of the sensing of the sample, the projection data from the distribution of the X-ray attenuation coefficient are measured.

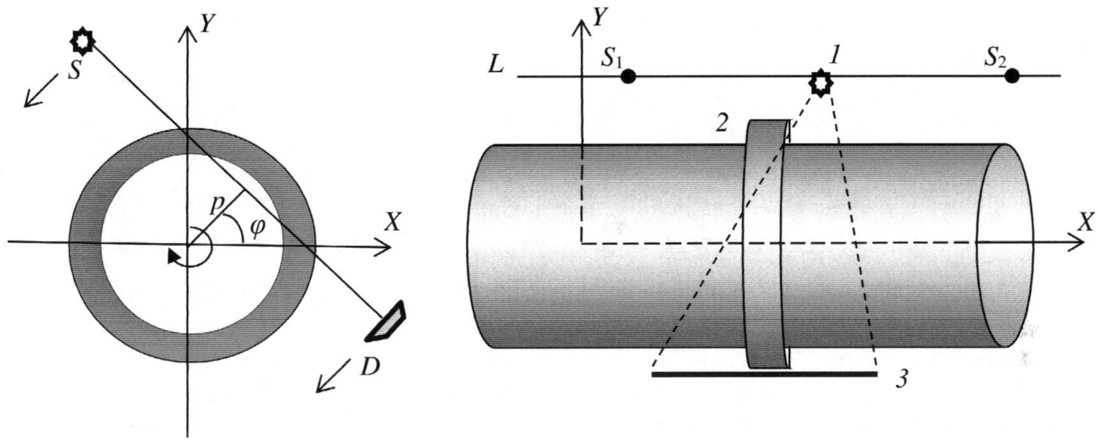

Figure 1. Left: Full set of projection data measuring system. Right: Planar registration scheme: 1 – X-ray source; 2 – object under study; 3 – detector; L – the line along which the source moves between the points S_1 and S_2

The Radon transform (1.1) is invertible. This means that the function $g(x,y)$ can be reconstructed (restored) from the projection data $f(p,\varphi)$. In particular, if there is a complete set of one-dimensional parallel projections (i.e. a set of projections registered by all angles), then the following inversion formula takes place:

$$g(x,y) = \frac{1}{8\pi^2}\int_0^{2\pi}\int_{-\infty}^{\infty}|\omega|\tilde{f}_\varphi(\omega)\exp\left(i\omega(x\cos\varphi + y\sin\varphi)\right)\mathrm{d}\omega\mathrm{d}\varphi. \tag{1.3}$$

In formula (1.3) and further the tilde sign above means the Fourier transform. As can be seen from (1.3), in the process of reconstruction the Fourier image of each projection is multiplied by the frequency module (this procedure is called ramp filtering). After that, the integration of the angle is performed.

In principle, a complete set of one-dimensional projections can be measured in the registration system schematically presented in Fig. 1, on the left. The radiation source and detector move synchronously in the direction indicated by the arrows, scanning the entire object. After that, the system rotates around the Z axis, and the next projection is measured at the following angle, and so on. Of course, in reality, a complete set of data cannot be obtained. However, as practice shows, to get a high-quality tomogram, it is sufficient to register projections in the range of angles from 0° to 360° in steps of 0.5° – 1°. Unfortunately, in the considered case such registration scheme often cannot be implemented. Access to many types of pipelines, in particular, in heat exchange devices is limited. Therefore, the registration system cannot make a complete turnover around the pipe. Moreover, sometimes a turn in this direction is impossible.

IOP Conf. Series: Materials Science and Engineering **681** (2019) 012027 doi:10.1088/1757-899X/681/1/012027

A different scanning geometry, called planar geometry, was developed for pipeline inspection. It is presented in Fig. 1, right. The X-ray source *1* moves along the line segment *L*, bounded by points S_1 and S_2. The radiation transmitted through object *2* is recorded by a fixed line or matrix of detectors *3*. In the latter case, the distribution of the attenuation coefficient can be obtained in some volume, which significantly increases the information content of the diagnosis. But this raises the problem of scattered radiation, which distorts the projection data, which in turn leads to incorrect reconstruction results. The use of a line of detectors weakens the influence of the scattering process, but the image of only one section is reconstructed in the plane passing through the line of detectors and the source. In particular, [9] describes a portable equipment that implements the described registration scheme.

For planar geometry of projection data collection, the inversion formula (1.3) cannot be used directly. In [9] its following modification is offered:

$$g(x,y)=\frac{1}{2\pi}\int_{S_2}^{S_1}\left\{\int_{-\infty}^{\infty}|\omega|\tilde{f}_s(\omega)\exp\left(i\omega\left(r_d\left(s-x\right)/y\right)\right)\mathrm{d}\omega\right\}\mathrm{d}s\,. \tag{1.4}$$

Here it is assumed that the line *L* coincides with the abscissa axis; r_d is the distance between the line *L* and detectors, *s* is the current coordinate of the source. The function $\tilde{f}_s(\omega)$ is the Fourier transform from the function $f_s(x)$ representing the fan projection, i.e. the data measured by the line of detectors at a certain position of the source.

Projection data obtained by planar registration, or measured during the rotation of the recording system around the object, in the range of angles less than 180°, is called incomplete. The papers [10 – 12] are devoted to tomography from such kind of data. There special procedures are developed to improve the quality of reconstruction. The present study also relates to this area. It takes into account the features of the diagnosed object and the specifics of the measurement process.

2. Deconvolution of projection data

2.1 Overview of previous results

It is known that in the images reconstructed from incomplete data, there are characteristic distortions: the structures are stretched in the directions from which the object is probed. In particular, in the planar registration scheme presented in Fig. 1, on the right, such a stretching will be observed mainly along the *Y* axis. Thus, cracks perpendicular to the pipe surface will be longer in the image than they actually are. Such erroneous diagnostics can lead to the dismantling of the pipeline section suitable for operation.

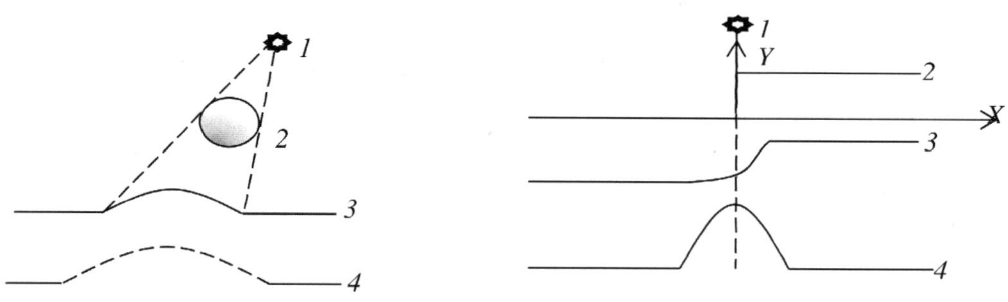

Figure 2. Left: The effect of increasing the projection of a small object: 1 – source; 2 – object; 3 – ray projection; 4 – projection observed in real measurements. Right: Explanation of the method of experimental determination of the hardware function: 1 – source, 2 – object, 3 – projection, 4 – derivative of the projection

In [9] it was shown that the stretching of the structure element is stronger the larger its size on the projection. It was also found that for small elements, this size on real projections is several times greater than the size they have on model projections calculated under the assumption that the equation (1.2) is satisfied; see Fig. 2, on the left. This anomaly is due to the imperfect properties of the recording system: the finite size of the focal spot of the X-ray source, the presence of scattered radiation, the inertia of the detectors, etc. We denote by $f_s^{(0)}(x)$ a model fan projection, which is also referred to hereinafter ray projection. We will consider all effects related to the effect of the equipment in a linear approximation. Then the registered projection can be expressed as follows:

$$f_s(x) = \int_{-\infty}^{+\infty} f_s^{(0)}(x')K(x-x')\mathrm{d}x' + \xi(x) . \tag{2.1}$$

The first term in the right part (2.1) is a convolution of the ray projection with the so-called hardware function $K(x)$, which is determined by the properties of the equipment used, but does not depend on the projection itself. The second term is a random function describing uncontrolled distortions – noises – always present in the results of real measurements.

If the hardware function $K(x)$ is known, the ray projection can be approximated from (2.1). According to [13], the following estimation takes place:

$$f_s^{(0)}(x) \approx \frac{1}{2\pi}\int_{-\infty}^{+\infty} \frac{\tilde{f}_s(\omega)}{\tilde{K}(\omega)+\alpha\omega^2}\exp(i\omega x)\mathrm{d}\omega . \tag{2.2}$$

Here α is the regularization parameter, which is generally determined from the statistical properties of the noise component $\xi(x)$. In practice, the function $K(x)$ can be determined experimentally. In particular, in [9] a method was proposed, the essence of which is explained in Fig. 2 (right side). Suppose that sample *2*, made of a homogeneous material, is a step, and the X-ray source *1* is located exactly above its cut. When this occurs, the projection *3* is measured. As a mathematical model of the attenuation coefficient distribution in the *XY* plane, we take the θ function. Then the measured projection in the absence of noise will be determined by the equation

$$f_s(x) = \int_{-\infty}^{+\infty} \theta(x')K(x-x')\mathrm{d}x' . \tag{2.3}$$

Differentiating the expression (2.3), and given that the generalized derivative of θ function is δ function we come to expression $K(x) = \mathrm{d}f_s(x)/\mathrm{d}x$. Thus, differentiating certain projections it is possible to obtain an estimation of the hardware function of the recording system. According to the results of [9] it turned out that it is well approximated by the Gauss function. Numerical simulations, as well as experiments carried out in [9], showed that the utilization of the deconvolution procedure to projection data reduces the error in determining the crack length by 1.5 – 3.5 times depending on its width, viewing angle, noise level and other factors.

2.2. Investigation of dependence on the value of the regularization parameter

The most important point of the theory of solving ill-posed problems is the choice of the value of the regularization parameter. For the deconvolution problem there are a number of methods for determining the value of α in the formula (2.2), see [13]. Their implementation requires nontrivial a priori information on the noise power spectral density $W(\omega)$ as well as on the analytical properties of the Fourier transform of convolution kernel $\tilde{K}(\omega)$. In the case of X-ray diagnostics, the available estimates of the statistical characteristics of noise and the hardware function of the equipment do not provide the accuracy required for the effective operation of the known methods for calculating the regularization parameter. In addition, the solution of the convolution equation (2.1) is only a part of the problem under study, therefore, the optimal value of α obtained for it may be suboptimal, or even unacceptable, in terms of reconstruction of the image of the attenuation distribution. In this situation, it seems appropriate to use numerical modeling to find the optimal values of α by brute force.

The paper deals with the planar registration scheme. The two-dimensional section of the sample with a crack is modeled by a rectangular area. The sides of the rectangle are parallel to the coordinate axes, its size along the *X* axis is much larger than the distance $|S_1S_2|$, see Fig. 1, right. A crack is simulated also as a rectangle with a large aspect ratio *a*:*b*, located symmetrically relative to axis *Y*. It is assumed that it starts from the surface of the material facing the source. In Fig. 4,*a* the model of the cross section of the sample is presented. In the selected conventional units $a=0.06$, $b=0.002$. The size of the shown area is 0.2×0.2, the sampling of the image is 1025×1025 pixels.

The error of the crack length estimation is calculated by the formula $\delta_l=(a_r-a)/a\times100\%$, where a_r – the length of the crack in the reconstructed image, a – its real length (i.e. the length of the rectangle modeling the crack). As in previous studies, the value of a_r defined as the length of the axis *Y* segment bounded by the points separating the regions where the attenuation coefficient is higher and lower than $g_{\min}+k(\overline{g}-g_{\min})$. Here $g_{\min}$ is the minimum value of attenuation coefficient in the reconstructed profile along the Y axis; $\overline{g}$ is the average value of the restored image; the coefficient *k* takes values from the range from 0 to 1. For the results below, *k* is 0.8.

The number of projections (source positions) is 500. The ray projections are convolved with the kernel $K(x)=\exp\left(-x^2\ln 2/\beta^2\right)$, which effective width β is 0.004, i.e. twice the width of the crack. Uncorrelated centered Gaussian noise with constant variance $\sigma^2=\xi^2\overline{f}^2$ (here ξ is a positive number and $\overline{f}$ is the average value of the projection data in the absence of noise) is added to each projection. Tomographic reconstruction was carried out according to the formula (1.4). The value $\varphi_0=\arctan\left(|S_1S_2|/r_d\right)$ is taken as the angle of view of the study area. In the process of modeling the parameters φ_0 and ξ are varied.

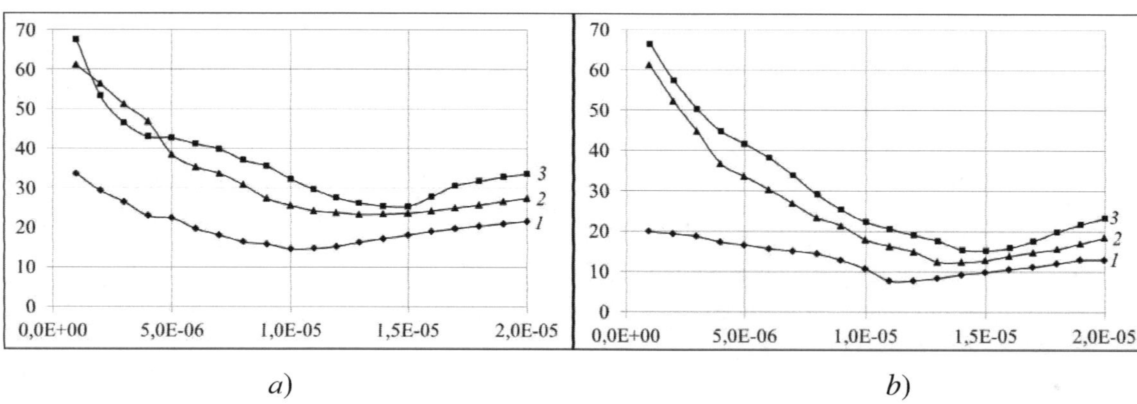

Figure 3. Dependences of the δ_l on the regularization parameter α. (a) Viewing angle 30°; (b) viewing angle 60°. Curves 1 - ξ=0 (no noise); curves 2 - ξ=0.01; curves 3 - ξ=0.02.

In Fig. 3 the dependences of the error δ_l on the value of the regularization parameter α in the formula (2.2) are presented. For Fig. 3,*a* the viewing angle is 30°, and for Fig. 3,*b* it is 60°. In Fig. 3,*a*,*b* for curves *1*, *2*, and *3* the values of the parameter ξ, which determines noise level, are 0 (noise is missing), 0.01 and 0.02, respectively. You can see that all the curves in Fig. 3 have a minimum. Its presence in the dependences of errors on a regularization parameter is typical for solving ill-posed problems by regularizing algorithms [13]. In the case under consideration, this is explained by the fact that for small α, the deconvolution procedure, even in the absence of noise in the original data, generates high-frequency distortions in the estimation of ray projections, which are then amplified by the reconstruction algorithm. On the other hand, the growth of α not only suppresses noise, but also blurs the projections of the crack, which, as mentioned above, leads to an increase in the length of the crack in the reconstructed image. From the analysis of the results presented in Fig. 4, the following conclusion was obtained: the values of the regularization parameters providing the least standard deviation for the estimation of the ray projection data were on average 1.2 – 1.4 times lower than those that provide the least error in determining the crack length. This is explained by the fact that the reconstruction of the image from projections is also an ill-posed problem, the solution of which, generally speaking, also needs to be regularized.

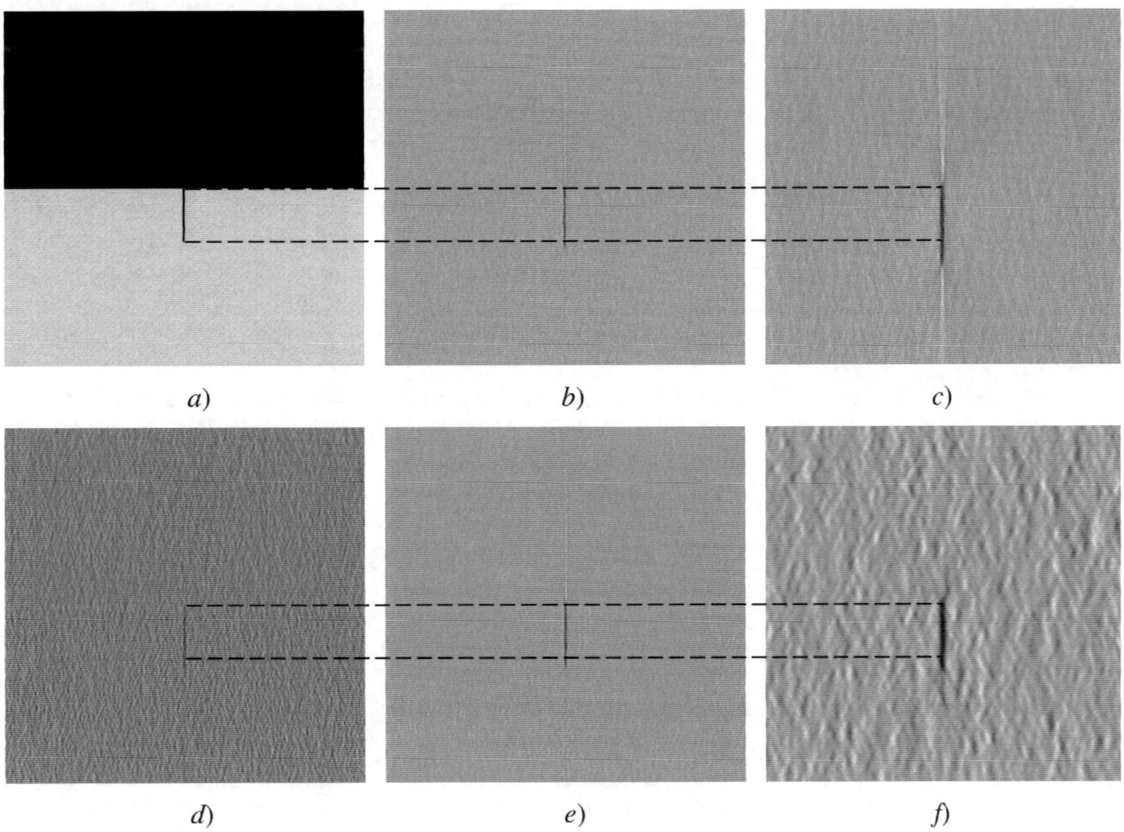

Figure 4. (a) Mathematical phantom modeling a crack in the material; (b) – (f) tomograms: (b) the ray projection data; (c) projections convoluted with a Gaussian kernel; (d) – (f) results after deconvolution of the data: (d) α=9·10^{-6}; (e) α=1.4·10^{-5}; (f) α=2·10^{-5}

Examples of the obtained tomograms are shown in Fig. 4,*b* – *f*. The noise in the projection data corresponds to the parameter $\xi = 0.01$. This means that the average signal to noise ratio is 100. The viewing angle is 60°. In Fig. 4,*b* the reconstruction result from the ray data is presented. In this case $\delta_l = 4.5\%$. The other tomograms are reconstructed from the projections convoluted with the Gaussian kernel. For the tomogram in Fig. 4,*c* deconvolution was not carried out. For Fig. 4,*d* – *f* data deconvolution was carried out by the formula (2.2) with the values of the parameter α equal to $9\cdot10^{-6}$,

10^{-5} and $2{\cdot}10^{-5}$, respectively. It is seen that, using deconvolution, it is possible with the correct choice of the value of the regularization parameter to significantly improve the estimation of the crack length. In particular, for Fig. 4,*e* $\delta_l = 11.2\%$ vs 27.4% for Fig. 4,*c*. Thus, the error decreased by almost 2.5 times. According to the tomogram in Fig. 4,*d* can be concluded that the value $\alpha = 9\cdot10^{-6}$ is underestimated: the noise level in the image is too high, which prevents a good visualization of the crack. Value $\alpha = 2\cdot10^{-5}$, by contrast, seems overestimated, the cracks are too blurry, so determining its true parameters is difficult, see Fig. 4,*f*. However, for Fig. 4,*d* and Fig. 4,*f* error values are 18.1% and 18.9%, respectively, i.e. less than for Fig. 4,*c*. Thus, the utilization of the deconvolution procedure was expedient even with the nonoptimal choice of α.

3. Conclusion

The actual problem of determining the size of the crack in the weld by X-ray tomography was considered. In previous studies, the authors found that the size of the crack in the image exceeds its true size, and this excess is much larger than its estimate, which is obtained within the framework of the ray approximation. The presumable reason for this is the distortion of projection data by the hardware function of the recording system, which in the linear approximation can be represented by a convolution operation. To improve the estimation of the crack depth, the procedure of projection data deconvolution was used. Since deconvolution, as well as tomographic reconstruction, is an ill-posed problem, the influence of the regularization parameter value on the estimation accuracy by means of numerical modeling was investigated in this paper. The dependences of the error in determining of the crack depth on the regularization parameter were obtained for different values of the parameters characterizing the registration conditions.

References

[1] Barrett H H and Swindell W 1996 *Radiological Imaging* vol 2 (New York: Academic Press)

[2] Redmer B, Ewert U, Likhachov A V, Pickalov V V, Onel Y, Li Z and Baranov V A 2000 *Proc. 15th World Conference on Non-Destructive Testing* IDN370

[3] Hanaor D A H, Hu L, Kan W H, Proust G, Foley M, Karaman I and Radovic M 2016 *Materials Science and Engineering,* A **672** 247

[4] Macovski A 1983 *Proc. IEEE* **71** 3730

[5] Kak A C and Slanley M 1988 *Principles of computerized tomographic imaging* (New York: IEEE Press)

[6] Kalender W A 2009 *Computed tomography: fundamentals, system technology. Image quality* (New York: Wiley)

[7] Herman G T 1980 *Image Reconstruction from Projections: The Fundamentals of Computerized Tomography* (New York: Academic Press)

[8] Natterer F 1986 *The Mathematics of Computerized Tomography* (New York: Wiley)

[9] Likhachov A V, Pickalov V V, Ewert U and Redmer B 2001 *Proc. 2nd World Congress on Industrial Process Tomography* (Hannover: NTH Press) p 711

[10] Hanson K M and Wecksung G W 1983 *Journal of the Optical Society of America* **73** 1501

[11] Boyd J 1995 *Journal of Nondestructive Evaluation* **14** 61

[12] Quinto E T 1998 *Inverse Problems* **14** 339

[13] Tikhonov A N and Arsenin V Ya 1977 *Solution of Ill-Posed Problems* (New York: Wiley)

IOP Conf. Series: Materials Science and Engineering **681** (2019) 012028 doi:10.1088/1757-899X/681/1/012028

Dependence of the structure and characteristics of a Russian alternative for AISI 304 stainless steel powder on the parameters of their laser cladding on substrates from low-carbon and structural steels

A B Lyukhter[1], P A Palkin[1], A V Zavitkov[1], D M Kononov[2] and A V Kireev[2]

[1] Research and educational center for implementation of laser technologies in machine building, Vladimir State University, 600000, Russia
[2] Institute of machine building and automobile design, Vladimir State University, 600000, Russia

E-mail: palkin@laser33.ru

Abstract. The work studies X8 CrNiSiMg 17-8-6 steel powder, a Russian alternative for AISI 304 steel, cladded by ytterbium fiber laser on two types of substrates: heat-resistant steel and low-carbon steel. The results demonstrate the effect of laser radiation power on the macrostructure of single cladded beads and their fusion with the substrate. The selected regimes allow forming solid cladded layers. The paper also studies the microstructure of metallographic sections of cladded specimens, the phase composition of cladded layers and initial powder. The deliverables include the data on wear resistance, resistance to intercrystalline corrosion and strength of cladded layer fusion with the substrate.

1. Problem Formulation

Implementation of fiber lasers for cladding layers of stainless and heat-resistant steels enables production of more wear resistant, higher-quality and denser coatings compared to conventional cladding methods. This method heats the part less and provides stronger fusion with the substrate [1]. High cooling rate of the cladded layers induces a characteristic microstructure that appreciably affects final mechanical and chemical properties of the coatings [2, 3]. Inaccurate selection of cladding regimes inevitably leads to defects in the layers, either inside them or inside the fusion with the substrate. Such strongly negative effects can be avoided in two ways: a) modeling or selection of optimal heat input and focal distance of the laser; b) implementation of a special multi-channel nozzle forming the flux of sputtered powder as a purely technological solution to make the cladded layers more sinterable. The work is to study the properties of Russian X8 CrNiSiMg 17-8-6 powder sputtered in an inert gas (argon) through the multi-channel nozzle directed by ytterbium fiber laser on a steel substrate under varied power and laser velocity, to discover and control the properties of such coating.

2. Laser cladding technology

The stainless-steel powder was cladded on two types of low-carbon steel: X12 CrNiTi 18-10 austenitic heat-resistant alloyed steel and X9 MgSi 2 structural ferrite low-alloyed steel. The cladding powder was represented by X8 CrNiSiMg 17-8-6 steel, a Russian alternative for AISI 304 stainless steel.

Laser cladding was performed by LRC laser robotic complex composed of LS-5 ytterbium fiber laser, Fanuc six-axis robot with coaxial powder feeding nozzle, two-axis positioning device and AT-1200 powder feeder.

2.1 Selecting optimal laser-cladding regime

This work considered ten regimes of laser cladding sorted according to laser radiation power. The main parameters of laser cladding (radiation power, produced layer hardness, substrate melting and maximum depth of heat-affected zone) are presented in Table 1. The laser velocity was selected according to the substrate heat capacity.

Table 1. Main cladding parameters

No.	Power (W)	Hardness	Substrate melting	Heat affected area (mm)
1	600	90-140 HB	no	0.6
3	800	150	no	0.6
4	1000	150–160 HB	yes	0.8
5	1250	120-150 HB	no	1
6	1500	150-170 HB	yes	1
7	1600	150-250 HB	yes	1.3
8	1700	150-270 HB	yes	1.3
9	1800	100-150 HB	yes	1
10	1900	150 HB	yes	1
11	2000	150–200HB	yes	1.3

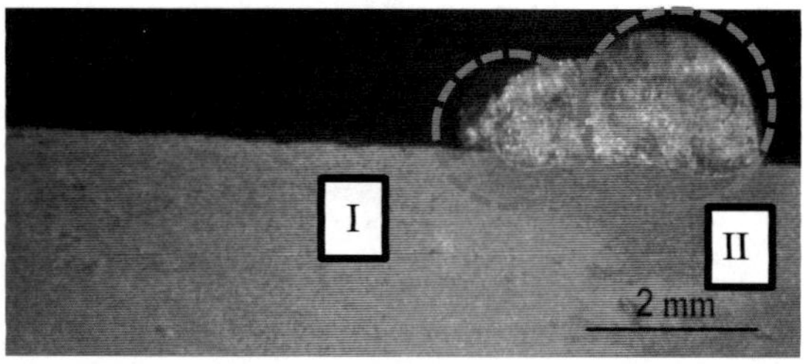

Figure 1. End polished section of single cladded track at low power

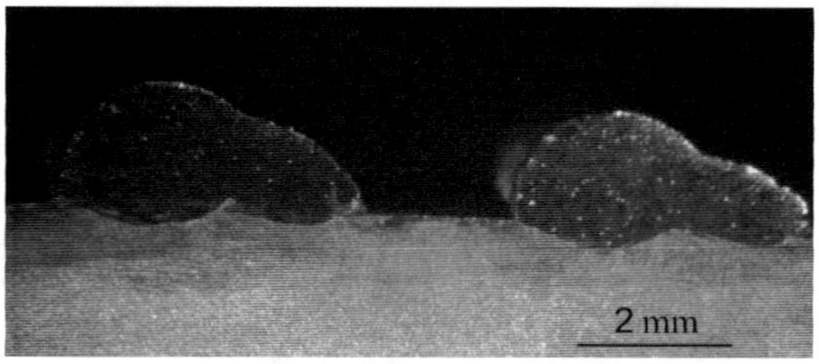

Figure 2. End polished section of multiple cladded tracks at high power

The specimens were cross-cut to produce metallographic polished samples that were studied by optical microscopy (Figures 1 and 2). Obviously, there are separate beads represented by a surface with two components: main body (Figure 1-I) and metal outflow from the main pool (Figure 1-II) formed after powder melting under the action of the shielding gas jet providing total crossing of the beads during formation of solid cladded layer.

3. Structure and mechanical properties of cladded layers

After trying several regimes of laser cladding, three regimes of laser beam energy input were selected. Figures 3–5 demonstrate edge polished samples of beads cladded in three regimes, correspondingly. Their main parameters are presented in Table 2.

Table 2. Main parameters

No.	Energy per unit length (W/m)	Hardness	Substrate melting	Heat affected area (mm)
Regime 1	240<	~ 30 HRC	yes	~ 2.5
Regime 2	>500	~ 30 HRC	yes	~ 2.5
Regime 3	240<x>500	~ 30 HRC	yes	~ 1.5

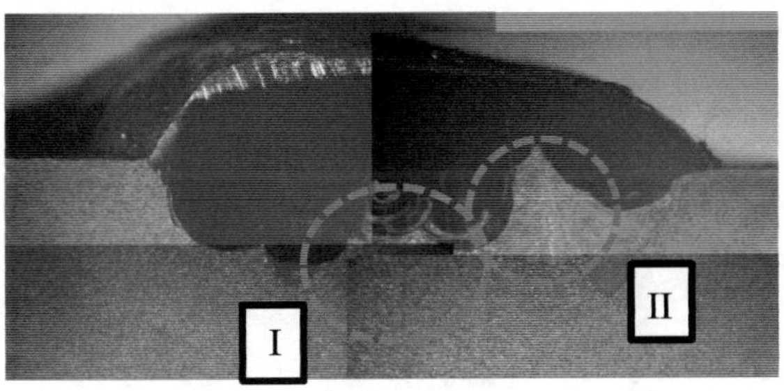

Figure 3. End polished sample cladded in regime 1

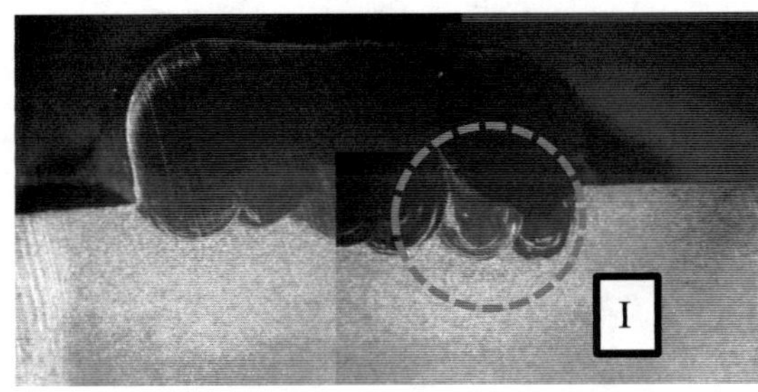

Figure 4. End polished sample cladded in regime 2

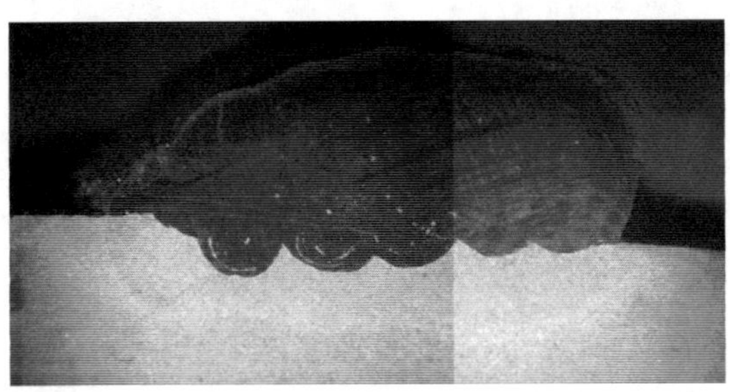

Figure 5. End polished sample cladded in regime 3

The metallographic slices demonstrate solid cladded layers: in regime 1, the substrate is not fully penetrated during formation of first beads (Figure 5-I) and on consequent incomplete mixing of cladded metal and base substrate metal (Figure 5-II); in regime 2, there is excessive penetration with incomplete mixing of the substrate metal. Regime 3 demonstrates the most optimal formation of cladded layers with partial penetration of the substrate, minimum formation of separate micropores and absence of substantial negative effects, including cracks and lacks of penetration.

4. Tribological properties

The cladded layers were used to produce samples for studying their tribological properties on a versatile friction machine. A hardened spherical tip made of high-carbon alloyed steel was used to make spherical friction tracks under load and without lubrication on the surface of the cladded layers to reveal the behavior of the friction coefficient during wear and to establish qualitative dependence of the wear resistance as the specimen mass loss on the path, load and velocity (Tables 3 and 4).

Table 3. Measurement parameters

Substrate	Velocity (m/s)	Load (N)	Distance (m)	Mass loss (g)
09G2S (European A 516-55 steel)	0.2	10	1000	0.042

Tribological test results of stainless steel laser cladding on the 09G2S substrate are presented as plots in Figures 6 and 7.

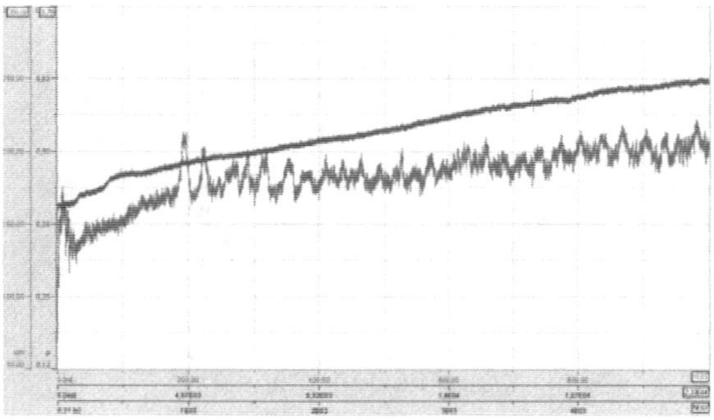

Figure 6. Dependence of friction coefficient on distance and load

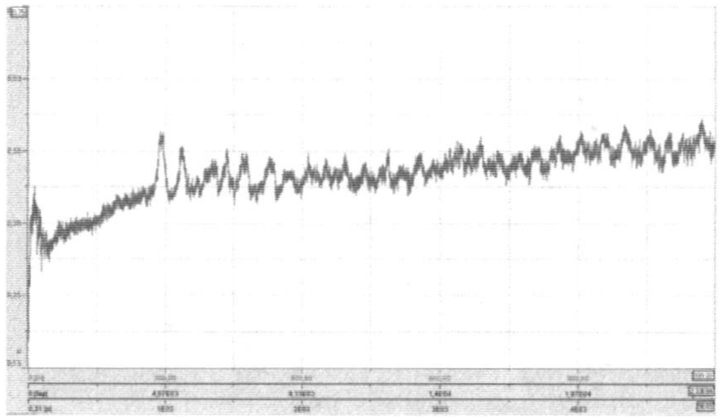

Figure 7. Dependence of friction coefficient on distance and load

Tribological test parameters for stainless steel laser cladding on X12 CrNiTi 18-10 substrate are presented in Table 2. The results are presented in Figures 8 and 9.

Table 4. Measurement parameters

Substrate	Velocity (m/s)	Load (N)	Distance (m)	Mass loss (g)
X12 CrNiTi 18-10	0.2	10	1000	0.05

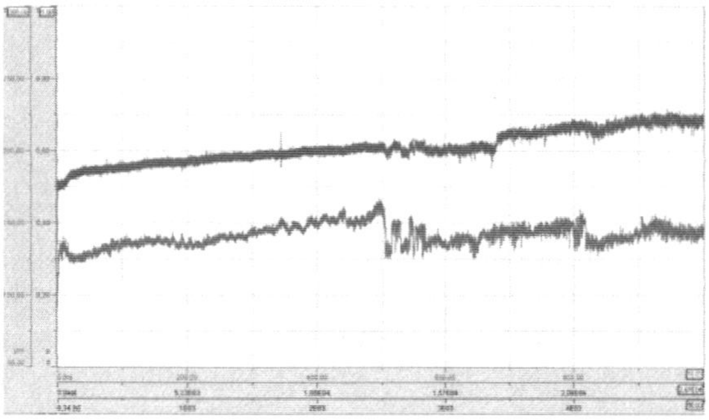

Figure 8. Dependence of friction coefficient on distance and load

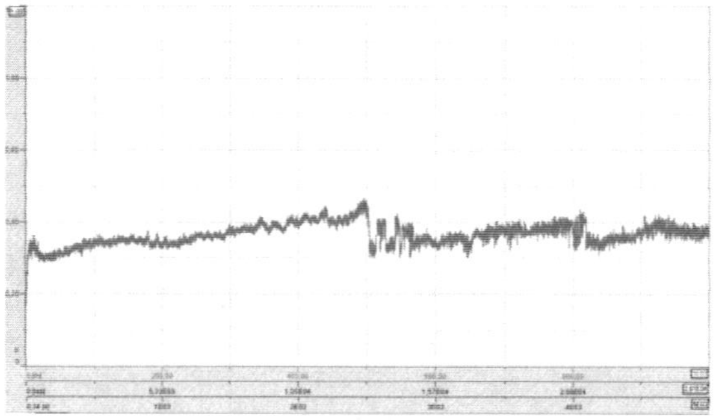

Figure 9. Dependence of friction coefficient on distance and load

The results of tribological tests have demonstrated similar mass loss for both substrate types (Tables 3 and 4). The cladded material wear on the stainless steel substrate (Figures 7 and 8) is characterized by a longer uniform section, unlike the layers applied on low-carbon steel where the plot is nonuniform starting from the beginning of the experiment, which is probably conditioned by different structure of cladded layers, depending on heat sink through the substrate material.

5. Phase analysis

X-ray diffraction (XRD) analysis was performed on a D8 ADVANCE X-ray diffractometer. The structure of the iron component of initial powder is represented by austenite and ferrite with prevailing ferrite component (Figures 10 and 11) [4]. Cladding on austenitic steel surface allows obtaining primarily austenitic structure (up to 95% of austenite, Figures 12 and 13), unlike cladded layers of ferrite steel surface, which structure remains austenitic-ferritic with prevailing fraction of austenite up to 79.21% (Figures 14 and 15).

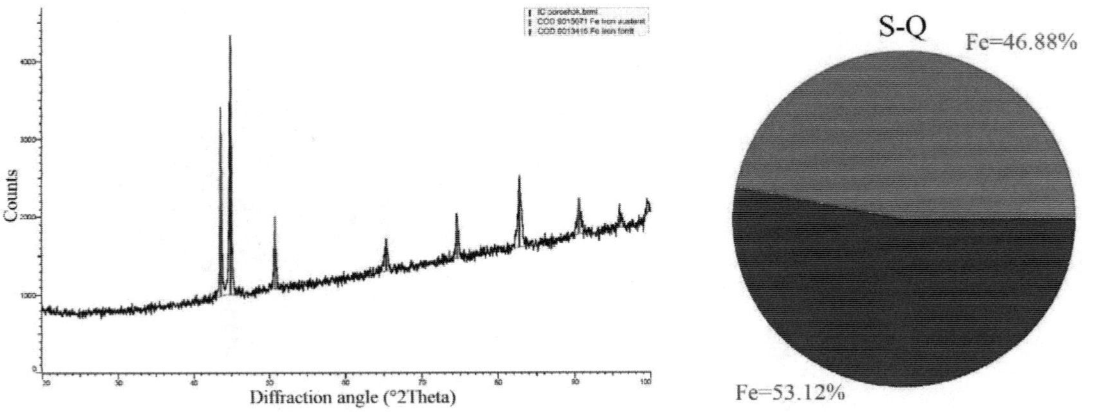

Figure 10. Dependence of intensity on 2θ angle

Show	Icon	Color	Index	Name	Parent	Scan	Pattern #	Compound Name
Yes			1	COD 9015071	Pattern List #5	IC poroshok.brml #1	COD 9015071	Iron
Yes			2	COD 9013415	Pattern List #5	IC poroshok.brml #1	COD 9013415	Iron

Formula	Y-Scale	I/Ic DB	I/Ic User	S-Q	Added Reference	d x by	Scan WL	Wavelength
Fe	62.4308 %	9.740		46.88 %		1.0000	Yes	1.54060
Fe	96.1740 %	13.240		53.12 %		1.0000	Yes	1.54060

System	Space Group	a	b	c	alpha	beta	gamma	Z	Volume	Density	Cell Tuned	F (N)
Cubic	F m -3 m (225)	3.59500							46.46		No	
Cubic	I m -3 m (229)	2.85500							23.27		No	

Figure 11. Results of phase composition investigation of initial powder

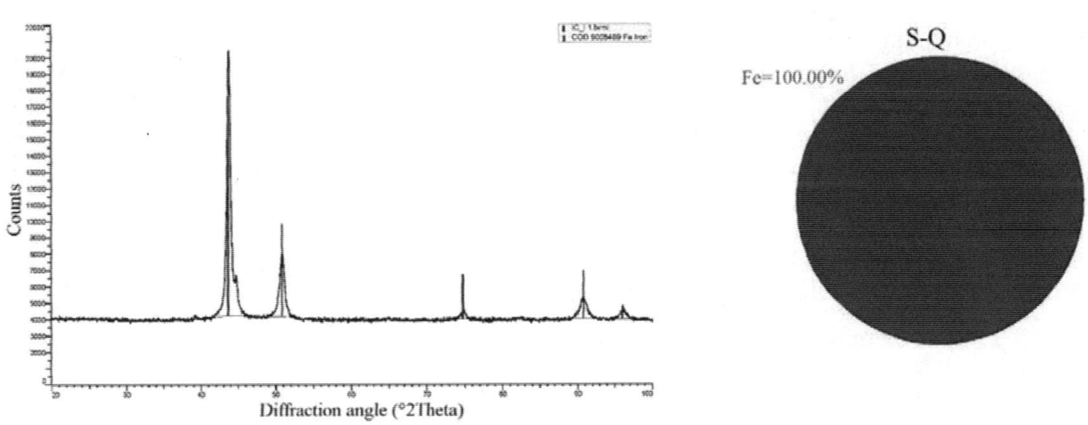

Figure 12. Dependence of intensity on 2θ angle

IOP Conf. Series: Materials Science and Engineering **681** (2019) 012028 doi:10.1088/1757-899X/681/1/012028

Show	Icon	Color	Index	Name	Parent	Scan	Pattern #	Compound Name
Yes			1	COD 9008469	Pattern List #9	IC_I 1.brml #1	COD 9008469	Iron

Formula	Y-Scale	I/Ic DB	I/Ic User	S-Q	Added Reference	d x by	Scan WL	Wavelength
Fe	78.4595 %	9.730		100.00 %		1.0000	Yes	1.54060

System	Space Group	a	b	c	alpha	beta	gamma	Z	Volume	Density	Cell Tuned	F (N)
Cubic	F m -3 m (225)	3.59100							46.31		No	

Figure 13. Results of phase composition investigation of cladding applied on austenitic steel substrate

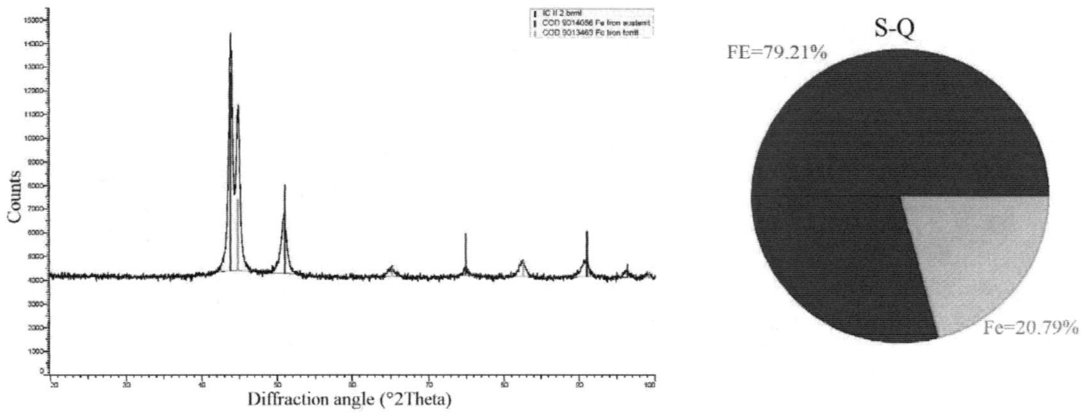

Figure 14. Dependence of intensity on 2θ angle

Show	Icon	Color	Index	Name	Parent	Scan	Pattern #	Compound Name
Yes			1	COD 9014056	Pattern List #3	IC II 2.brml #1	COD 9014056	Iron
Yes			2	COD 9013463	Pattern List #3	IC II 2.brml #1	COD 9013463	Iron

Formula	Y-Scale	I/Ic DB	I/Ic User	S-Q	Added Reference	d x by	Scan WL	Wavelength
Fe	83.2200 %	9.720		79.21 %		1.0000	Yes	1.54060
Fe	29.8000 %	13.260		20.79 %		1.0000	Yes	1.54060

System	Space Group	a	b	c	alpha	beta	gamma	Z	Volume	Density	Cell Tuned	F (N)
Cubic	F m -3 m (225)	3.58100							45.92		No	
Cubic	I m -3 m (229)	2.86040							23.40		No	

Figure 15. Results of phase composition investigation of cladding applied on ferrite steel substrate

6. Microstructure analysis

The microstructure of cladded layers was studied in the optical range using a metallographic microscope. The heat affected zone of the substrate metals primarily demonstrates medium-grain structure, which is connected with the grain growth under the thermal effect conditioned by heat sink through the substrate in the presence of small-grain component (Figure 16a). The cladding structure, a separate bead in particular (Figure 16b), demonstrates appreciably ramified dendrite structure formed along the heat sink lines during crystallization. The microstructure of the cladded layer is predominantly a solid solution of alloying elements in iron of austenitic type with presence of intermetallides on grain boundaries (Figure 16c).

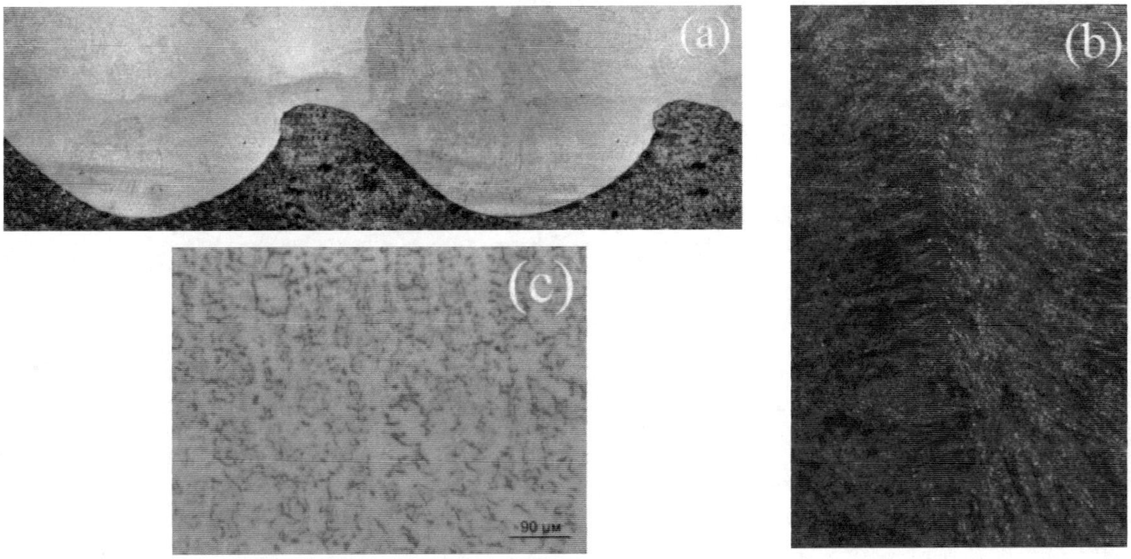

Figure 16. Metallographic images of cladding

7. Conclusions

High cooling rates during laser cladding of stainless steels have a considerable impact on the microstructure and phase composition; hence, affect their functional properties [5]. With high energy additivity, the formation of layers from powder materials by high-power fiber laser beam is complicated by a considerable problem of balance of velocity and energy density of the laser providing sufficient sintering of cladded layers with minimum formation of pores and burn-off of alloying elements. In the work, the laser beam power was selected to avoid cracking in pre-heated substrate. Cladding was carried out by a four-channel power feeding unit that was forming the deposited layer with set continuity and selected optimal laser input power. As a result, the replacement of a conventional single-axis coaxial nozzle with a four-channel nozzle for cladding by ytterbium fiber laser using LRC-5 at VLSU provides predominant formation from mainly ferrite powders of cladded austenitic layers with dendrite structure and minimum number of separate micropores with a surface hardness of 25–30 HRC without unacceptable defects, including cracks. This will allow using the method for cladding working surfaces suffering simultaneously loads and corrosion, for instance, pipeline valves.

References

[1] Alam M Edrisy A Urbanic J and Pineault J 2017 Microhardness and Stress Analysis of Laser-Cladded AISI 420 Martensitic Stainless Steel *Journal of Materials Engineering and Performance* **26** 3 1084
[2] David S A 1987 Effect of rapid solidification on stainless steel weld metal microstructures and its implications on the Schaeffler diagram *Oak Ridge National Lab* (USA) 10487
[3] Kurz W Trivedi R 1994 Rapid solidification processing and microstructure formation *Materials Science and Engineering* **179** 46–51
[4] Blawert C Weisheit A Mordike BL Knoop RM 1996 *Surf Coat Technol* **85** 15
[5] Hemmati I Ocelı´k V Hosson J 2011 Microstructural characterization of AISI 431 martensitic stainless steel laser-deposited coatings *J Mater Sci* **46** 3405–3414

Laser welding of aluminium alloys for the aircraft industry

A Malikov[1], A Orishich[1], E Karpov[1,2] and I Vitoshkin[1]

[1] Khristianovich Institute of Theoretical and Applied Mechanics, Siberian Branch of Russian Academy of Sciences, 4/1, Institutskaya street, Novosibirsk, Russia, 630090
[2] Lavrent'ev Institute of Hydrodynamics, Siberian Branch of Russian Academy of Sciences, 15, Lavrentyeva street, Novosibirsk, Russia, 630090

E-mail: smalik@ngs.ru

Abstract. The results of laser welding of an Al-Mg-Mn and two Al-Mg-Li alloys with different lithium contents which used in the aircraft industry are presented in this paper. Microstructure, phase composition and mechanical properties of the welds were investigated. A fundamental difference in the processes of formation of structural-phase compositions in the weld metal of these alloys was found.

1. Introduction

Aluminum alloys of Al-Mg (e.g., Russian AMg5, AMg6) and Al-Cu-Mg (e.g., Russian D16 and 1163) systems are widely used for production of aerospace vehicles [1–3]. At the same time, progress in the aircraft industry is associated with using of contemporary high-strength aluminum alloys as structural material. To date, new aluminum-lithium alloys have been developed [4–6]. Alloying with lithium lowers metal density in comparison with traditional aluminum alloys. Examples are heat-treatable wrought alloys such as Al-Mg-Li-*X* (*X* = Zn, Mn, Zr, Sc), Al-Cu-Mg-Li, and Al-Cu-Li-*X* (*X* =Mg, Zn, Mn, Zr, Sc); Al-Cu-Li-*X* high-strength alloys as well as Al-Mg-Li-*X* average-strength ultra-light corrosion-resistant alloys.

Al-Mg-Li corrosion-resistant alloys (e.g., Russian 1420 and 1424) are the most promising candidates for replacing traditional Al-Mg alloys (e.g., Russian AMg6 and AMg5) in the aircraft industry. This is due to the fact that they have higher strength and fatigue characteristics. However, detailed studies of the properties of these alloys are necessary. Also, heat treatment (hardening, aging and annealing) and cold working procedures should be developed to improve structural-phase composition and mechanical properties. Different welding procedures (friction stir, gas tungsten arc, electron-beam, as well as laser-beam with and without filler wire) have been developed to replace riveted joints and, as a result, reduce weight of structures. However, welded joints of these Al-Li alloys made by fusion welding have low mechanical properties. It was shown in reviews on laser welding of aluminum alloys [7–9] that tensile strength of welded joints of Al-Mg-Li alloys is only about 60…85% of the base metal strength. The main reasons are burning-out of alloying elements with low melting point (magnesium, lithium, zinc, and manganese) as well as porosity and hot cracks in the weld metal. Optimization of welding parameters or using of filler wires was recommended by different authors to reduce the influence of these factors. The authors of papers [10, 11] presented microstructure and strength properties of the 2A97 alloy laser welds with and without filler wire. The weld strength was 84% of the base metal strength. In [12], the process of T-joint formation of 2060-T8/2099-T83 Al-Li alloys using laser welding with filler wire was studied; phase composition and mechanical characteristics of the welded joint were shown. In [13], laser welding of the 5A90

Al-Li alloy was investigated; microstructure and microhardness of the welds and the base metal were presented. However, it was not taken into account that these alloys are cold work- and precipitation-hardenable. Due to this fact their strength properties are determined by formation of hardening phases as a result of these treatments. Aluminum alloys of the Al-Mg-Li system are characterized by a complex phase composition which changes during heat treatment. This is due to the large number of alloying elements compared to traditional alloys. For example, Mg, Li, Zr, Sc, and Zn are present in the 1424 alloy. Alloying elements react chemically with aluminum. Therefore, in addition to the main strengthening δ-Al_3Li phase, double and triple intermetallic phases are formed in these alloys (for example, $S_1(Al_2MgLi)$ [14, 15]) which have a significant effect on strength and ductility of the metal.

The aim of the paper is to study regularity of changing in mechanical properties, microstructure and phase composition of AMg6, 1420, and 1424 aluminum alloys used in the aircraft industry during laser welding.

2. Materials and experimental procedure

The chemical compositions of the investigated aluminum alloys, which are a solid solution of α-Al with alloying elements, are given in Table 1. The AMg6 alloy also includes dispersed particles of Al_3Mg_2 (β-phase) and Al_6Mn, which are the products of the solid solution decomposed during crystallization. The 1420 alloy contains hardening phases of Al_3Li (δ′-phase) and Al_2MgLi (S_1-phase) as well as dispersed particles of $Al_3(Zr)$ phase. The 1424 alloy distinguishes from 1420 a lower lithium concentration. It also contains hardening phases Al_3Li (δ′-phase) and Al_2MgLi (S_1-phase) but other dispersed particles (Al_3Sc and $Al_3(Sc, Zr)$ phases).

Data on ultimate tensile strength (σ_{UTS}), yield strength ($\sigma_{0.2}$), elongation (δ), and density ρ of these alloys are shown in Table 2.

Table 1. Chemical composition of aluminium alloys (%, wt.)

Alloy	Al	Mg	Mn	Zn	Li	Zr	Sc	Ti	Fe
AMg6	base	5.8…6.2	0.5…0.8	0.2	–	–	–	0.02…0.10	0.4
1420	base	5.8…6.2	0.10…0.25	0.05	2.0…2.2	0.01		–	–
1424	base	4.7…5.2	0.05…0.25	0.4…0.8	1.61	0.09	0.07	–	–

Table 2. Mechanical properties of aluminium alloys

Alloy	ρ, g/cm^3	σ_{UTS}, MPa	σ_{YS}, MPa	δ, %
AMg6	2.65	315…375	190	10.0
1420	2.50	440…450	280	18.5
1424	2.54	450…500	322	8.6

Butt joints were welded using a 'Siberia' automated laser technological facility (Figure 1, a) which includes a continuous wave CO_2 laser with a power of up to 8 kW [16]. It was developed at the Khristianovich Institute of Theoretical and Applied Mechanics, Siberian Branch of Russian Academy of Sciences. Scheme of the laser welding process is shown in Figure 1, b. The laser beam was focused on the alloy surface using ZnSe lens with a focal distance of 254 mm. The beam diameter on the lens was equal to 25 mm. The diameter of the focused beam in its waist was evaluated as a sum of the diffraction diameter and diameter of the dissipation spot resulting from the spherical aberration. The calculated total diameter was 250 μm for the CO_2 laser.

An inert gas (helium) was used to protect the welded joint. The gas was fed through a nozzle tilted at an angle of 45° to the welded plates. The weld root was also protected with helium. The protective gas flow rate was 5 L/min.

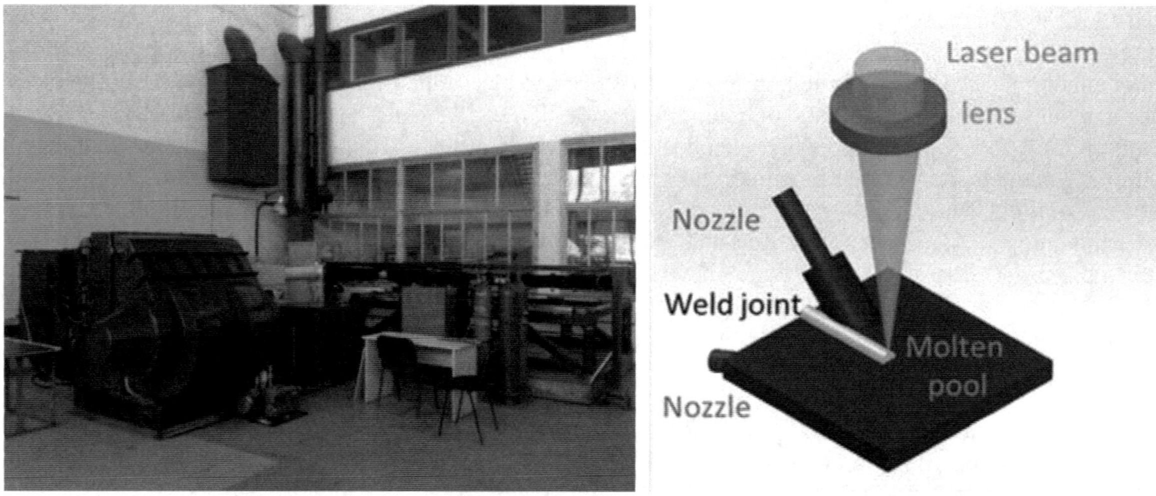

a b

Figure 1. Photograph of the 'Siberia' automated laser technological facility (a) and scheme of the laser welding process (b).

The oxide film was removed by means of chemical milling to a depth of 0.15…0.20 mm using 20% NaOH solution. The sample edges were scraped bright with a metallic scraper immediately before welding.

Firstly, the keyhole laser welding process was optimized in terms of speed, laser power, diameter and location of the focal spot, as well as shielding gas flow rate. Quality criteria were sound welds without any discontinuities such as pores, lack of fusion, undercuts and hot cracks. The optimal laser welding parameters are presented in Table 3.

Table 3. Optimal laser welding parameters

Alloy description	Thickness, mm	Laser power, kW	Welding speed, m/min	Focus position, mm
AMg6	1.5	2.6	4	-3
1420	1.4	2.7	4	-3
1424	1.6	2.9	4	-3

3. Results and discussion

Optical micrographs of the cross-sections of the weld metal, heat affected zone and the base metal are presented in Figure 2. The figure shows that all base alloys had a typical recrystallized structure. No dendritic structure was observed. Agglomerated particles in a wide range of sizes from tenths of a micrometer up to 10 μm were inside the metal matrix of a solid solution. The initial structure was transformed during laser welding and a typical granular structure was formed after crystallization. Dark agglomerates in the weld metal located mainly at boundaries of dendritic grains, making them very contrasting. The amount of dark agglomerates sharply increased in the welds of the studied aluminum alloys in comparison with the base metal. Their size was about 1…5 μm in most cases.

Images of the cross-sections of the etched welds and the base metal made using a scanning electron microscope (SEM) in the mode of backscattered electrons at different magnifications are presented in Figures 3 and 4. It follows from their comparison with Figure 2 that changes in microstructure of the welds and the base metal captured using electronic and optical microscopes coincide at the micron level. Uneven etching of the surface of the cross-sections indicates the uneven distribution of the main alloying elements in dendritic cells. Two types of inclusions can be distinguished on the basis of obvious differences in morphology and color after etching: light and dark which are often combined into complex agglomerates with dimensions of 5…10 μm. The amount of the dark agglomerates in the welds is greater than in the base metal. The results of EDX analysis show an increase in Mg content in the dark agglomerates compared to the solid solution for the welds of the 1420 and 1424 alloys from 4.6 up to 5.6% and from 3.2 up to 5.5%, respectively; Mg diffused to the dendrite boundaries. Diffusion of Mg into the dark agglomerates did not occur in the AMg6 alloy.

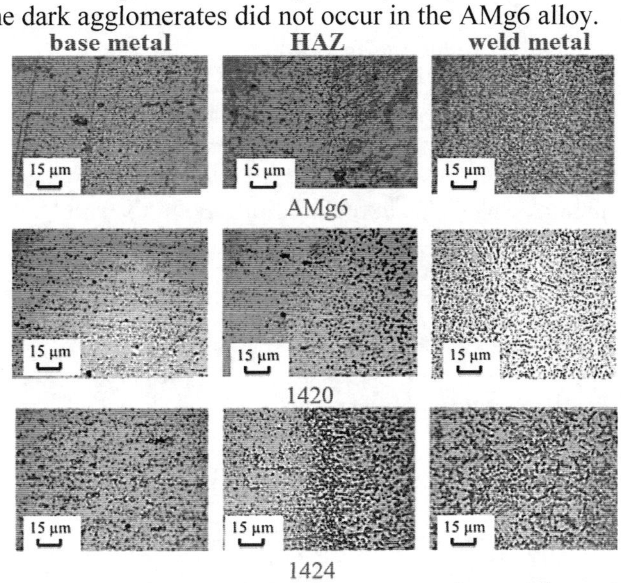

Figure 2. Optical micrographs of the cross-sections of the weld metal, heat affected zone and the base metal.

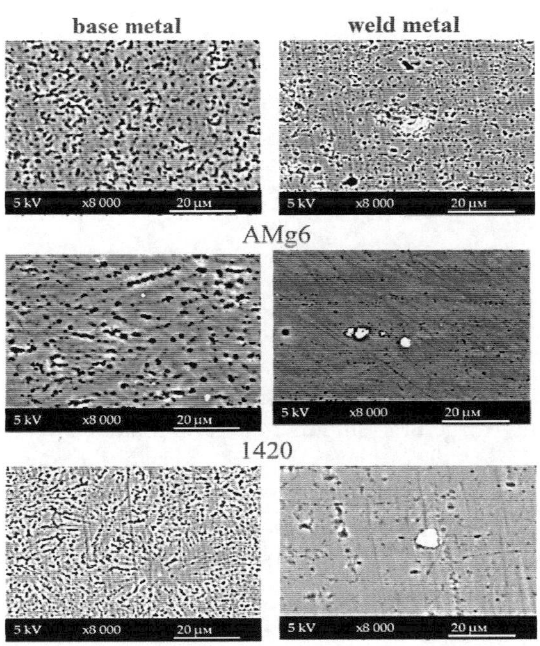

Figure 3. SEM images of the cross-sections of the welds and the base metal.

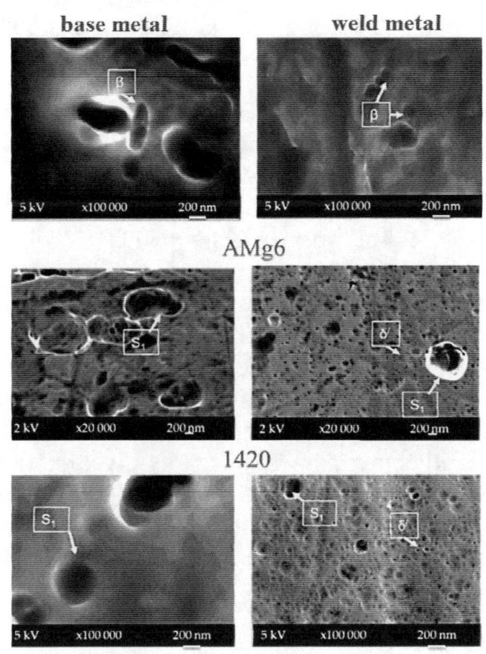

Figure 4. SEM images of the cross-sections of the welds and the base metal.

In the AMg6 alloy, β-phase (Al_3Mg_2) particles with dimensions of 200 nm were observed at high magnification both in the base metal and in the weld. In the 1424 and 1420 alloys, two types of particles with sizes of 50 nm and 200 nm particles were found in the base metal. In the welds of these alloys, the dark particles with a size of 50 nm (apparently δ′ (Al_3Li)) were absent but the amount of the dark particles with a size of 200 nm (S_1 (Al_2MgLi)) increased.

Figure 5 shows diffraction patterns of the base metal which included intense reflections of the main α-Al phase as well as low-intensity reflections related to intermetallic phases. The AMg6 alloy had a number of reflections that could be Al_3Mg_2, Al_6Mn, and $Al_5Mg_{11}Zn$ phases. Wide reflections of the Al_3Li phase were observed in the 1424 alloy. Perhaps, based on published data [15], they indicated presence of S_1 phase (Al_2MgLi). Intermetallides in the base metals were presented in very low concentrations. Intensity of the reflections of the intermetallic phases drops significantly in both alloys after exposure to laser radiation. This is due to their dissolution in the weld pool.

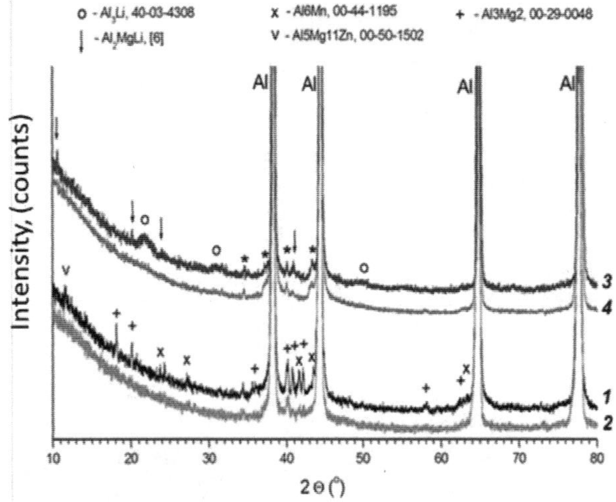

Figure 5. XRD patterns of the base metal and the welds: 1 – AMg6 alloy, 2 – welded joint of AMg6 alloy, 3 – 1424 alloy, 4 – welded joint of 1424 alloy. The numbers indicate a quantity of cards in the PDF-4 powder database (2011).

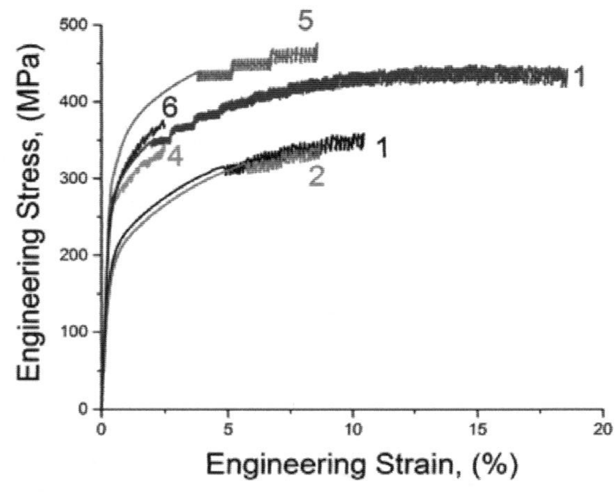

Figure 6. Stress-strain curves: 1 – AMg6 alloy, 2 – welded joint of AMg6 alloy, 3 – 1420 alloy, 4 – welded joint of 1420 alloy, 5 – 1424 alloy, 6 – welded joint of 1424 alloy.

Figure 6 shows stress vs. strain dependences for the studied alloys and the welded joints.Tensile strength of the AMg6 alloy welds was almost 98% of the base metal strength, elongation was 86% of the base metal elongation. Tensile strength of the welds of the 1420 and 1424 alloys was 78% of the base metal strength; their elongation was about 13…28% of the base metal elongation.

It can be concluded based on the results of microstructural studies and XRD analysis that the strengthening phases were dissolved and Mg diffused from the solid solution to the dendrite boundaries in the 1420 and 1424 alloys due to the laser welding process. Only metastable S_1 (Al_2MgLi) was in the solid solution. This caused embrittlement of the welds. Magnesium diffusion was not observed in the AMg6 alloy weld metal, and the β-phase (Al_3Mg_2) was in the solid solution of both the base metal and the weld.

4. Conclusion

Comprehensive studies of laser welding of the AMg6, 1420, and 1424 aluminum alloys were performed. Microstructure, phase composition, and mechanical properties of the welded joints were investigated. The obtained results made it possible to reveal the fundamental features and differences in the crystallization processes of the studied aluminum alloys. Hardening phases formed in the 1420 and 1424 alloys due to cold work and heat treatment were dissolved in the weld pool under the influence of laser radiation. This caused embrittlement of the welds.

Acknowledgments
This study was funded by RFBR and region Novosibirsk according to the research project No.19-48-540008.

References

[1] Campbell F C 2006 *Manufacturing Technology for Aerospace Structural Materials* (London: Elsevier)
[2] Belov N A, Eskin D G, Aksenov A A 2005 Alloys with Lithium *Multicomponent Phase Diagrams* (London: Elsevier) chapter 8 pp 257–286
[3] Fridlyander I, Grushko O, Shamrai V, Klochkov G 2007 *Met. Sci. Heat. Treat.* **49** 279–283
[4] Rioja R J, Liu J 2012 *Metall. Mater. Trans. A Phys. Metall. Mater. Sci.* **43** 3325–3337
[5] Abd El-Aty A, Xu Y, Guo X, Zhang S-H, Ma Y, Chen D 2018 *J. Adv. Res.* **10** 49–67
[6] Khokhlatova L B, Kolobnev N I, Oglodkov M S, Mikhaylov E D 2012 *Metallurgist* **56** 336–341
[7] Kashaev N, Ventzke V, Cam G 2018 *J. Manuf. Process.* **36** 571–600
[8] Xiao R, Zhang X 2014 *J. Manuf. Process.* **16** 166–175
[9] Oladimeji O O, Taban E 2016 *Weld. World.* **60** 415–457
[10] Ning J, Zhang L J, Bai Q L, Yin X Q, Niu J, Zhang J X 2017 *Mater. Des.* **120** 144–56
[11] Fu B, Qin G, Meng X, Ji Y, Zou Y, Lei Z 2014 r *Mater. Sci. Eng. A.* **617** 1–11
[12] Han B, Tao W, Chen Y, Li H 2017 *Opt. Laser Technol.* **93** 99–108
[13] Cui L, Li X, He D, Chen L, Gong S 2012 *Mater. Charact.* **71** 95–102
[14] Fridlyander I N, Khokhlatova L B, Kolobnev N I, Rendiks K, Tempus G 2002 *Met. Sci. Heat Treat.* **44** 2002 3–8
[15] Alekseev A A, Lukina E A, Zaytsev D V, Fridlyander I N 2006 *Mater. Sci. Forum.* **519–521** 259–264
[16] Malikov A, Orishich A, Golyshev A, Karpov E 2019 *J. Manuf. Process.* **41** 101–110

IOP Conf. Series: Materials Science and Engineering **681** (2019) 012030 doi:10.1088/1757-899X/681/1/012030

Analysis and simulation of kinetics of elasto-plastic weld failure in structures at cryogenic temperatures

N A Makhutov[1], I V Makarenko[1], and L V Makarenko[1]

[1] Blagonravov Mechanical Engineering Research Institute, Russian Academy of Science, 4, Bardin street, Moscow, Russia

E-mail: I.V.Makarenko@yandex.ru

Abstract. The article presents the results of solving one of the fundamental problems in the field of increasing strength, service life, survivability, reliability and safety of welded joints in critical structures. A model for simulation of elasto-plastic welds failure in steel structures at cryogenic temperatures was developed. It proposed clarifying calculation and experimental methods for analysis and simulation of failure mechanisms taking into account types of forming defects. In addition, effect of heating and cooling cycles, kinetics of elasto-plastic deformation as well as fields of residual stresses and strains were used to predict the results.

1. Introduction

Technologies of welding parts with inhomogeneous physical and mechanical properties require development of theoretical approaches to their analysis and simulation using the general theory of elastic-plasticity. This will increase strength, service life, survivability, reliability and safety of welded joints of critical structures. In particular, one of the urgent problems is development of algorithms for calculating strength and survivability taking into account multidirectional kinetics of fracture of welded structures at cryogenic temperatures. To solve it, prediction of crack resistance of welds should be done considering initial differently oriented defects such as semi-elliptic cracks. It is also necessary to simulate effects of heating and cooling cycle, kinetics of elastic-plastic deformation under loading, fields of residual stresses and strains, as well as dependences of these parameters on temperature. Inability to obtain accurate analytical solutions on distribution of physical and mechanical properties over all volume of welded joints necessitates experimental studies.

2. Materials and experimental procedure

The studies were carried out on two types of welded plates along the weld (WS) and fusion line (FL). Schemes of the samples are shown in Figure 1. The samples were tested for eccentric tension according to current standards. Width of the welds was 20…100 mm, the temperature range was 77…295 K.

The plates were welded using gas tungsten arc welding with filler wire. The number of discontinuities formed using this method is much less than when using shielded metal arc welding. This is especially important for the operation of structures in high vacuum and low temperature conditions. In addition, the mechanical properties of the weld metal are more stable. In turn, this ensures stable values of the maximum level of stress intensity factor (K_C) for a material of a given thickness.

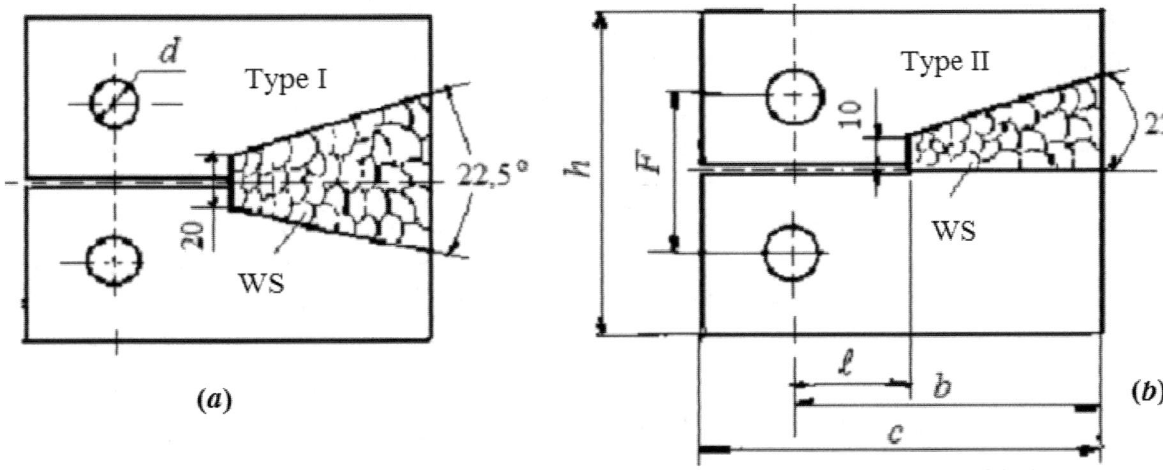

Figure 1. Scheme of welded plates with crack-like discontinuities such as lack of fusion: a – samples with a symmetrical weld; b – samples with an asymmetric weld.

Due to the large cross-section of the welds, multilayer welding was done. The cross-section of each bead was 15…40 mm^2 (depending on the welding mode and dimensions of the samples). The number of layers in the welds was up to 100…150. Heat input (q/v), determined using the equation $q/v = J\cdot U\cdot \eta_0$ (where J is welding current; U is arc voltage; η_0=0.75), was in the range of 756…1133 mm.

Repeated thermo-deformation effect of welding cycles on steels such as 18%Cr-8%Ni type caused precipitation of chromium carbides. This process most intensively occurs in the temperature range of 873…1123 K. The reason for this is the fact that the filler wire does not contain titanium which accelerates carbide formation. After 3…4 multiple exposures, the effect of thermal welding cycles on welded metal at cryogenic temperatures decrease below critical level [1]. Therefore, each subsequent weld bead was deposited after cooling the previous one to 373 K.

3. Statement and solution of the problem

A crack begins to propagate when stress intensity factor reaches a certain critical value. The deformation criterion of failure is determined by the following equation [2–3]:

$$\overline{K}_{\alpha e} = \overline{K}_{\alpha c e}, \qquad (1)$$

where $\overline{K}_{\alpha e}$ is the relative coefficient of intensity of elasto-plastic deformations; $\overline{K}_{\alpha c e}$ is the relative critical coefficient of intensity of deformations; $\alpha = I, II, III$ is the type of crack (normal, transverse or longitudinal).

According to [2-3], the relationship with the force criterion is governed by the dependence $\overline{K}_{\alpha e} = \overline{K}_{\alpha}^{P_{ke}}$, where $\overline{K}_{\alpha}^{P_{ke}}$ is the stress intensity factor and P_{ke} is the parameter of deformation fracture criteria. It is obvious that the force fracture criterion for mixed-type cracks is represented by the function $F(K_\alpha, K_{\alpha c}) = 0$, where $K_{\alpha c}$ is the critical stress intensity factor. The Irwin's force fracture criterion and the critical value criterion for the invariant J-integral $\int_C \frac{1}{2}\sigma_{ij}\varepsilon_{ij}dx_2 - T_i\frac{\partial u_i}{\partial x_1}ds = J_c$ are equivalent, since the relations of the deformation theory of plasticity are the same as the relations of nonlinear elasticity (T_i are the stresses applied to the crack faces; u_i is the displacements; σ_{ij} and ε_{ij} are the components of the stress and strain tensor).

As a result of the studies, a functional change in the average integral values of the yield strength of the welded joints σ_{Tj} was determined as dependences on temperature. They are described by the

IOP Conf. Series: Materials Science and Engineering **681** (2019) 012030 doi:10.1088/1757-899X/681/1/012030

obtained empirical linear functions. Yield strength of the weld $\sigma_{TW} = f_1\ (T)$ is represented by equation (2):

$$\sigma_{TW} = h_1 + h_2 T\ . \tag{2}$$

Yield strength of the fusion line $\sigma_{TL} = f_2\ (T)$ is represented by equation (3):

$$\sigma_{TL} = h_3 + h_4 T\ , \tag{3}$$

where $h_1 = 583 \cdot MPa$, $h_2 = -0{,}8 \cdot MPa/K$, $h_3 = 557 \cdot MPa$, $h_4 = 0{,}76 \cdot MPa/K$.

Based on the experimental results, dependences of the relative critical stress intensity factors $\overline{K}_{Cj} = K_C / \sigma_{Tj} = f(T,t)$ as functions of temperature T and weld thickness t were obtained. They are determined by equation (4) for the weld, equation (5) for the fusion line.

$$\overline{K}_{CW} = (k_1 + k_2 \cdot T) - (k_3 + k_4 T)(t - k_5)^2 , \tag{4}$$

where $k_1 = 0{,}086 \cdot \sqrt{m}$, $k_2 = 4{,}4 \cdot 10^{-4} \cdot \sqrt{m}/K$, $k_3 = 18{,}53 \cdot \sqrt[4]{m}$, $k_4 = 0{,}123 \cdot \sqrt[4]{m}/K$, $k_5 = 0{,}07 \cdot m$.

$$\overline{K}_{CL} = (k_6 + k_7 \cdot T) - (k_8 + k_9 T)(t - k_5)^2 , \tag{5}$$

where $k_6 = 0{,}053 \cdot \sqrt{m}$, $k_7 = 5{,}1 \cdot 10^{-4} \cdot \sqrt{m}/K$, $k_8 = -14{,}1 \cdot \cdot \sqrt[4]{m}$, $k_9 = 0{,}22 \cdot \sqrt[4]{m}/K$.

Figure 2a shows the results of $\overline{K}_{Cj}$ for the welded joint of the 18%Cr-9%Ni steel. Figure 2b gives the limit of fracture function $f_F = f(\overline{K}_{CW}, T, t)$ for a wide class of welded joints of austenitic stainless steels such as 18%Cr-9%Ni type. Dots located below this fracture surface provide operational parameters for the corresponding structural parts. The coefficients of equations (4) and (5) k_i for $i = 1, 2, 3, \ldots 9$ depend on the type of welding and the parameters of the mode:

$$f_F = f(\overline{K}_{CW}, T, t) = \overline{K}_{CW} - (k_1 + k_2 \cdot T) - (k_3 + k_4 T)(t - k_5)^2 = 0 \tag{6}$$

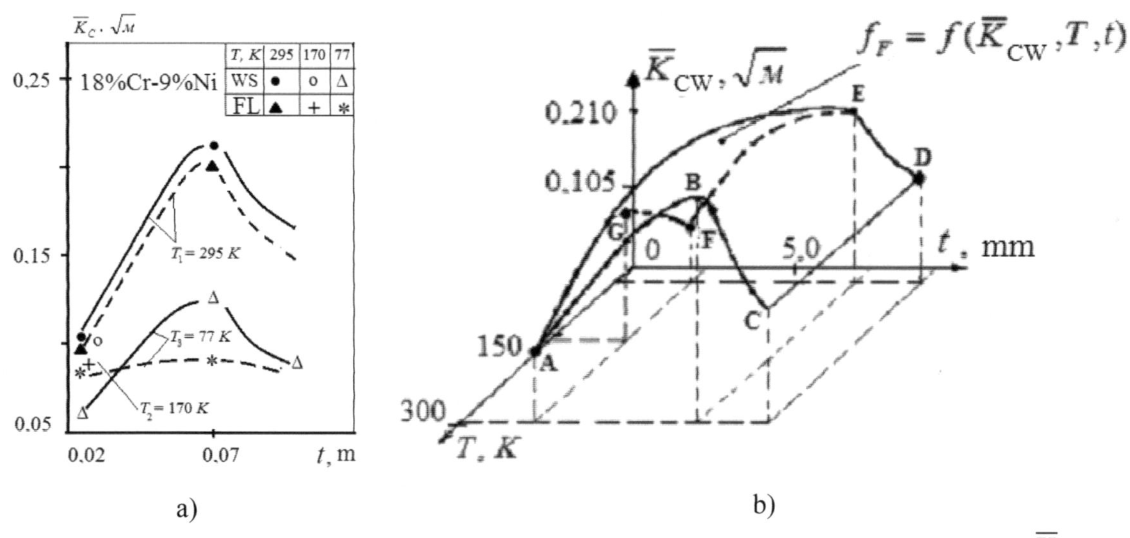

Figure 2. Experimental and calculated according to equations (1) and (2) dependences of $\overline{K}_{Cj}$ on thickness t of welded joint WS and FL in the range of cryogenic temperatures (a) and limit of a fracture function for welds of steels such as 18%Cr-8%Ni type

It follows from the analysis of the data presented in Figure 2 that the weld has higher $\overline{K}_{Cj}$ values compared to the fusion line. There is a decrease in the level of $\overline{K}_{Cj} = \overline{K}_{Cj}(t)$ with a decrease in temperature from 295 K to 77 K. In the graph presented, an extreme thickness range can be observed after which a monotonous decrease in $\overline{K}_{Cj}$ occurs. This parameter tends to values that make it possible to form the condition for the maximum reduction in plastic deformations during failure by normal breakaway:

$$\overline{K}_C \rightarrow \overline{K}_{IC}. \tag{7}$$

Discontinuities in welds (both initial and emerged during operation) can be modeled as surface semi-elliptical cracks in the metal. When calculating the stress-strain state near their contour, it is important to take into account the anisotropy of physical and mechanical properties of the material and kinetics of residual stresses [4–7]. This was done by the finite element method using ANSYS MECHANICAL software [8]. The relative stress factors along the contour of the cracks were determined. Their maximum dimensions are regulated according to criterion (8), if condition (9) is fulfilled based on equations (4), (5) and (6).

$$\overline{K}_{\alpha i} \angle \overline{K}_{\alpha iC}. \tag{8}$$

$$f(\overline{K}_{\alpha i}, T, t) \cdot \angle \cdot f(\overline{K}_{CW}, T, t). \tag{9}$$

Figure 3a shows the function of the intensity of the relative deformations vs. relative distance from the crack contour $\overline{e}_i = f(r/T)$.

Based on the deformation criteria of failure [2, 3] and taking into account [5–7], the calculation of the relative strain intensity factors for misoriented surface semi-elliptic cracks can be obtained by equation (10):

$$\overline{K}^*_{ie} = \frac{(2\pi r^*)^{p_{ire}} \left[\overline{\sigma}_{ni}\right]^{\frac{1-m_i}{m_i\,(1-m_i)}}}{\overline{e}_{ni}\left[f(r^*/l^*_{ij})\right]^{p^{(k)}_{ike}}} \overline{e}^*_i \sin^2 \beta^*. \tag{10}$$

The relative coefficient of strain intensity for the first fracture model $\overline{K}_{1e} = \overline{K}^*_{ie}$ was calculated taking into account the relative elasto-plastic strains $\overline{e}^*_i = \overline{e}_x$ directed along the axis of the first principal stress. In this case, $\overline{\sigma}_{ni}$ and $\overline{e}_{ni}$ are relative nominal stresses and elastic-plastic deformation in the weld; β^* is the angle of inclination of the crack plane to the direction of the first principal stress.

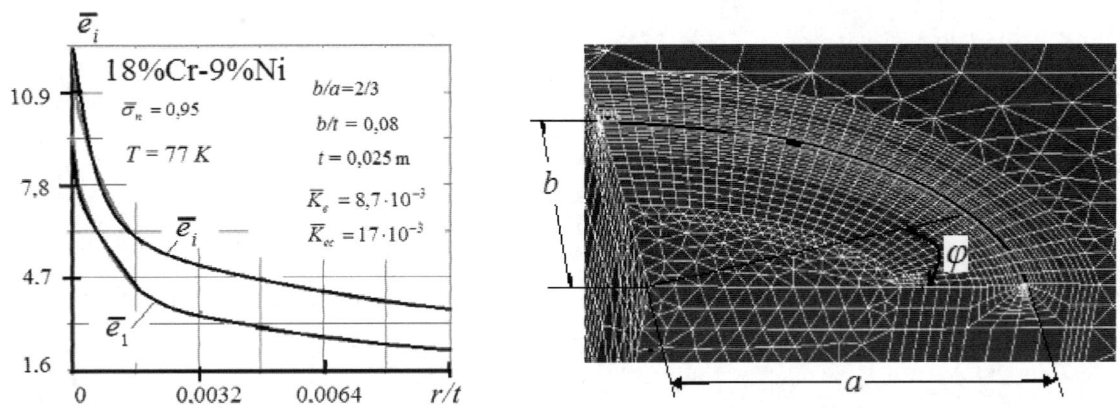

a) b)

Figure 3. Distribution of relative strains near the contour of a semi-elliptical crack vs. relative distance (a) and finite element model of the crack (b)

As a result of solving the elasto-plastic problem using the deformation criteria of fracture [2-3], the relative intensity coefficient of deformations was obtained for the crack with the parameters a = 0.001 m, b = 0.0015 m, b/a = 2/3, t = 0.025 m. On the surface, they are significantly larger than the critical coefficient of strain intensity $\overline{K}_{ec} = 0{,}017$ at T = 77 K. The results of this study are consistent with [2–18].

4. Conclusion

The paper presents a refinement methodology for prediction allowable critical dimensions of initial and operational semi-elliptical inclined surface cracks. It is based on the above mentioned factors of nonlinear boundary loading conditions. In addition, fracture toughness dependences of the studied steels were presented as a limit of a function depending on cryogenic temperatures and thickness of welded joints. Basic concepts and a general methodology for the refinement calculation of strength and survivability of welded joints of critical structures were formulated on the basis of the deformation criterion of failure. A model for simulation of these processes was also proposed for determining an equivalent coefficient of strain intensity along a front of cracks during inelastic deformation.

References

[1] Makarov I I, Grudzinsky B V 1975 Effect of thermal diffusion welding cycle on the ductility of welds at cryogenic temperatures *Automatic welding* **9** 38–42 (in Russian)
[2] Makhutov N A 1981 *Deformation Criteria of Failure and Strength Analysis* (Moscow: Mechanical engineering) (in Russian)
[3] Makhutov N A 2005 *Structural strength, Service Life and Technogenic Safety* (Novosibirsk, Science) (in Russian)
[4] Makhutov N A , Makarenko I V, Makarenko L V 2017 Calculation and experimental analysis of the stress-strain state for inclined semi-elliptical surface cracks *Inorganic Materials* **53** 1502–05
[5] Makhutov N A , Makarenko I V, Makarenko L V 1999 Kinetics of residual stress fields in inhomogeneous austenitic steels during elastoplastic deformation *Factory laboratory* **65** 40–4 (in Russian)
[6] Makhutov N A , Makarenko I V, Makarenko L V 2004 Investigation of spatial mechanical heterogeneity of welded joints of austenitic stainless steels *Factory laboratory* **70** 39–49 (in Russian)
[7] Makhutov N A , Makarenko I V, Makarenko L 2004 Effect of anisotropy of physical and mechanical properties on kinetics of cracks in austenitic steels *Strength issues* **1** 113–9 (in Russian)
[8] 2010 *ANSYS. Structural Analysis Guide* (Canonsburg: SAS IP, Inc)
[9] Lach R, Grellmann W 2017 Mixed mode fracture mechanics behaviour of PMMA *Macromol Symp* **373** 1–6
[10] Wei Z, Deng X, Sutton M A, et al. 2011 Modeling of mixed-mode crack growth in ductile thin sheets under combined in-plane and out-of-plane loading *Eng. Fract. Mech.* **78** 3082–101
[11] Chen X, Deng X, Sutton MA, et al. 2014 An inverse analysis of cohesive zone model parameter values for ductile crack growth simulations *Int. J. Mech. Sci.* **79** 206–15
[12] Wang Y J, Ru C Q 2016 Determination of two key parameters of a cohesive zone model for pipeline steels based on uniaxial stress-strain curve Eng. Fract. Mech. **163** 55–65
[13] Zhao G H, Zhao L, Zhang Y X, et al. 2018 Finite element analysis of dynamic fracture behaviour of drill pipe under various impact loads *Mechanika* **24** 404–11
[14] Li W, Siegmund T 2002 An analysis of crack growth in thin-sheet metal via a cohesive zone model *Eng. Fract. Mech.* **69** 2073–93
[15] Wang Y J, Ru C Q 2016 Determination of two key parameters of a cohesive zone model for pipeline steels based on uniaxial stress-strain curve *Eng. Fract. Mech.* **163** 55–65

[16] Daimon R, Okada H 2014 Mixed-mode stress intensity factor evaluation by interaction integral method for quadratic tetrahedral finite element with correction terms *Eng. Fract. Mech.* **115** 22–42

IOP Conf. Series: Materials Science and Engineering **681** (2019) 012031 doi:10.1088/1757-899X/681/1/012031

Regularities of the influence of temperature and pressure on the grain size in the synthesized intermetallic compound Ni_3Al

V E Ovcharenko [1], K O Akimov [1]

[1]Institute of Strength Physics and Materials Science SB RAS, Tomsk, Russia

E-mail: ove45@mail.ru, kibaarg@mail.ru

Abstract. The intermetallic compound Ni_3Al (γ'-phase, ordered solid solution) is the main strengthening phase of nickel superalloys is, the content of which in modern superalloys reaches up to 0.89. The efficiency of the intermetallic phase as a heat-resistant component in large part limited by low ductility and by strength in a wide temperature range accordingly. An increase in strength limit of the intermetallic component is possible with a decrease in grain size less than a critical value, less than 10 microns. Application of the known methods of plastic deformation for refining an intermetallic grain is almost impossible, but it is physically justified in the period of the grain structure nucleation under non-equilibrium conditions of the exothermic reaction of an intermetallic compound formation in a powder mixture of nickel and aluminum. The retention of low-dimensional grain structure in the synthesized intermetallic compound is possible with combining the processes of crystallization and compaction of the high-temperature synthesis product. This paper presents the results of investigation of the influence of high-temperature synthesis product deformation on the formation of a grain structure in a Ni_3Al intermetallic compound synthesized under pressure.

1. Introduction

The conventional technology for producing intermetallic compounds and their alloys is based mainly on melting and casting methods. However, the difficulties of this technology associated with dendritic and zonal segregation in ingots, which determine the relatively low values of strength properties in a cast state led to the development of powder metallurgy technologies. The most significant results were obtained in the development of technologies for obtaining the Ni_3Al intermetallic compound with high values of thermal conductivity and the ratio of strength to the specific weight [1], and high resistance to high-temperature oxidation [2]. These technologies include the methods of vacuum sintering under pressure [3], high-temperature impact compaction of a powder intermetallic compact [4], mechanical doping [5], and electric spark plasma sintering [6] of the initial powder materials.

The key parameter determining the strength of the sintered intermetallic is the grain size. It is theoretically shown that for ordered intermetallic compounds, which are characterized by low ductility under tension as a result of fracture along the grain boundaries, there is a critical grain size below which an increase in the crack resistance of the polycrystal occurs (the stress intensity coefficient K_{1c} increases, the intermetallic compound ductility increases) [7]. Using the example of Ni_3Al intermetallic samples obtained from foils with a thickness of 300–500 μm with a grain size of 1 to 83 μm, it is shown that refinement of a grain of the γ' phase is accompanied by an intensive increase in yield strength and tensile strength — when the grain size decreases from 18 to 1.34 μm, yield and tensile strengths increase from 280 to 1254 MPa and from 350 to 1757 MPa, respectively [8]. It is noted that the most significant increase in tensile strength with a simultaneous decrease in the plasticity of the intermetallic compound is observed when the grain size in the intermetallic compound is less than 10 μm. The above refers to the Ni_3Al intermetallic compound samples obtained either by the method of directional solidification followed by cold rolling, or by the methods of controlled deformation of the cast intermetallic in the form of thin foils for use in the manufacture of microcircuits, microsensors, parts of thermal evaporators, chemical reactors, etc. [9]. To obtain bulk samples of high-strength Ni_3Al intermetallic compound of structural purpose with a low-dimensional grain structure, the issue of technological support remains open.

Solving the problem of forming a low-dimensional grain structure in bulk samples of an intermetallic compound is possible by plastic deformation methods under the conditions of nucleation and crystallization of the intermetallic phase in the exothermic reaction process of an intermetallic compound formation in a compact from a powder mixture of initial elements [10, 11]. The thermophysical conditions of the volumetric reaction of an intermetallic compound formation in the powder mixture of initial elements ensure synchronicity of passing of the phase transformations simultaneously in the entire volume of the powder billet, which makes it possible to consolidate individual structural fragments of the high-temperature synthesis product when the intermetallic phase of a given composition reaches crystallization. The mathematical calculations carried out using the example of the Ni_3Al intermetallic compound showed that the application of pressure to the thermally reactive powder mixture of nickel with aluminum during the thermal explosion mode reduces the grain size in the synthesized intermetallic by an order [12, 13].

This paper presents the research results of the influence of deformation during the high-temperature synthesis process under pressure on the dimension of the grain structure of the synthesized Ni_3Al intermetallic compound.

2. Materials and methods

High-temperature synthesis of the Ni_3Al intermetallic compound under pressure was carried out on experimental bench consist of hydraulic press equipped with a high-frequency generator for heating the steel die-mold prior to self-ignition of the compact from 3Ni + Al powder mixture (nickel particle size was 1 ÷ 3 μm, aluminum - 5 ÷ 10 um). Additionally, the experimental stand equipped with a temperature recorder of heating the powder compact, a digital manometer for recording the pressure in the hydraulic system of the main cylinder of the press and the wake-up timer of the working pressure application in the main cylinder of the press (Fig. 1a). The stoichiometric 3Ni+Al powder mixture was placed in a cylindrical steel die-mold with an internal diameter of 58 mm heated by high-frequency currents. Initiation of the expansion stroke of the hydraulic press and compacting of the high-temperature synthesis product occurred automatically according to the specified program of the technological cycle. In order to further deformation the product of high-temperature synthesis in the working space of the die-mold, the product was compacted with partial extrusion through holes of different diameters in the lower part of the mold (Fig. 1b). In place to change the magnitude of the additional deformation, the synthesis product was extruded through holes of various diameters - 3 and 4 mm

The grain structure of intermetallic samples was studied by optical metallography (Neophot 32) on lamellar specimens cut from the central part of the synthesized compacts (Fig. 2).

Metallographic specimens were prepared by mechanical polishing with a gradual decrease in the size of the diamond abrasive particles of 1 micron. Grain structure was detected by argon ion etching at an accelerating voltage of 0.6 kV. Grain size determined by the method of random secants with averaging over 150 measurements. The phase composition of the samples was investigated by X-ray diffraction analysis using a DRON-7 X-ray diffractometer in CoKα radiation at an accelerating voltage of 40 kV and a current of 40 mA.

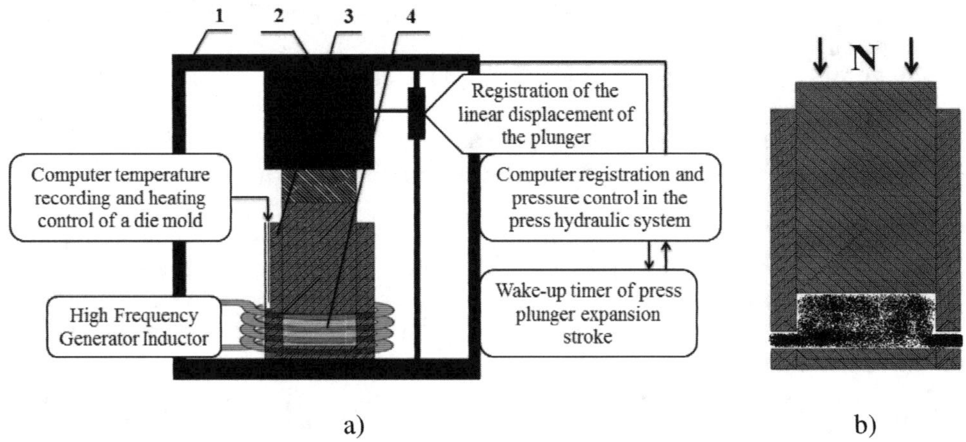

Figure 1. Block diagram of an experimental stand for high-temperature synthesis of an intermetallic compound under pressure during continuous heating of the powder mixture of initial elements in a steel die-mold: working space of a hydraulic press, 2 — press master cylinder, 3 — steel die-mold, 4 — powder compact (a), a die-mold scheme for high-temperature synthesis of an intermetallic compound under pressure with partial extrusion of a synthesis product (b)

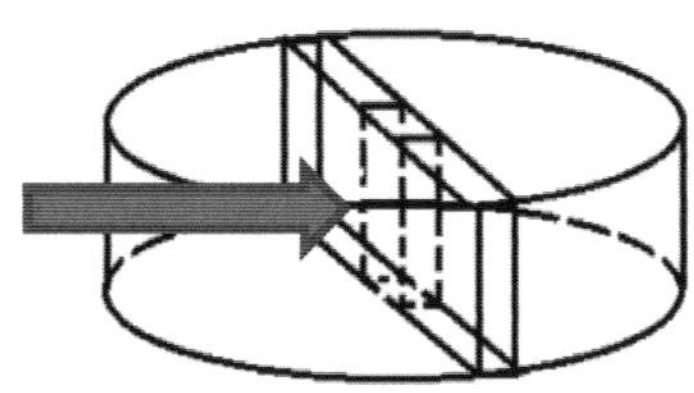

Figure 2. Scheme of intermetallic samples preparation for statistical evaluation of grain size. The arrow shows the sampling site for statistical evaluation of the grain size

3. Results and discussion

In Fig. 3 shows the dependences of the pressure applied to the powder compact of initial elements on the time of the high-temperature synthesis of the Ni_3Al intermetallic compound under pressure, including with partial extrusion of the synthesis product.

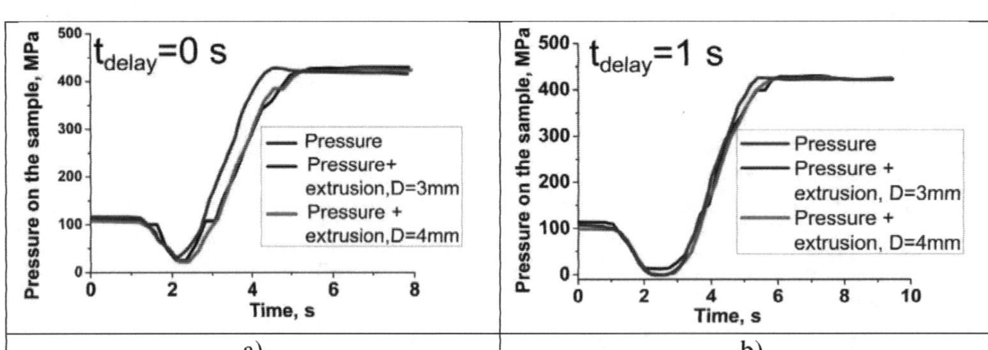

Figure 3. Dependencies of the pressure applied to the thermoreactive mixture of the initial elements on the time of the synthesis process of Ni_3Al intermetallic compound with a delay time of pressure application to synthesis product of 0 s (a) and 1 s (b)

Presented on Fig. 3 dependencies demonstrate identical character. There is high repeatability of nature of the change in the pressure on the thermoreactive powder mixture throughout the phase transformations during the volumetric exothermic reaction, including with extrusion of the synthesis product The above means that in each investigated variant, including extrusion of the synthesis product, the reaction of the formation of intermetallic in the powder compact of the initial elements takes place in full (Fig. 4). The latter is confirmed by the results of X-ray diffraction analysis presented in Fig. 4 in the form of diffraction patterns obtained from a reference, synthesized under pressure and synthesized under pressure with extrusion of the high-temperature synthesis product of the Ni_3Al intermetallic compound.

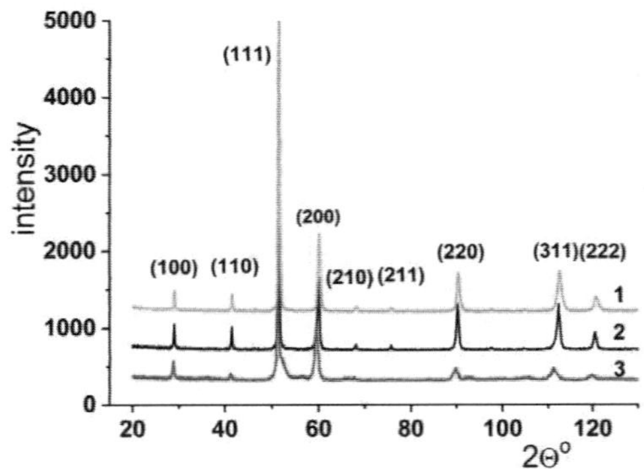

Figure 4. Diffraction patterns from the reference (1), synthesized under pressure (2) and from the synthesized under pressure with partial extrusion of the product of high-temperature synthesis (3) of the Ni_3Al intermetallic compound

The main difference of presented dependencies of the pressure value on the time of the synthesis is the increase in the residence time of the thermoreactive system in the melt state (with the minimum pressure value in the die-mold) at an increase in the delay time of applying the compaction pressure to the synthesis product. Herein the increase in the delay time has a significant effect on the final grain size in the intermetallic compound specimens synthesized under pressure. In Fig. 5 shows the depèndences of the grain size synthesized under pressure samples of the Ni_3Al intermetallic compound including with the extrusion of the synthesis product on the time delay of applying pressure to the high-temperature synthesis product. We can state the following main features of the presented dependencies:

- the initial increasing in the delay time up to 0.3 s leads to an increase in the grain size in the synthesized intermetallic samples,
- regardless of the presence or absence of extrusion of the synthesis product, with a further increase in the delay time,
- the largest grain size is formed in intermetallic samples synthesized under pressure without extrusion of the synthesis product, a decrease in the grain size in the synthesized intermetallic compound occurs
- deformation of the high-temperature synthesis product in the die-mold with its partial extrusion leads to a fold decrease in the grain size in the intermetallic samples synthesized under pressure,
- an increase in the deformation of the high-temperature synthesis product with an increase in the proportion of extruded synthesis product through an opening in the die-mold of increased diameter further reduces the grain size.

In the intermetallic synthesized under pressure: an increase in the diameter of the holes in the mold from 3 to 4 mm leads to a decrease in the grain size in the synthesized intermetallic, in comparison with the samples synthesized under pressure without extrusion, ~ 1.8 times and ~ 2.5 times, respectively.

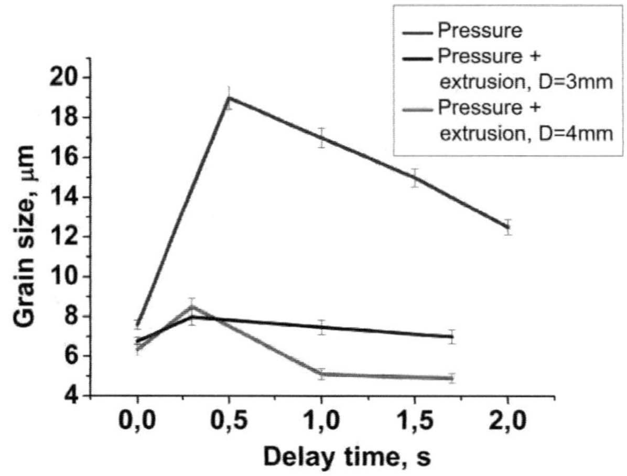

Figure 5. Dependences of grain size in synthesized under pressure and synthesized under pressure with partial extrusion of synthesis product samples of intermetallic compound Ni_3Al on the delay time of compaction pressure application to the product of high-temperature synthesis

4. Conclusion

The presented results make it possible to substantiate the promising application of intense plastic deformation to form a low-dimensional grain structure in the intermetallic compound Ni_3Al under the conditions of combining the exothermic reaction of the intermetallic compound formation in the powder mixture of the initial elements with the deformation of the high-temperature synthesis product. Deformation processes play a key role in the formation of a low-dimensional grain structure and so that a greater extent, that the higher the degree of the synthesis product deformation. It is necessary to conduct comprehensive studies of the intermetallic compound structural-phase states at all stages of the high-temperature synthesis process under pressure with deformation of the synthesis product to detail the patterns of a low-dimensional grain structure formation.

Acknowledgments

The work was carried out as part of the Basic Research Program of the Russian Academy of Sciences for 2013–2020 (project No. 23.2.2.).

References

[1] Stoloff N S et. al. 2000 *Intermetallics* **8 (9-11)** 1313-1320
[2] Sikka V K et. al. 2000 *Intermetallics* **8 (9-11)** 1329-1337
[3] Nazarian-Samani M and Kamali A 2010 *J. Alloy. Compd.* **492** 196-200
[4] Hadjiafxenti A et. al. 2013 *Vacuum* **96** 73-78
[5] Antolak-Dudka A et. al. 2013 *Intermetallics* **42** 41-44
[6] Fragea N et. al. 2018 *Intermetallics* **102** 26-33
[7] Chan K S 1990 *Scripta Metall. Mater.* **24** 1725-1750
[8] Jozwik P and Bojar Z 2007 *Arch. Metall. Mater.* **52 (issue 2)** 322-327
[9] Polkowski W et. al. 2015 *Mater. Lett.* **139** 46-49
[10] Mukasyan A S and Shuk C E 2017 *Int. J. Self Propag. High Temp. Synth.* **26 (3)** 145-165
[11] Ovcharenko V E and Lapshin O V 2007 *Physics and chemistry of materials treatment* **3** 76-83
[12] Ovcharenko V E et. al. 2005 *Phys. Mesomech.* **8** 65-68
[13] Ovcharenko V E et. al. 2006 *Combust. Expl. Shock Waves* **42 (3)** 64-70

Laser welding of thin metal filaments from titanium nickelide

A I Potekaev[1,2], S V Gunter[1], E A Ibragimov[2], A A Klopotov[1,3] and S V Galsanov[1]

[1]NationalResearchTomskStateUniversity, Tomsk, Russia
[2]Yurga Institute of Technology (branch) of the National Research Tomsk Polytechnic University, Yurga
[3]Tomsk State University of Architecture and Building, Tomsk, Russia

E-mail: guntersv@inbox.ru.

Abstract. The paper presents the results of a study of welded joints of thin filaments from the TiNi(MoFe) alloy. Welding was carried out in order to produce compounds from thin filaments used in medicine as a suture material and for manufacture of metal fabrics. The experiments have shown the perspectivity of using pulsed laser welding for producing welded joints of thin filaments. Formation features of the welding pool as a result of the interaction of a laser beam with a composite material, which are thin filaments from the TiNi (MoFe) alloy, have been revealed using scanning electron microscopy.

1. Introduction

The used in medicine alloys based on titanium nickelide TiNi(MoFe) exhibit a set of unique properties: superelasticity and shape-memory combined with a high level of chemical resistance under conditions of changes in stress, temperature, and deformation. At present, thin and superfine filaments (up to 40 μm) made of TiNi(MoFe) alloys are beginning to gain wide application in various fields of medicine [1]. Thin filaments are used to produce suture material, metal fabric, and knitted nets for various purposes, which are successfully used in medicine (Fig. 1) [1-3]. Practical application of superfine filaments from TiNi(Mo, Fe) raises the necessity to perform welding of these filaments. In this regard, the task of developing a technology for producing welded joints from thin filaments from alloys based on titanium nickelide is currently relevant.

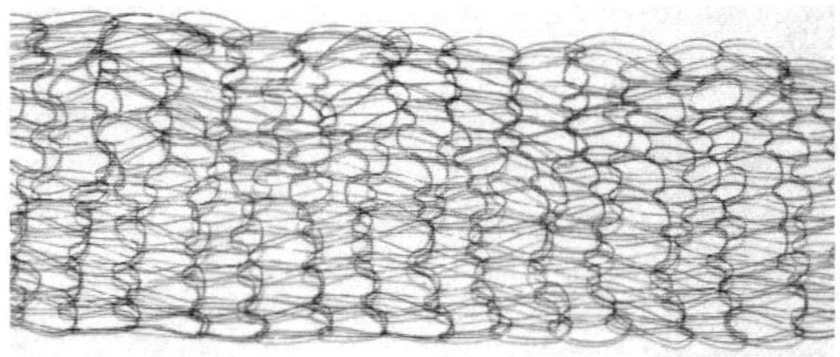

Figure 1. Knitted material from titanium nickelide, produced according to the knitting technology from superelastic thin-profile filament 40÷90 μm.

According to literature data, welding of alloys based on titanium nickelide is carried out using different methods. For example, in [4] the effect of diffusion welding on the microstructure of an alloy based on TiNi was studied. As a result, it has been found that it is necessary to use alloys with a minimum content of particles of the secondary phases Ti_2Ni and Ti_4Ni_2O. This is conditioned upon the fact that the presence of secondary phases contributes to a significant decrease in physicomechanical properties. Diffusion welding is lengthy and, as a result, formation of a liquid phase under reverse peritectoid flow reaction TiNi + $Ti_2Ni \rightarrow$ L + TiNi is possible. Laser welding does not have such a disadvantage. The use of laser radiation energy for welding is characterized by several advantages: the local nature of the thermal effect, minimal thermal deformation, wide range of regulating energy characteristics of the laser beam, possibility of making high-quality welding seams on a wide class of materials [5-7].

The locality of the laser radiation effect during welding of titanium nickelide alloys determines special properties of welding joints [8-11]. Based on the analysis of literature data, it has been concluded that the use of laser welding to produce welding joints of elements from TiNi-based alloys is promising.

It should be noted that production of thin filaments from the TiNi(Mo, Fe) alloy is associated with a long thermo-mechanical treatment of the alloy (multiple drawing through dies). This treatment results in the fact that thin filaments from the alloy based on titanium nickelide become a multicomponent composite material: a core and a shell. The core of a thin filament consists of the TiNi(Mo, Fe) alloy that possesses superelastic properties and shape-memory effects [12,13]. The filament shell has a complex structural-phase composition: titanium oxide TiO, TiO_2 and their derivatives, and oxycarbonitrides. [3,14]. The physicomechanical properties of the core and the shell are significantly different from each other. An important feature of such a composite material is the dependence of the physicomechanical properties on the filament diameter. This phenomenon is associated with an increase in the volume fraction of the oxide layer with a decrease in the filament diameter [3.14]. The interaction of the laser beam with such a composite material is quite complicated and manifests itself in the occurrence of different processes simultaneously. We shall note two main ones. The first process is associated with transition of laser electromagnetic radiation into thermal energy, which in a local area leads to a phase transition "solid – liquid". The second process is associated with chemical reactions as a result of instant heating and cooling.

The purpose of the paper is to present the results of structural studies of the welding joint of thin filaments from the TiNi(Mo, Fe) alloy obtained using laser welding.

2. Materials and research methods

Welding joints were performed on thin filaments of the TiNi(Mo, Fe) alloy with a cross section of 160 μm, produced by repeated drawing through dies with a certain step and subsequent short-term heating from 500 to 800 °C [3].

The structure of welding seams was examined using a VEGA3 TESCAN scanning electron microscope. Welding of thin filaments from titanium nickelide was carried out under two variants. The first variant was butt welding, when edges of the two ends of the wire are welded (Fig. 2).

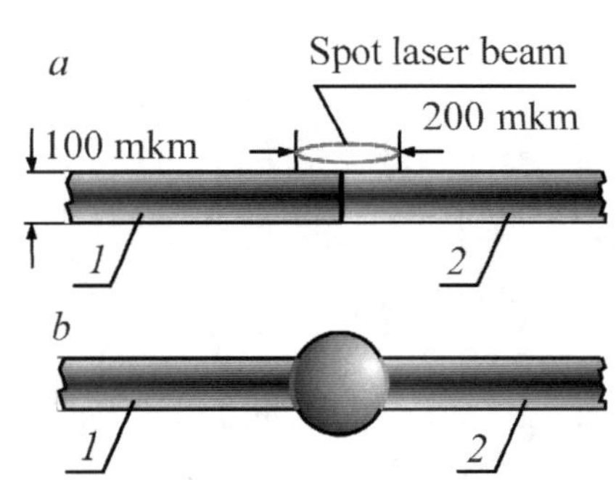

Figure 2. Butt weld scheme of two thin filaments *1* and *2* of the TiNi(Mo, Fe) alloy: *a* is the arrangement of filaments before welding; *b*is after welding

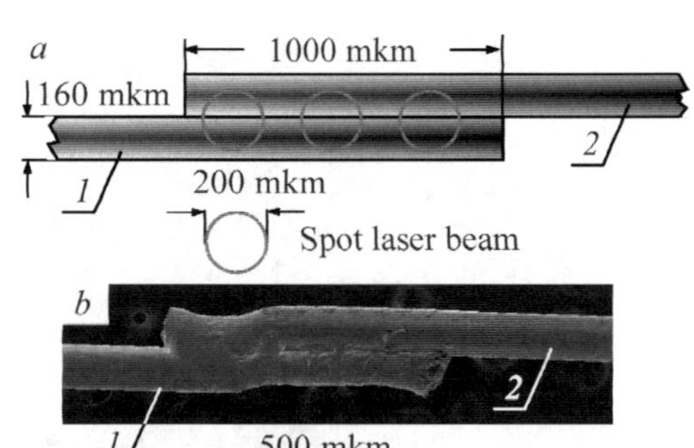

Figure 3. Overlapping weld scheme of two thin filaments*1* and *2* of the TiNi(Mo, Fe) alloy: *a* is the arrangement of filaments before welding; *b*is the overlapping weld micrograph

A laser beam from an ytterbium fiber laser LK-100-V was used for welding. Welded thin wires belong to small-sized elements. Therefore, in this case, welding of thin wires was carried out in a mode providing melting of the alloy without its intensive evaporation. A pulsed radiation mode was used to perform welding of thin wires. The laser beam parameters are as follows: beam pulse duration 30 μs, number of pulses 3, pulse repetition rate 0.3 s^{-1}, beam power 10 W, laser spot diameter $d_{l.i.}$=100 μm.

The second variant is overlapping welding of two thin filaments. This option includes welding of two ends of the wire according to the scheme shown in Fig. 3.

3. Results and discussion

3.1. Surface morphology of thin filaments from the TiNi(Mo, Fe) alloy

Studies of the morphology of the surface areas of the thin filament from the TiNi(Mo, Fe) alloy were carried out using methods of scanning electron microscopy. The results of the study are presented in Fig. 4.

According to the published data [3, 14], the oxide layer of the wire from the TiNi(Mo, Fe) alloy consists mainly of titanium oxides, such as TiO, TiO_2, and their derivatives. In the process of manufacturing the wire, and then thin filaments, formation of oxycarbonitrides, in addition to titanium oxides, occurs on the surface of the wire. Oxycarbonitrides inevitably remain on the surface layer of the wire and are the result of high temperature exposure to the lubricant during drawing through the die.

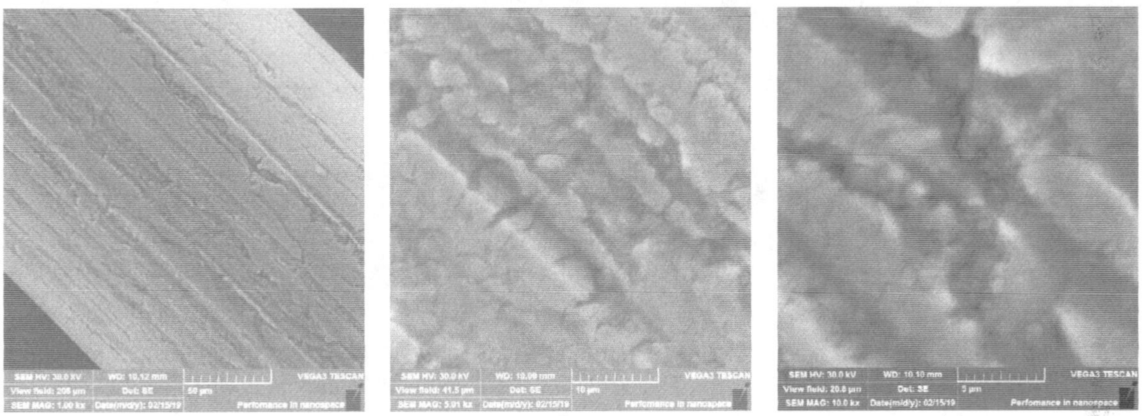

(a) (b) (c)

Figure 4 (a-c). Microphotographs of surface areas of the thin wire 220 μm from the TiNi(Mo, Fe) alloy with different zoom

The morphology of surface high-temperature oxide phases on the filament surface is characterized by complex shapes. The surface, which is based on oxide and oxycarbonitride phases, is formed as a result of the thermomechanical deformation of the material upon drawing of thin alloy filaments in a hot state through the spinnerets. The microphotographs show gaps and dents (Fig. 4 a). Gaps and dents on the surface are elongated along the filament surface. These gaps and dents were formed as a result of producing the wire during drawing through the dies in a hot state. In the transverse direction, these gaps have dimensions from 1 to 4 μm. In addition, smaller gaps of the oxide layer with characteristic sizes less than 1 μm are visible on the surface (Fig. 4b). Moreover, microcracks are clearly visible at high magnification (Fig. 4,c).

The above mentioned surface morphology of thin filaments has been obtained as a result of formation of surface oxide and oxycarbonitride phases during drawing of the alloy based on titanium nickelide in a hot state through spinnerets. The formation process of oxide and oxycarbonitride phases on the surface of thin filaments is heterogeneous and multi-stage, because the process of filament production includes repeated and short-term heating at temperatures from 500 to 800 °C. Such thermomechanical treatment is accompanied by significant mass transfer under conditions of intense diffusion processes on the surface along interstitial sites of the lattice and along extended defects of atoms of main alloying elements and segregant atoms [15-17]. The mass transfer of oxygen and carbon over the surface of filaments is facilitated by loosening the surface structure. The interaction of aggregate atoms on the filament surface is carried out in local areas with an increased concentration of defects, where the interatomic bonds are weakened.

Studies of the elemental composition of the oxide layer and the core of filaments from the TiNi alloy using X-ray spectral microanalysis have revealed that the composition of the core does not change with a decrease in the filament diameter. In this case, the oxide layer contains oxygen in addition to main alloy-forming components Ti and Ni. The oxygen content in the oxide layer of the wire with a diameter of 80 microns is 1.5 times greater than in the surface oxide layer of the wire with a diameter of 40 microns [3.14].

Thus, the data presented indicate that thin filaments produced as a result of repeated drawing through dies at high temperatures are a composite material.

3.2. Butt welding

Mastering of the technology for welding of thin filaments from the TiNi(Mo, Fe) alloy using laser irradiation was carried out on a large number of samples. The results of this work are shown in microphotographs in Fig. 5. On the above micrographs it can be seen that exposure to the laser beam has led to formation of spherical-shaped balls in the welding joint. For filaments with a diameter of 60 μm, the diameter of the welded ball reaches values of the order of 180 μm. The welded ball, connecting filaments with a diameter of 100 microns, has a diameter of about 200 μm. It has been found that the welded ball has a discoloration. This phenomenon reflects the presence of a large amount of oxides on the surface.

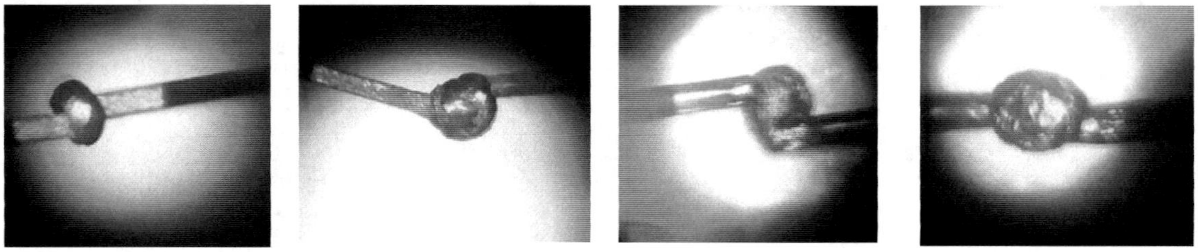

(a) (b) (c) (d)

Figure 5 (a-d). Photographs of welding joints of thin wire with a diameter of 60 μm (*a, b*) and 100 μm (*c, d*) produced by laser welding

Thus, it has been shown that high efficiency of energy input under exposure to a laser beam during welding of thin filaments from the TiNi(Mo, Fe) alloy leads to formation of a welding ball that ensures the connection of two ends of the thin filament. In this case, rapid cooling of the melt should lead to formation of a specific (in the structure and the composition) state in the welded ball and in the zones of thermal influence [11].

3.3. Overlapping welding

The use of overlapping welding has allowed obtaining a good welding joint. This is visually visible in microphotographs presented in Fig. 6. An important point should also be noted. The compositional structure of thin filaments (oxide shell and core made of the superelastic alloy TiNi(Mo, Fe)) has manifested in structural features of the weld pool [3, 14]. This fact is conditioned upon the fact that several different processes occur simultaneously in the process of interaction of the laser beam with the composite material.

We shall note two main processes. The first one occurs as a result of the thermal energy transfer during the interaction of laser radiation with the material and its transfer in the local area to a liquid state with formation of a multiphase liquid. The second process is associated with chemical reactions that occur as a result of instant heating and equally rapid cooling. In microphotographs, it is possible to distinguish characteristic interaction zones of laser radiation with the filament material (Fig. 6). The zone in the center of the laser crater, which was formed as a result of melting and then crystallization of a narrow layer of the molten matrix alloy, is clearly visible. The other zone, on the edge of the laser crater, clearly shows the boundary "liquid phase - solid material" and reflects the interaction process of the titanium oxide shell with the main molten metal of the filament.

Fig. 6*a–f* shows microphotographs obtained under the exposure to a laser beam in the region of the end of the 2nd filament (Fig. 3). It can be seen that the thermal study of the laser beam in the region of the end of the 2nd filament has led to formation of a melt from the main alloy, which manifested itself in the folded surface. The interaction of the molten alloy from the 2nd filament with the oxide surface of the 1st filament has manifested in formation of elongated growths in the shape of tentacles on the surface of the oxide film of the filament.

Fig. 6 g–k shows micrographs illustrating the result of the interaction of the molten metal in the central part of the contact of two filaments after exposure to a laser beam. The micrographs clearly show that the molten metal connects two filaments in the local area. Moreover, microphotographs in Fig. 6 g– i show that there is no oxide film in the central part of the weld pool. Micropores with sizes up to 2 μm are found in the joints of the weld pool from the molten and non-molten metal (Fig. 6 j– l). The presence of pores indicates formation of gas bubbles which have led to formation of micropores. Formation of gas bubbles reflects the process of involving substances from the oxide film into the molten metal and formation of closed cavities in the molten metal where oxygen is released. The quantity and sizes of pores depend on the time of ascent and removal of gases from the melt, i.e. the lifetime of the liquid phase or bubble degassing of the weld pool.

Moreover, a thermally activated process of microplastic deformation of metal occurs in the area of interaction between the weld pool and the non-molten metal. In addition to the diffusion welding zone, a warping zone of the oxide film is formed before the front of the molten bath (Fig. 6 *j– l*).

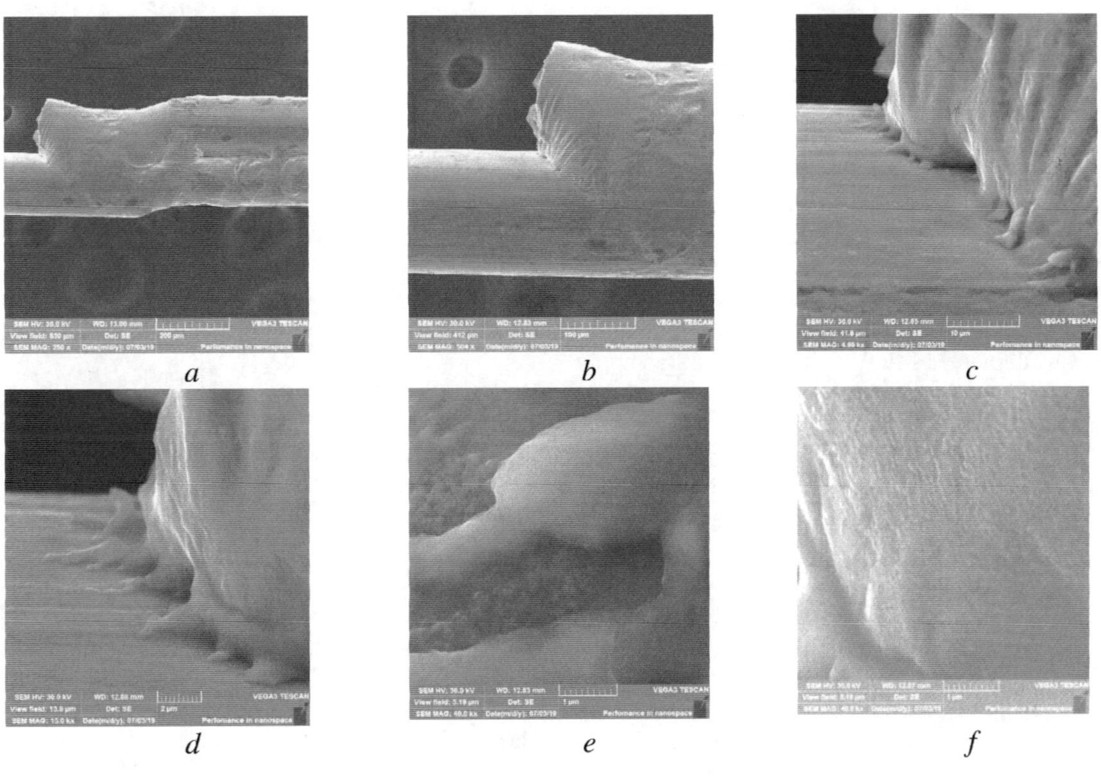

a *b* *c*

d *e* *f*

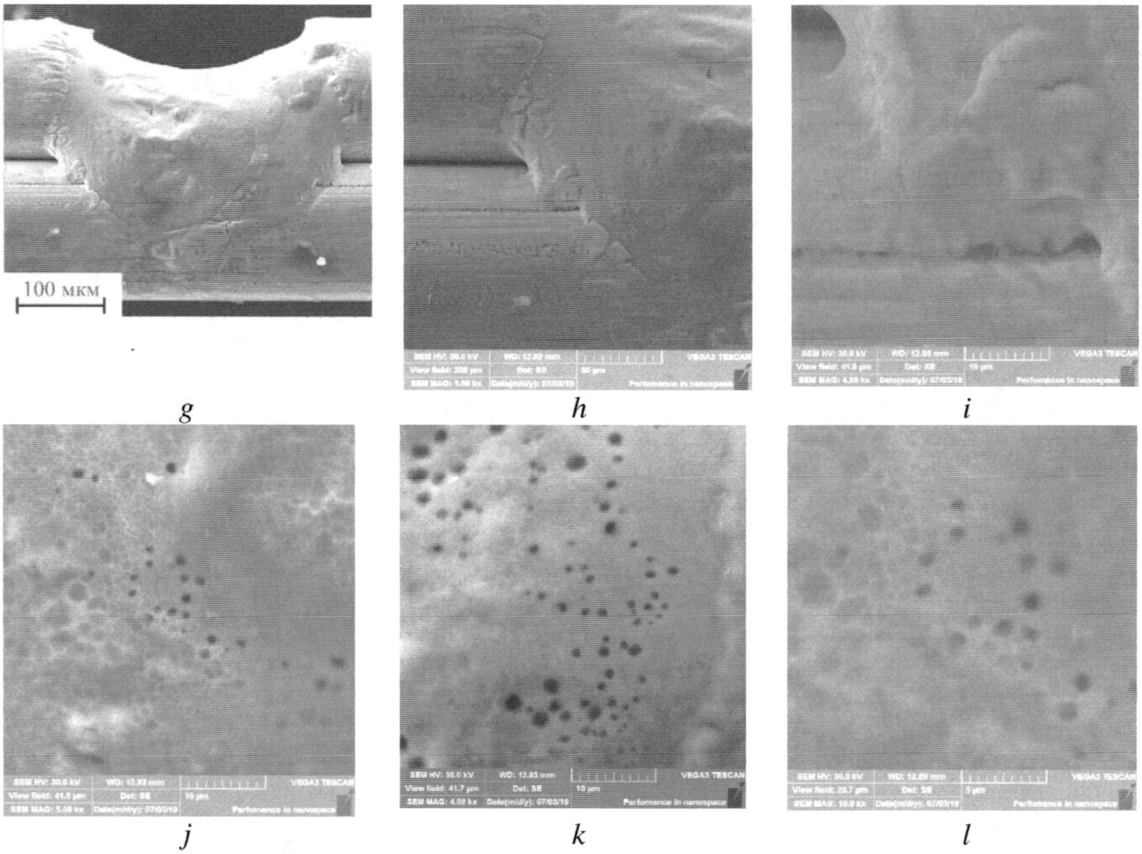

Figure 6 (a-l). Microphotographs of the fusion areas of two filaments in the area of exposure to a laser beam

4. Conclusion

It has been established that overlapping and butt welding with a laser beam of thin filaments from the TiNi(Mo, Fe) alloy result in formation of weld pools in local areas, which create a physical contact between thin filaments. The presence of micropores with sizes up to 2 μm at the joints of the weld pool from the molten and non-molten metal has been revealed.

Experimental studies have shown that pulsed laser welding allows obtaining reliable welding joints from thin filaments of the TiNi(Mo, Fe) alloy.

References

[1] V. E. Gunther, A. A. Radkevich, S. B. Kang, T. L. Chekalkin, E.S. Marchenko, S.V. Gyunter, Biomedical Physics & Engineering Express 5, 1-9 (2019).

[2] V. E. Gyunter, E. S. Marchenko, S. V. Gyunter, G. A. Baigonakova, Tech. Phys. Lett. 44, 811–813 (2018).

[3] S. V. Gunther, O. V. Kokorev, T. L. Chekalkin, V. N. Hodorenko, G. Ts. Dambaev, Adv. Mater. Lett. 6, 774-77 8 (2015),

[4] K. S. Senkevich, Science and Heat Treatment, 55, 685-670 (2014).

[5] T. D. B. Kannan, T. Ramesh, P. Sathiya "A Review of Similar and Dissimilar Micro-joining of Nitinol", JOM 68, 1227-1243 (2016).

[6] S. T. Auwal, S. Ramesh, F. Yusof, S. M. Manladan, The International Journal of Advanced Manufacturing Technology 96, 475-490 (2018).

[7] L. Liu, H. Wang, Metall and Mat Trans A 42, 1044-1051 (2011).
[8] Y. T. Hsu, Y. R. Wang, S. K. Wu, C. Chen, Metall and Mat Trans A, 32 569-578 (2001).
[9] P. Sevilla, F. Martorell, C. J. A. Libenson, F. Planell, J. Gil, J. Mater Sci: Mater Med., 19 525-529 (2008).
[10] A. S. Zocram, A. A. Akbari, F. Mohsenitar, Weld World, 60 11-19 (2016).
[11] M. Mehrpouya, A. Gisario, G. B. Broggiato, M. Puopolo, S. Vesco, Int. J. Adv. Manuf. Technol. 103 1593-1601 (2019).
[12] V. E. Gyunter, V. V. Ovcharenko, A. A. Klopotov, Tech. Phys. Lett. 28 182-183 (2000).
[13] V. Ë .Gunter, T. Yu. Maletkina, A. A. Klopotov, J. Appl. Mech. Tech. Phys. 39, 312-318 (1998).
[14] D. Zaworonkow, M. Chekan, K. Kusnierz, Biomed. Phys. & Eng. Exp. 4 (2018).
[15] K. Fukuda, Y .Kado, T. Yoshikawa, K. Oishi, Y Mae, J. Mat. Eng. Perform 11, 544-551 (2002).
[16] A. Martinez-Villafane , F. H. Stott, J. G. Chacon-Nava, G. C. Wood, Oxidation of Metals. 57, 267-280 (2002).
[17] C. L. Zeng, M. C. Li., G. Q. Liu, W. T. Wu, Oxidation of Metals. 58, 171-185 (2002).

Non-vacuum electron-beam surfacing of coatings with the ultrafine structure

I M Poletika[1] **and M V Perovskaya**[1]

[1] Institute of Strength Physics and Materials Science of SB RAS, Tomsk, Russia

E-mail: mv_perovskaya@inbox.ru

Abstract. The paper employs an ELV-6 accelerator to surface low-carbon steel with a tungsten carbide powder thus obtaining steel coatings with high hardness but low wear resistance. The abrasive wear resistance is considerably increased by changing the composition of the surfacing mixture due to the introduction of nickel as an austenite stabilizer and titanium carbide as a modifier. Besides, the paper deals with surfacing with a powder mixture of chromium carbide, chromium, and titanium carbide to produce bifunctional coatings with both high wear and corrosion resistance, which is explained by the presence of chromium in the layer. The methods of physical metallurgy are used to study mechanisms responsible for property changes. A substantial increase in wear resistance is shown to be related to the ductile austenitic structure formed in the deposited layer matrix when nickel is introduced and to rapid structure refinement in the presence of multiple centers of crystallization represented by dispersed precipitates of TiC.

1. Introduction

Abrasive wear of the surfaces of machine parts and mechanisms considerably reduces their service life. A widely-used method for restoration and hardening of parts is electric arc surfacing. It is analogous to electron-beam surfacing performed with the electron accelerator, which was developed by BINP SB RAS in collaboration with ISPMS SB RAS [1, 2]. The accelerator is characterized by a high quality of formed coatings and ease of injection of alloying elements. Electric arc surfacing with tungsten carbide gives high hardness and satisfactory resistance to abrasive wear. This carbide can also be used for electron beam surfacing. To improve the wear resistance of coatings, tungsten carbide can be mixed with nickel allowing for structure austenization and titanium carbide as a modifier forming a homogeneous dispersed structure. A common and relatively inexpensive system for electric arc surfacing that ensures high resistance to abrasive wear is Cr-C. Hardening is achieved by precipitation of chromium carbides in the deposited layer. Chromium injected into surfacing mixtures during processing can improve corrosion resistance of coatings. Formed coatings will protect products against both wear and corrosion. An additional increase in wear resistance can be achieved by the structure refinement of deposited metal due to the introduction of modifiers.

The aim of the work is to determine the mechanical properties of the composites obtained by the method of non-vacuum electron-beam cladding of the powder mixture (Cr_3C_2 + TiC) on low-carbon steel. In this study, the ultimate strength and deflection of composite with coating were determined depending on the deposition modes.

2. Experimental

Surfacing is performed with an ELV-6 electron accelerator at the radiation energy of 1.4 MeV. Low-carbon steel is initially surfaced with a mixture of WC, Ni and TiC powders in the ratio of 6:3:1. The substrate were 5x10 mm plates cut from sheets (14 mm) of S235 low-carbon steel. The beam current I equaled 40 mA and the deposition velocity V was varied from 1.45 to 2.0 cm/s. A 1-mm powder layer was applied to the specimen surface before processing. The packed density was 0.45 g/cm^2. The surfacing mixtures was injected with flux MgF_2 to protect against environmental influences. To increase the processed surface area the beam was scanning with the magnetic sweep width l = 5 cm. The second series of experiments presented surfacing with a powder mixture of chromium carbide Cr_3C_2, chromium, and titanium carbide TiC in the ratio of 6:3:1. At the beam current 40 mA the translational velocity of a specimen under the beam, V, varied in the range of 0.5–1.2 cm/s. In both cases the radiation energy density W at the specimen surface ranged from 5.6 to 7.72 kJ/cm^2. The deposited metal structure is investigated by the Neophot metallographic microscope. The X-ray phase analysis was performed with a DRON-2M diffractometer. The PMT-3 device was employed to measure the microhardness distribution in the deposited layers. The average microhardness $H\mu^{ave}$ in the layer was measured by 100–150 measurements. The abrasive-wear tests were conducted at loosely fixed abrasive particles fed into the friction zone and pressed against the specimen by a rotating rubber roller (GOST 23.208-79). The corrosion resistance was determined in nitric-acid tests by measuring the mass wear of the specimen.

3. Results and Discussion

When surfacing with a mixture of tungsten carbide, titanium carbide and nickel powders, the deposited layer thickness L varied from 1.5 to 3 mm (Fig. 1a).

Coating microhardness (Fig. 1b) and coefficient of wear resistance (Fig. 1c) had high average values and however they decrease with increasing radiation energy density W. At W = 5.6 – 6.8 kJ/cm^2, the deposited layer contained a significant amount of γ-iron, a small amount of α-iron, tungsten carbide Fe_3W_3C, and titanium carbide TiC.

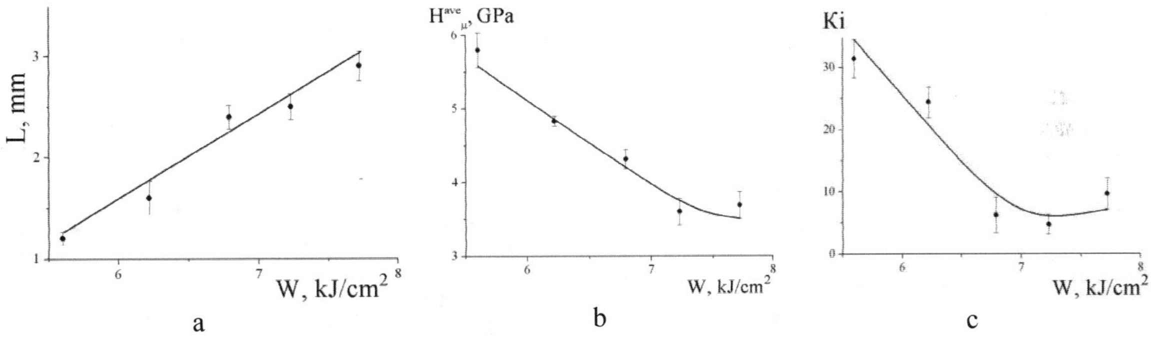

Figure 1. Thickness L (a), average microhardness $H\mu^{ave}$ (b), and coefficient of wear resistance Ki of the deposited layer (c) versus specific surface energy W for the surfacing mixture WC+Ni+TiC.

Moreover, the intensity of γ-ion lines was much higher than that of α-iron lines, which is indicative of the formation of the austenitic-martensitic structure with a large volume fraction of the austenitic phase and a small volume fraction of the martensitic phase. A solid solution becomes enriched with nickel.

According to the X-ray diffraction patterns, with increasing energy density W, the intensities of lines of γ-iron, tungsten carbide, and titanium reduce, while those of α-iron lines (martensite) increase. This bars evidence of a decreasing volume fraction of the wear-resistant austenitic phase. The observed changes are related to an increasing thickness of the deposited layer (Fig. 1a) and decreasing average concentration of alloying elements in it, including nickel that stabilizes the austenite.

Coatings formed are peculiar in high structure refinement. The micrographs of the deposited layers exhibit a significant reduction in grain size (up to 1–5 μm, Figs. 2a and 2b).

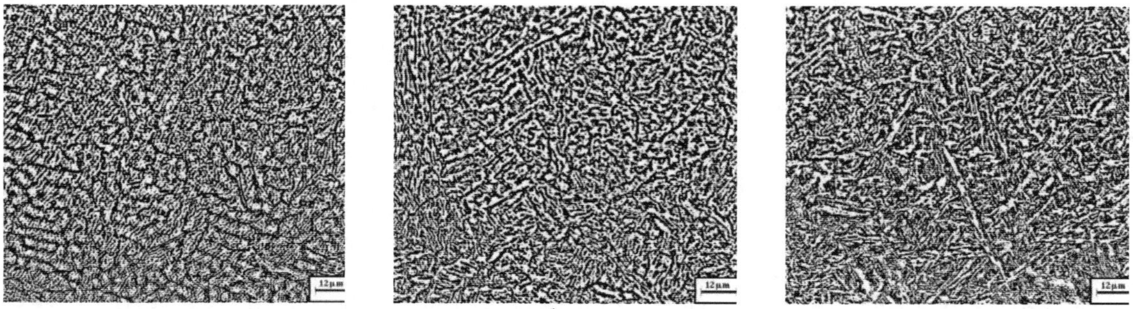

a b c

Figure 2. Structure of the layer surfaced with the mixture WC+Ni+TiC: W = 5.60 (a); 6.79 (b), and 7.72 kJ/cm^2 (c).

This is due to the large number of dispersed particles TiC precipitated in the molten layer (T_{melt} = 3180 °C). They are the first to precipitate from the melt presenting multiple centers of crystallization and hindering the austenite grain growth. Besides, high quenching rates applied to the solid phase induce the formation of the ultrafine-grained lath martensite. With increasing radiation energy austenitic-martensitic grains and martensite packets grow in size (Fig. 2c) and consequently hardness and wear resistance drop (Figs. 1b and 1c).

At the next experimental stage S235 steel was surfaced with a mixture of Cr_3C_2, Cr and TiC powders. The layers formed varied in thickness from 1 to 3 mm (Fig. 3a).

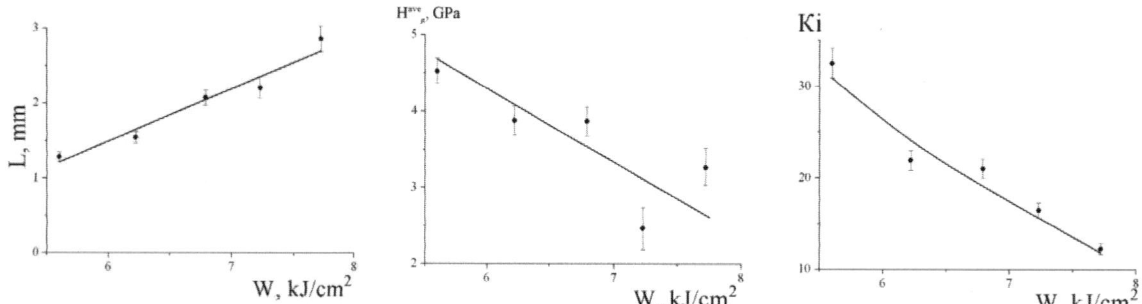

a b c

Figure 3. Thickness L (a), average microhardness $H\mu^{ave}$ (b), and coefficient of wear resistance Ki of the deposited layer (c) versus specific surface energy W for the surfacing mixture Cr_3C_2+Cr+TiC.

The coating hardness was 4.5 GPa at the specific surface energy W = 5.6 kJ/cm^2 (Fig. 3b). Coatings surfaced with this mixture had the abrasive wear resistance of 31.4 (Fig. 3c). With increasing depth of the remelted layer the structure changes as follows. At low specific surface energy the ultrafine hypoeutectic dendritic structure with the grain size 2–6 μm is formed (Figs. 4a and 4b). High wear resistance is associated with considerable structure refinement in the presence of dispersed titanium carbides as well as with precipitation of chromium carbides and wear-resistant austenitic phase in the coating. Moreover, TiC added to the surfacing mixture makes the structure more equiaxial so that wear is not accompanied by brittle fracture of long dendrite branches.

According to the X-ray phase analysis, for a thin deposited layer the radiographs show intense lines of chromium carbide Cr_7C_3 which compounds enter the austenitic-martensitic matrix with the predominant austenitic phase. The X-ray phase analysis reveals no compounds of titanium carbide, apparently, due to their high dispersability. With increasing thickness of the layer the intensity of chromium carbide lines reduces so that at the layer thickness 5–6 mm these lines are almost indistinguishable against the background. The intensity of austenitic phase lines also decreases. The α-iron (martensite) prevails in the structure.

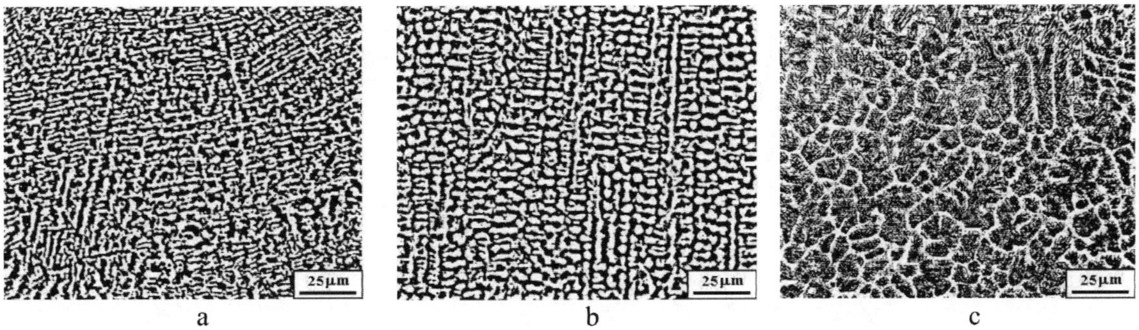

Figure 4. Structure of the layer surfaced with the mixture Cr_3C_2+Cr +TiC: W = 5.60 (a), 6.79 (b), and 7.72 kJ/cm^2 (c).

Figure 5 demonstrates measurement results for the corrosion resistance of the deposited coatings at dissolution in concentrated nitric acid and comparison of St3 and 12Kh18N10T steel specimens. In all tests the corrosion resistance of coatings is shown to achieve high values and approach the corrosion resistance of stainless steel 12Kh18N10T.

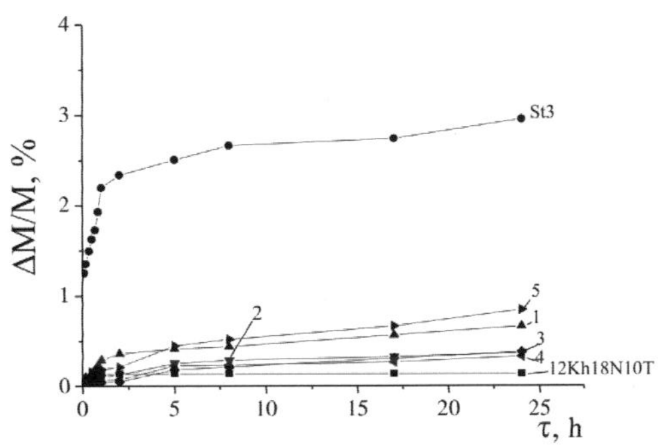

Figure 5. Relative mass wear of steel specimens and coatings surfaced with the mixture Cr_3C_2+Cr+TiC versus time in nitric acid: W =5.60 (1), 6.22 (2), 6.79 (3), 7.23 (4), 7.72 kJ/cm^2 (5).

The specific position of the curves in Fig. 5 is obviously dependent on the chromium content in a solid solution, from which it passes to the metal surface and forms a protective oxide film. With multiple chromium carbides precipitated in the layer corrosion losses increase slightly, which is related to the decreasing chromium content in a solid solution and propagation of interphase corrosion.

4. Conclusion

Deposition of tungsten carbide mixed with nickel and titanium carbide on low-carbon steel yields coatings with high hardness and wear resistance. The found effect is related to the nickel possibility to stabilize the wear-resistant phase, i.e. austenite, and to the precipitation of multiple dispersed particles TiC causing the formation of the homogeneous highly dispersed structure. Non-vacuum electron-beam surfacing of a mixture of chromium carbide, chromium and titanium carbide powders on low-carbon steel St3 provides bifunctional coatings with high wear resistance and corrosion resistance approaching that of stainless steel.

Acknowledgments
The work was carried out in the framework of the Basic Research Program of the State Academies of Sciences for 2013–2020 (Project No. 23.2.1).

References

[1] Poletika I M, Golkovski M G, Borisov M D, Salimov, R.A., and Perovskaya, M V 2005 *Physics and Chemistry of Materials Treatment* **5**: 29–41.
[2] Poletika I M, Golkovskii M G, Krylova T A and Perovskaya M V 2009 *Metal Science and Heat Treatment* **51**: 115-122.

IOP Conf. Series: Materials Science and Engineering **681** (2019) 012034 doi:10.1088/1757-899X/681/1/012034

EBSD analysis of the structure of a weld joining a chromium-nickel steel sheet and a titanium alloy sheet through a copper insert

N B Pugacheva[1], N S Michurov[1], and E I Senaeva[1,2]

[1]Institute of Engineering Science, Ural Branch of the Russian Academy of Sciences, 34 Komsomolskaya St., 620049, Ekaterinburg, Russia,
[2]B.N. Yeltsin Ural Federal University, 19 Mira St., Ekaterinburg, 620002, Russia

E-mail: nat@imach.uran.ru

Abstract. The structure of the different zones of a CO2 laser welded joint between a chromium-nickel steel sheet and a titanium alloy sheet, with a copper insert between them, is studied by EBSD analysis. The phase compositions of all the zones have been determined. The weld material has been found to consist of a Cu-based solid solution and unevenly distributed intermetallic particles of different stoichiometric compositions. The heat-affected zones consist of supersaturated iron- and titanium-based solid solutions. After laser welding, the microdeformations of the phase lattices are distributed unevenly over the cross section of the welded joint; namely, the maximum microdeformations and the resulting microstresses are concentrated in the heat-affected zone at the boundary between the weld and the titanium alloy.

1. Introduction

The application of the EBSD technique to analyzing the structural state of materials is becoming increasingly popular. This technique is mainly applied to the analysis of deformed alloys and less often to the study of cast alloys and welds. It is promising to use recrystallization and stress maps, as well as the analysis of the textures of the zones of welded joints produced by high-energy radiation sources, e.g. laser and electron-beam ones [1, 2]. When these heat sources are used, high solidification rates, together with convective mixing of the melt in the weld pool, may cause the formation of significant residual stresses and microstrains in welds, particularly if dissimilar materials are to be joined, when heterophase structures are formed.

The aim of this study is to investigate the structural state of the different zones of a laser-welded joint between a chromium-nickel steel and a titanium alloy with an intermediate copper insert.

2. Materials and research methods

We studied a weld produced by laser welding of 3 mm thick sheets of the 12Kh18N10T chromium-nickel steel and the VT1-0 titanium alloy with an intermediate copper insert (99.9 wt% Cu). The welding was performed with a continuous CO_2 laser in the Khristianovich Institute of Theoretical and applied Mechanics, SB RAS (Novosibirsk). A mixture of carbon dioxide and air was used to protect the surface of the weld pool and the overheated heat-affected zones. The welding speed was $V = 1.0$ m/min, the laser radiation power was $W = 2.3$ kW, and the focal distance was $F = -2$ mm. The phase composition of the weld and the heat-affected zones was determined with the application of the procedure of electron backscatter diffraction (EBSD) analysis using the Aztec software and a

IOP Conf. Series: Materials Science and Engineering **681** (2019) 012034 doi:10.1088/1757-899X/681/1/012034

Tescan Vega II XMU scanning electron microscope equipped with an Oxford NordlysMax^2 device. The electron-probe microanalysis was performed using an INCA Energy 450 device for energy-dispersive microanalysis with an ADD detector.

3. Results and discussion

Due to the rapid convective mixing of the melt in the weld pool under the action of a laser beam [1, 2], there occurs mutual penetration of the steel, copper insert, and titanium alloy elements in the entire volume of the pool. As the melt cools, first the intermetallics solidify and then the copper-based solid solution does. According to the data of the electron-probe microanalysis, the copper-based solid solution is significantly supersaturated with iron, chromium, and titanium [2, 3]. High cooling rates, about 1000 °/s, prevent titanium, iron, and copper from binding completely into intermetallics; therefore, the copper-based solid solution contains up to 10 wt% of iron and titanium. The phase EBSD analysis has demonstrated that the intermetallics in the weld material have different stoichiometric compositions, see Fig.1. Near the boundary between the weld and the titanium alloy there is a practically continuous chain of Ti_2Cu, Ti_3Cu_4 and TiFe intermetallics. The presence of a 100 to 150 μm thick intermetallic layer in the weld on the side of the titanium alloy was reported earlier in [4].

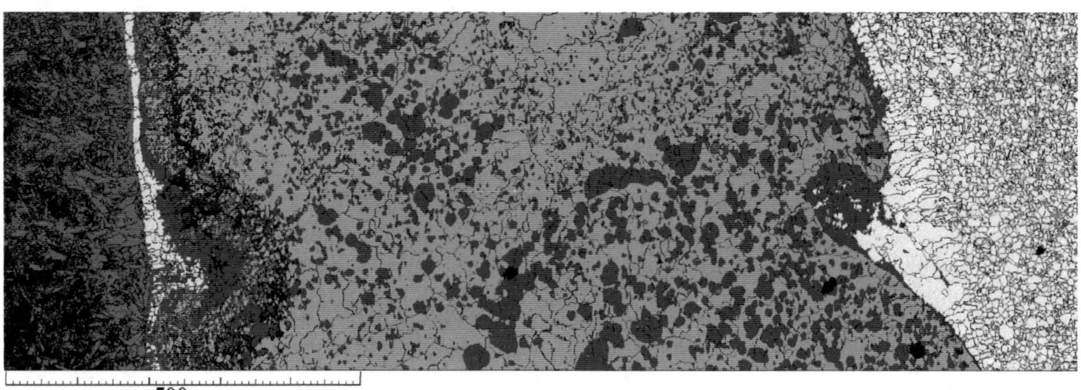

Figure 1. The phase composition of the weld: titanium alloy (green), Ti_2Cu (turquoise), Ti_3Cu_4 (pink), TiCu (brown), Ti_2Fe (red), TiFe (blue), steel (yellow), and copper-based solid solution (orange).

The thickness of the intermetallic layer found in the weld ranged between 70 and 180 μm. The intermetallic particles are arranged in layers as follows: right on the boundary with the titanium alloy there is a layer of $CuTi_2$ intermetallic particles, and then there come the TiFe layer and the layer consisting of a mixture of Ti_3Cu_4 and TiCu intermetallic particles; the intermetallic layer is followed by the weld material consisting of TiFe, Ti_2Fe intermetallics uniformly distributed in the Cu-based solid solution, Fig. 1.

The titanium alloy material is constituted by polyhedral grains of the α-phase, Fig. 2a. The analysis of the recrystallization maps has shown that recrystallized grains prevail in it (blue in Fig. 2c) and that there is a small amount of the subgrain and deformed structure (yellow and red, respectively). In the heat-affected zone the structure of the titanium alloy consists of packet and polyhedral grains of the α-phase, Fig. 2b. Due to high heating and cooling rates, as well as owing to the polymorphic $\beta \rightarrow \alpha$ transformation under cooling after welding, there are fewer recrystallized grains in this zone, grains with a subgrain structure (yellow in Fig. 2d) and deformed ones (red in Fig.2d) prevail.

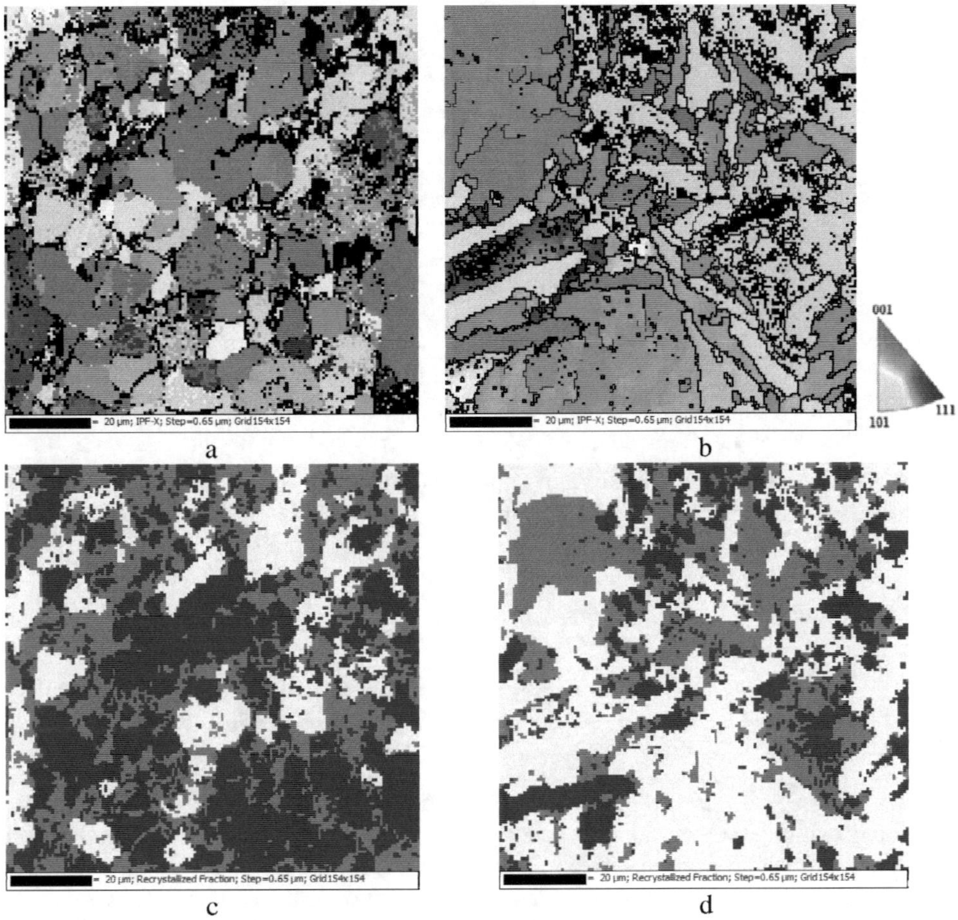

Figure 2. The microstructure (a, b) and recrystallization maps (c, d) of the titanium alloy in the initial state (a, c) and in the heat-affected zone (b, d); deformed grains (red in c and d), recrystallized grains (blue), and subgrains (yellow).

The electron-probe microanalysis of the titanium alloy near the weld has shown the presence of such alloying elements as Fe, Cu, Cr, and Ni in the titanium solid solution (Fig.3 a, the green, violet, red, and yellow lines, respectively), which act as stabilizers of the β-phase in titanium (). To determine the possibility of β-Ti stabilization in the heat-affected zones, we used a molybdenum equivalent Mo_{eq} [4, 5]:

$$Mo_{eq}=\sum\frac{x_i\cdot C''_{Mo}}{C''_i}=\%Mo+\frac{\%Ta}{4.5}+\frac{\%Nb}{3.6}+\frac{\%W}{2.25}+\frac{\%V}{1.5}+\frac{\%Cr}{0.63}+\frac{\%Mn}{0.65}+\frac{\%Fe}{0.35}+\frac{\%Co}{0.7}+\frac{\%Ni}{0.9}$$

As a result, the value of the molybdenum equivalent Mo_{eq} of the heat-affected zone proves to be equal to 20.9%; this corresponds to pseudo-β-titanium alloys, for which a β-structure is fixed during hardening. The EBSD study of the region under study has verified the presence of a solid solution of the alloying elements in the β-titanium at the boundary between the titanium alloy and the weld, with uniformly arranged TiFe and Ti_2Cu intermetallics and inclusions of the Cu-based solid solution (Fig. 3b).

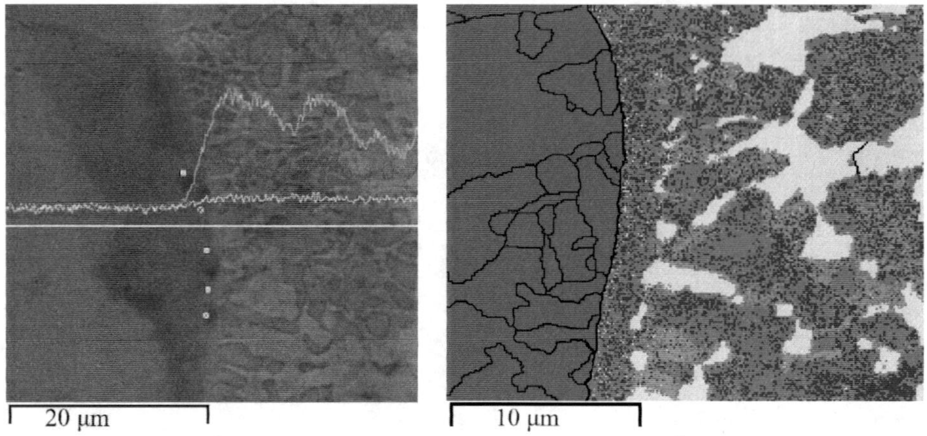

a b

Figure 3. The boundary between the titanium alloy and the weld: the distribution of the chemical elements along the line and the areas of local analysis (a); the phase composition of the boundary between the weld and the titanium alloy (green – α-Ti, red – β-Ti, orange – Cu, blue – TiFe, turquoise – Ti_2Cu) (b).

Table 1 – Averaged values of the content of the chemical elements in the titanium alloy at the boundary with the weld

Element	Si	Ti	Cr	Mn	Fe	Ni	Cu
wt%	0.1	74.4	1.9	0.1	5.9	0.8	16.8

In the central part of the weld the TiFe, TiCu, and Ti_3Cu_4 intermetallic particles (Fig. 4a) do not form clusters, they are distributed at some distance from each other in the Cu-based solid solution. The Cu-based solid solution grains are deformed after laser welding (red in Fig. 4b), whereas the intermetallics remain undeformed due to their high hardness and strength, and in the analysis of the recrystallization maps obtained by the EBSD technique they are qualified as recrystallized (blue in Fig. 4b). It was reported in [6] that the intermetallic particles chaotically arranged and distributed in the solid solution contribute to a decrease in the total level of strains and residual stresses due to their redistribution and dissipation in the weld volume. Thus, in the weld material there are no stress concentrators able to accelerate failure caused by external loads.

In the heat-affected zone on the side of the steel there prevails a subgrain structure. At the boundary between the weld and the steel the TiFe and TiCu intermetallic particles do not form continuous chains (Fig. 1). The analysis of the stress map has shown that no critical maximum stresses arise after welding, Fig. 5. Although the weld zone contains a considerable number of intermetallic particles, maximum microstrains after laser welding appear in the heat-affected zone on the side of the titanium alloy, Fig. 5.

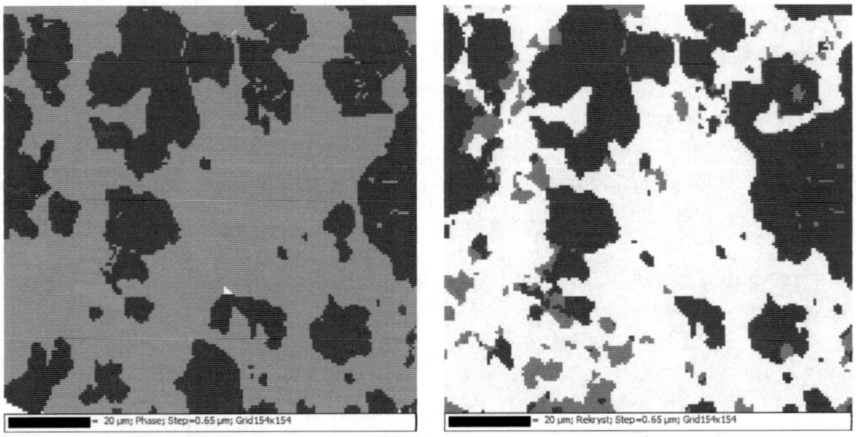

a b

Figure 4. The phase composition (a) and recrystallization maps (b) of the weld material: (a) – the Cu-based solid solution (orange), TiFe (blue), Ti_3Cu_4 (pink), and TiCu (brown); (b) – deformed grains (red), recrystallized grains (blue), and subgrains (yellow)

Figure 5. Stress maps for the weld joining chromium-nickel steel and titanium alloy sheets through a copper insert

4. Conclusion

The EBSD analysis has demonstrated that the weld joining the 12Kh18N10T and the VT1-0 alloy through a copper insert consists of a Cu-based supersaturated solid solution and Ti_2Cu, Ti_3Cu_4, TiCu, Ti_2Fe, and TiFe intermetallic particles. Near the boundary with the titanium alloy the intermetallic particles are arranged in layers, whereas in the weld and near the boundary with the steel they are distributed in the solid solution. The doping of the titanium alloy with Cu, Fe, Cr, and Ni transforms it from the α-alloy into a pseudo-β-alloy in the heat-affected zone near the weld. After the welding is completed, high cooling rates and the difference in the thermal coefficients of linear expansion of the phases in the welded joint cause the formation of deformed regions and residual stresses, whose maximum values are concentrated at the boundary between the weld and the titanium alloy.

5. Acknowledgments

The work was performed on the equipment installed in the Plastometriya collective use center of the IES UB RAS within state assignment, theme No. AAAA-A18-118020790145-0.

References

[1] W.W. Duley. Laser Processing an analysis of materials. Pienum Press, New York. 1983. 474.
[2] Cherepanov A.N., Orishich A.M., Shapeev V.P., and Pugacheva N.B. Thermophysics and aeromechanics 22 (2015) No. 2, 135–142.
[3] Veretennikova, I., Pugacheva, N., Smirnova, E., and Michurov, N. (2018). The laser-welded joint of an austenitic corrosion-resistant steel and a titanium alloy with an intermediate copper insert. Letters on Materials, 8(1), 42–47.
[4] Malcher, L. and Mamiya, E. N. (2014). An improved damage evolution law based on continuum damage mechanics and its dependence on both stress triaxiality and the third invariant. International Journal of Plasticity, 56, 232–261.
[5] Jia, Y. and Bai, Y. (2016). Ductile fracture prediction for metal sheets using all-strain-based anisotropic eMMC model. International Journal of Mechanical Sciences, 115–116, 516–531.
[6] N. S. Michurov, N. B. Pugacheva, I. A. Veretennikova, Yu. V. Khalevitsky, E. O. Smirnova, and A. S. Igumnov. Contribution of intermetallics to the formation of the stress-strain state of tensile welds on the AISI 321 steel and the grade 2 titanium alloy with a copper insert// AIP Conf. Proc. **2053**, 040058 (2018); doi: 10.1063/1.5084496.

IOP Conf. Series: Materials Science and Engineering **681** (2019) 012035 doi:10.1088/1757-899X/681/1/012035

A 120-kV electron welding gun with plasma cathode: vacuum and non-vacuum application

N G Rempe[1]

[1] Elion Ltd., ul. Sadovaya, 17, 634510, Tomsk, Russia
Tomsk State University of Control Systems and Radioelectronics, 40 Lenina Prospect, 634050, Tomsk, Russia

E-mail: remnik77@gmail.com

Abstract. The article studies the generation of small-diameter focused electron beams using electron-optical system of a welding gun with plasma cathode. The results show that the beams always form in the electron-optical system with plasma cathode when the gas in the region of beam emission, acceleration and transportation is present. The presence of the gas causes positive volumetric charge formation during the passage of electrons. This charge electrically neutralizes the beam, and in some cases can lead to overcompensation of its electron charge. The compensation of the volumetric charge in the region of beam formation and acceleration is the principal discrepancy of an electron-optical system with plasma cathode from a thermal cathode system. It is demonstrated that the brightness of the electron beam from a plasma cathode gun meets that of the beams from a thermal cathode gun. The paper describes the electron gun featuring voltage up to 120 kV and presents images of polished samples of weld penetration in stainless steel and titanium in vacuum. The gun served as the basis of a device for ejecting electron beam into the atmosphere. It is shown that the pressure rise up to the atmospheric one is provided through two gas-dynamic stages. The main characteristics of the atmospheric gun are given.

1. Introduction

Currently, plasma cathode electron guns are widely studied and implemented in practice to generate electron beams with various parameters. Electron beams produced by plasma guns can be used virtually in all modern electron-beam technologies, such as welding, cladding, material modification, etc. [1–4]. The guns are constantly enhanced; new information about their peculiarities and practical application regularly comes out.

The most complex to investigate and improve, from our perspective, are guns generating focused electron beams with high brightness. Such beams can be used in both traditional technologies and new fields that have been rapidly developing over the recent years. They first of all include additive technologies [2] and non-vacuum application that requires ejecting beam into a pressurized gas through a reliable and inexpensive system.

Such guns require much effort to investigate, because in addition to studying plasma as electron emitter, specific attention should be paid to the processes of beam formation [2–7]. Often, such studies are performed to achieve parameters that were already obtained by guns with solid state cathode. The primitive transfer of design principles from thermal cathode guns to plasma cathode guns is impossible.

IOP Conf. Series: Materials Science and Engineering **681** (2019) 012035 doi:10.1088/1757-899X/681/1/012035

This study exhibits the results obtained over the last several years of investigation on the specificities of the electron-optical system with plasma cathode aimed at formation of focused small-diameter electron beams.

2. Electron-optical system of guns with plasma cathode

Despite a vast diversity of plasma cathode guns available, focused electron beams can be produced only by using a limited range of discharge types. These include discharges with hollow cathode, high-voltage glowing discharge and some other [1]. Moreover, plasma beam industrial technologies implement plasma cathode guns of only one type based on reflective discharge with hollow cathode [3–7].

The electrons in an electron-optical system based on reflective discharge with hollow cathode are captured into the accelerating gap from the plasma surface. Plasma is generated in a discharge chamber schematized in Figure 1. The discharge chamber includes hollow cathode 1, anode 2 and emitter cathode 4 with a hole along its axis. Permanent magnet 3 creates a magnetic field to initiate and preserve stable discharge in the chamber. In the stationary regime, the discharge chamber contains plasma with nonuniform concentration distribution which maximum corresponds to the axial region of the discharge. In an electron-optical system with plasma cathode, four typical regions can be distinguished: region of primary beam formation, acceleration region (accelerating gap), transportation region (drift space), focusing region (focusing system).

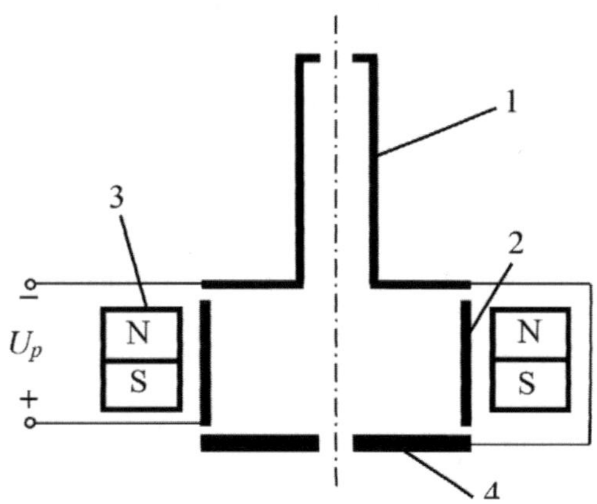

Figure 1. Scheme of discharge chamber of plasma cathode gun: 1) hollow cathode; 2) cylindrical anode; 3) permanent magnet; 4) emitter electrode.

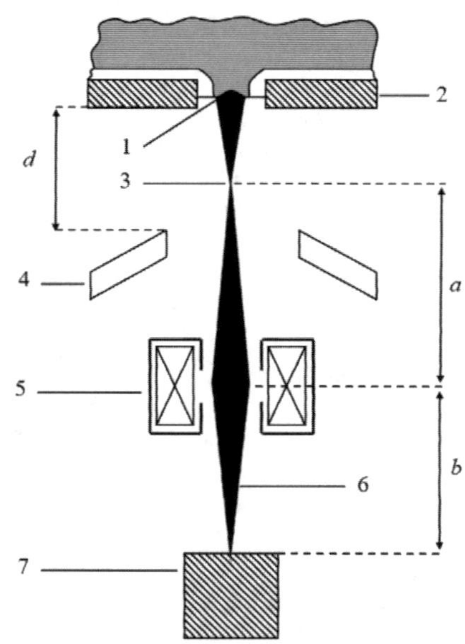

Figure 2. Electron-optical system with plasma cathode: 1) emitting surface of plasma; 2) emitter electrode; 3) crossover; 4) accelerating electrode; 5) focusing system; 6) electron beam; 7) electron collector; d is the accelerating gap; a is the distance from the crossover to the focusing system; b is the distance from the focusing lens to the electron collector.

The region of initial formation of the electron beam includes emitting surface of plasma 1 (Figure 2) and electric field near this surface. The emitting surface of plasma can form both in the cylindric channel inside emitter electrode 2 of the discharge chamber and outside of it.

The accelerating region (Figure 2) is the region between electrodes 2 and 4 where high-voltage electric field is concentrated which strength determines the electron energy in beam 6.

The transportation region (drift space) is an equipotential space between accelerating electrode 4 and electron collector (welded part) 7.

Focusing region, as a rule, is a part of transportation region where the magnetic field of focusing system 5 is localized.

Plasma, as an electron emitter, has a number of peculiarities that make difference between an electron-optical system with plasma cathode and a system with solid state cathode. The shape and position of the surface are not fixed and depend on plasma concentration, strength and shape of the electric field, pressure and type of gas and other factors [1]. When the electric field strength changes, the emitting surface displaces. The displacement continues until the equality of the electrostatic field pressure and gas-kinetic pressure of plasma.

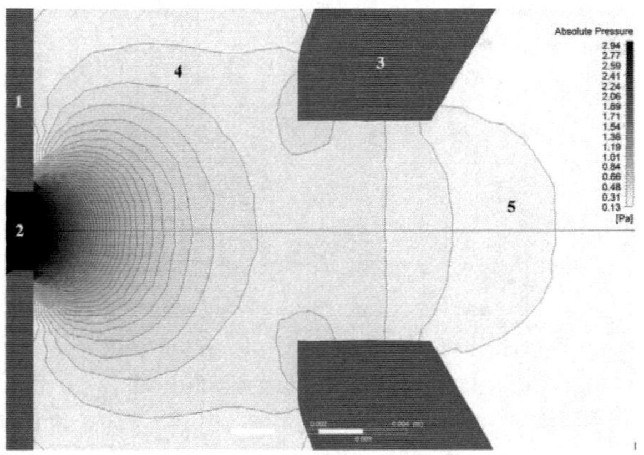

Figure 3. Illustration of pressure distribution in the accelerating gap of a plasma cathode gun: 1) emitter cathode; 2) emission channel; 3) extractor; 4) accelerating gap; 5) drift space. Black means the highest pressure, white is the lowest.

The displacement of the emitting surface can determine the type of emission parameters and electron-optical characteristics of the electron gun. This is why the development of plasma guns requires measures for stabilizing the emission surface of plasma. Typically, the parameters of electron gun are chosen for the emitting surface to be localized in the channel [6,7]. The emission channel sets the lateral dimensions of the emitting surface and limits the penetration of the electric field into the discharge chamber from the side of the accelerating gap.

According to the experimental results, the velocity distribution of the main fraction of electrons emitted from plasma can be considered to be similar to the Maxwell distribution. The electron temperature—depending on the discharge parameters and type—can amount to 5–15 eV [1,2], which is considerably higher than the temperature of thermoelectrons. It was believed that obtaining fine-focused beams in such beams is complicated due to high temperature of electrons in plasma. However, there is a range of factors specific for plasma cathode which action can have additional effect on focusing. One of such factors is permanence of gas in the acceleration region and electron drift space.

3. Gas in electron-optical system with plasma cathode

The plasma cathode guns are specific due to the necessity of certain gas pressure in the vicinity of plasma emitter, which in majority of cases is higher than that in the working chamber. A pressure of plasma gas in the discharge chamber is about $5 \cdot 10^{-2}$ torr. Such pressure is provided by controlled gas puffing with a rate of 10–20 cm^3atm/hr. The puffed gas, as a rule, is evacuated through the accelerating gap.

Electron emission from plasma occurs through the hole in the cathode in the discharge chamber into the region of the accelerating field between the cathode and grounded high-voltage anode (extractor), as shown in Figure 3. The figure shows the pressure distribution in the accelerating gap for constant pressure in the vacuum chamber of 10^{-3} torr during gas puffing into the discharge chamber with a rate of 10 cm^3atm/hr. Pressure distribution in the region of electron beam generation was modeled in ANSYS software.

The pressure distribution (Figure 4) in the accelerating gap of the plasma cathode gun was calculated for different combinations of the rate of gas puffing into the discharge chamber and pressure in the vacuum chamber. The distributions demonstrate that pressure gradient is always present in the accelerating gap. Near the emission channel and inside of it, the pressure can exceed that in the vacuum chamber by 1–2 orders of magnitude. This means that the formation of electron beam in plasma cathode gun occurs in appreciably different conditions from those of thermal cathode guns. In the accelerating gap, along with accelerated electrons, plasma electrons and ions constantly present that can substantially affect the motion of fast electrons in the beam.

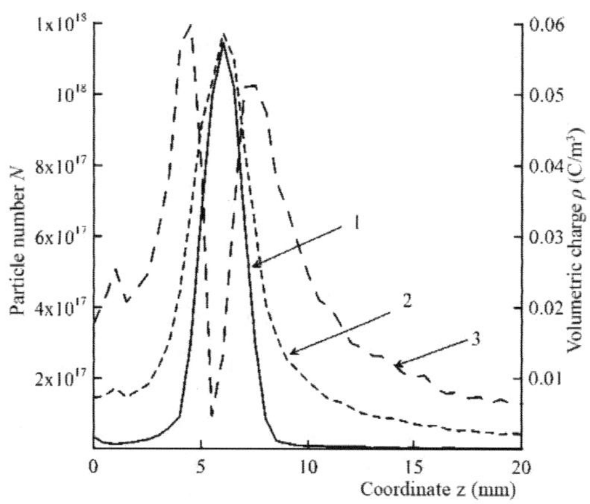

Figure 4. Distribution of electrons (1), ions (2) and volumetric charge (3) on the axis of the accelerating gap.

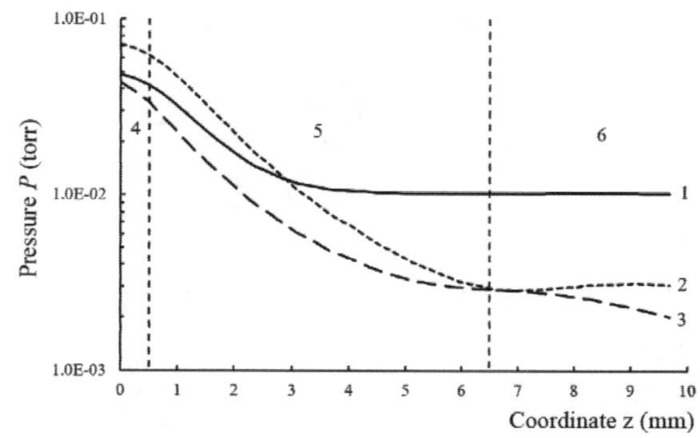

Figure 5. Pressure distribution in the accelerating gap of plasma cathode gun under different pressure in the vacuum chamber (torr) and rate of gas puffing into the hollow cathode (cm^3atm/hr): 1) 10^{-2}, 5; 2) 10^{-3}, 10; 3) 10^{-4}, 15. 4) boundary of the emission channel; 5) boundary of the accelerating gap; 6) channel region in the extractor.

Figure 4 presents axial (along coordinate z) distribution of electrons, ions born due to ionization and spatial charge in the accelerating gap. In calculations, the gap was filled with nitrogen with pressure distribution corresponding to dependence 1 in Figure 5. The maximum of electron distribution (dependence 1) corresponds to the position of the electron beam crossover that—according to the calculations—is localized in the acceleration space. The electron current density in the crossover is maximum. Following the calculations, the ion concentration is also maximum here (dependence 2) and is comparable with that of electrons. The equality of concentrations of electrons and ions in the region of crossover neutralizes the beam charge (dependence 3). In the remaining part of the calculation domain to the left and to the right of the crossover, the total charge is positive. Hence, the volumetric charge of the electron beam here is overcompensated. The simplest explanation exists for the overcompensation effect in the region to the right of the crossover. This region has the conditions for ion accumulation. These conditions express in low mobility of ions because they reside in the equipotential space.

The distributions demonstrate that gas ionizes along the whole accelerating gap. The majority of ions escape along the field into the region of plasma boundary. The ion current into plasma does not exceed 1% of the electron beam current for all pressure distributions in the accelerating gap and is maximum for conditions that correspond to the pressure distribution as per dependence 2 (Figure 4).

The probability of ionization is known to drop with increasing electron energy. This means that the main contribution to gas ionization and formation of ion current is made by electrons with low energy that corresponds to the maximum of the ionization function. Such conditions are created for electrons emitted from plasma near the emission surface. Gas flow from the discharge chamber leads to the formation of the increased pressure region localized in the accelerating gap near the emission channel. Therefore, the presence of gas in the accelerating gap of the plasma cathode gun causes the formation of positive volumetric ion charge after electron beam passage. This charge can not only electrically neutralize the beam, but in certain cases overcompensate its electron charge. The compensation of the volumetric charge in the region of beam formation and acceleration is the principal discrepancy of an electron-optical system with plasma cathode from a thermal cathode system. The compensation affects the electron trajectories of the beam, while the overcompensation can facilitate additional ion focusing of the beam. Beam overcompensation in the thermal cathode gun is also possible, but only in the drift space after electron beam passage in residual gases and vapors of metals from heated or melted metal. Thus, we can state that higher pressure in the accelerating gap and plasma cathode gun in the majority of cases promotes its higher quality.

This statement was checked by measurements made elsewhere [2]. The brightness of the electron beam was assessed by measuring the diameter and gun beam convergence angle at accelerating voltage of 60 kV and distance of more than 0.5 meters from the focusing lens.

To assess the brightness of the electron beam with Gauss law of current distribution along the section, the following expression was used:

$$B = \frac{i_b}{\pi^2 \cdot r_f{}^2 \cdot \theta^2},$$

where $\boldsymbol{i_b}$ is beam current; $\boldsymbol{r_f}$ is the effective radius of beam in the focal spot, $\boldsymbol{\theta}$ is convergence angle.

To measure the effective radius of beam and convergence angle, a measurement system was used developed by TWI Ltd. (Cambridge, UK). The measurements demonstrated that the brightness of the electron beam in plasma cathode guns is about $1 \cdot 10^{10}$ A·m^{-2}·sr^{-1}, which corresponds to the brightness of electron beams generated by thermal cathode guns with LaB_6 cathode [2].

4. Vacuum electron gun with high beam brightness.

The results of the study were used to design the plasma cathode gun schematically presented in Figure 6. The discharge forms in the discharge cell formed by inner surfaces of a hollow cathode 2, cylindrical anode 3 and emitter cathode 1 (Figure 6, highlighted region). Between the hollow and emitter cathodes an axially symmetric field is created by a permanent magnet, the field possessing induction on the axis of about 0.1 T. The discharge sustains continuously with the current from 0.1 to 1.5 A, which allows generating electron beam with the current from several to some hundreds of mA. Electron emission from plasma occurs through the hole in the emitter cathode in the discharge chamber into the region of the accelerating field between the cathode and grounded extractor. The electron beam formed in the accelerating gap is focused by the field of electromagnetic lens 4 and transported by beam guide 6 to the welded part.

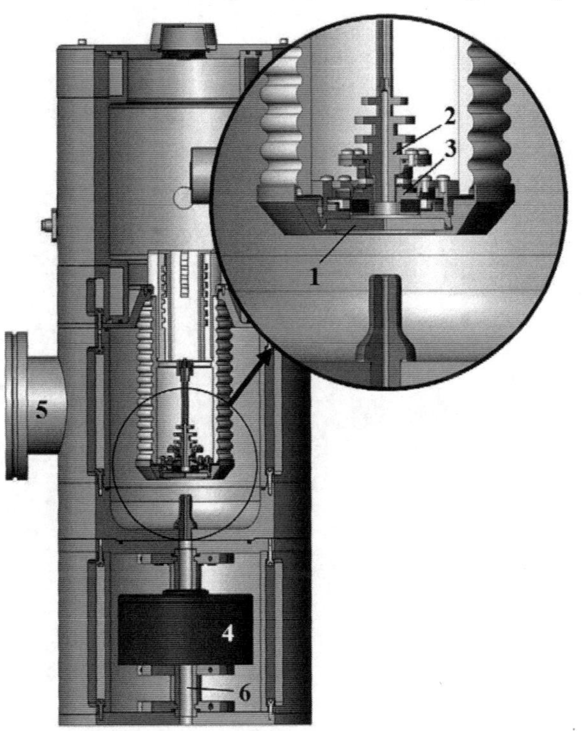

Figure 6. Scheme of discharge chamber of plasma cathode gun.

The gun body has pipe 5 with ISO100-KF flange enabling connection of a turbomolecular pump. The turbomolecular pump can evacuate the accelerating gap and beam guide 6 of the gun.

The gun forms an electron beam with an energy of up to 120 keV and power of up to 24 kW.

Welding capability of the gun was assessed by weld penetration of specimens from various metals. Figure 7 depicts polished weld penetrated specimens from AISI 321 stainless steel (a) and VT14 titanium (b). The weld penetrations were made with a power of the electron beam of 12 kW for stainless steel and 6 kW for titanium. The working distance (from the focusing lens to the specimens) was 300 and 670 mm. The beam was focused on the part surface. For the same rates of the welding manipulator moving rate, the shape and depth of penetrations made for different distances differed negligibly. The presented specimens are characterized by narrow and deep penetration profile with parallel walls, which is characteristic of the electron beam with an electron beam with small convergence angle.

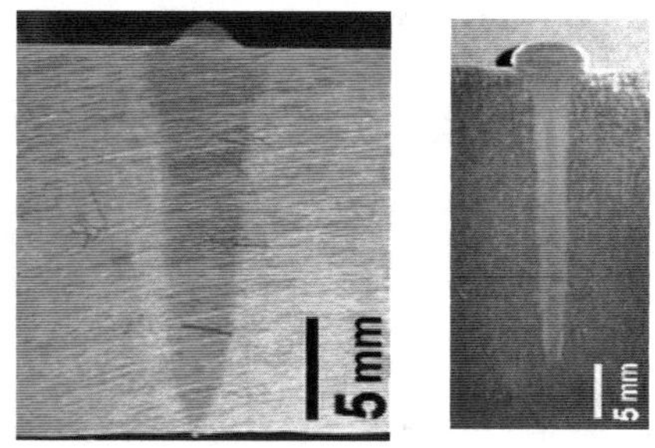

a b

Figure 7. Polished weld penetrated specimens from AISI 321 stainless steel (a) and VT14 titanium (b).

5. Non-vacuum electron-beam system based on plasma cathode gun

The implementation of plasma as a source of electrons allows largely reducing the requirements to residual atmosphere in the region of initial formation and acceleration of an electron beam. An electron beam can be generated by a plasma cathode gun under a pressure 2–3 orders of magnitude higher than that of thermal cathode guns. According to the above, higher pressure in the accelerating gap and plasma cathode gun in the majority of cases promotes its higher quality. The noted specificity allows assuming that plasma cathode guns can turn out to be effective both for vacuum applications and as a component of non-vacuum (atmospheric) electron-beam systems. Such work was accomplished in [7, 8]. The beam was transported into the atmosphere through gas-dynamic windows enabling step-wise increase of gas pressure up to atmospheric one. A scheme consisting of two windows was chosen and studied. Figure 8 shows the model of electron beam ejecting into the atmosphere obtained primarily by calculation methods. The ejection device is designed in a way that the accelerating gap of the gun (region between the emitter cathode surface and extractor) and beam guide separated by the extractor make up the second stage. The stages are counted starting from the atmosphere.

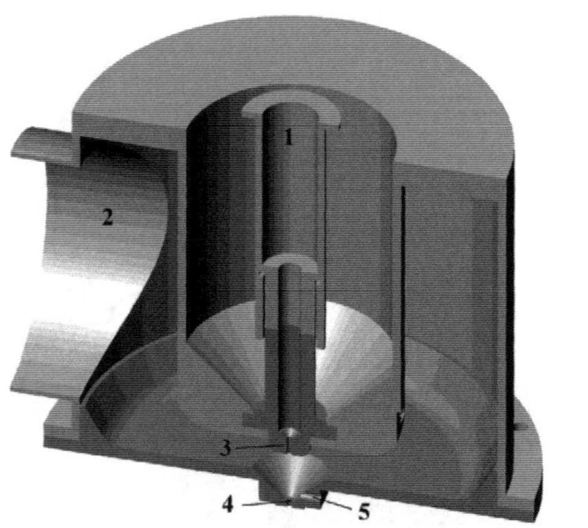

Figure 8. Three-dimensional model of the device ejecting electron beam into the atmosphere: 1) beam guide, 2) first-stage evacuation pipe, 3) second-stage diaphragm, 4) first-stage diaphragm, 5) cross-jet.

The beam guide tapers at the exit; the small taper base faces the first stage and transits into elongated cylindrical channel 3 with a length of 10 mm and diameter of 2.5 mm.

In the transportation channel, the electron beam is focused by the magnetic field in the plane of first-stage output diaphragm 4. The second stage is evacuated by a turbomolecular pump with a production rate of 800 l/s (nitrogen). The first stage has a 2-mm output diaphragm which ejects the electron beam immediately into the region with atmospheric pressure. The stage is evacuated with the rate of 500 l/s through pipe 2 by a Roots vacuum pump. In the first stage, there is a channel to create a cross-jet located in perpendicular to the output diaphragm axis.

The gun with the output device was mounted on a robot (Figure 9) that was performing welding, cutting and cladding in the atmosphere. The characteristics of equipment are presented in Table.

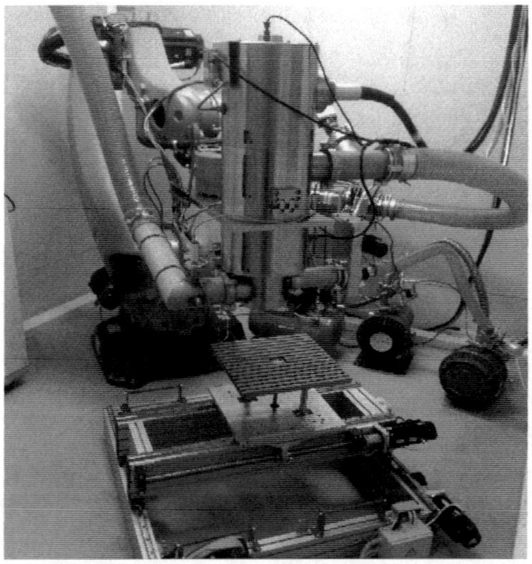

Figure 9. External view of gun and output device mounted on the robot.

Table 1. Characteristics of non-vacuum electron-beam system based on plasma cathode gun.

Parameter	Value
Accelerating voltage [kV]	120
Electron beam current [mA]	1-200
Number of evacuation stages	2
Pressure drop [torr]	from $5 \cdot 10^{-4}$ to 760
Coefficient of current passage, not less [%]	86
Working distance in atmosphere [mm]	5-15

Preliminary experiments on metal welding at atmospheric gas pressure were carried out. The welding joint was made between two 6-mm plates (from copper and aluminum) with 80 foils between them with average thickness of 15 μm each (terminals of accumulator batteries). The welding of such joint by electron beam with an energy of 120 keV and power of

4 kW (for copper terminal) and 2 kW (for aluminum terminal) was performed over 30 seconds. The length of the welding seam was 280 mm. The studies have demonstrated high mechanical strength of the joint. However, the presence of pores and defects in the joint disallowed recommending the technology for welding metal parts for vacuum-tight applications.

6. Conclusions

The presented characteristics in combination with the level of development of plasma cathode electron guns allow using them in the majority of traditional electron-beam vacuum technologies, in newly elaborated thermal processes (for instance, non-vacuum ones), in scientific experiments and other applications.

References

[1] Zavyalov M.A., Kreindel Yu.E., Novikov A.A. and Shanturin L.P., 1989 *Plasma processes in technological electronic guns*. Moscow: Energoatomizdat, 256 p.

[2] Hassel T, Rempe N, Kornilov S, Beniyash A, 2012 *Welding and Cutting*, V. 11, No. 2, P. 122-127.
[3] Kornilov S, Rempe N, Beniyash A, Murray N, Hassel T and Ribton C, 2013 *Technical Physics Letters*, V. 39, № 10, P. 843-846.
[4] Osipov I and Rempe N, 2000 *Review of Scientific Instruments*,V.71, №4, V.1638–41.
[5] Galansky V.L., Gruzdev V.A., Osipov I.V., Rempe N.G, 1992 *Russian Physics Journal*, V. 35, No 5, P. 28-33.
[6] Kornilov S, Rempe N, *15th International Symposium on High Current Electronics: Proceedings*. Tomsk: Publishing house of the IAO SB RAS. 19-24 September 2010, P. 76-79.
[7] Kornilov S and Rempe N, 2012 *Technical Physics,* V. 57, No. 2, P. 236-241.
[8] Aksenov A, Kornilov C, Motorin M and Rempe N, 2017 *Instruments and Experimental Techniques*, V. 60, No 2, P. 233-236.
[9] Borovik V, Shatravin V, Kornilov S Rempe N , Junusov I, Shidlovskiy S and Shashev D, 2016 MATEC Web of Conferences 7. Сер. "7th Scientific Conference with International Participation "Information-Measuring Equipment and Technologies", IME and T 2016", Article number 01034.

IOP Conf. Series: Materials Science and Engineering **681** (2019) 012036 doi:10.1088/1757-899X/681/1/012036

Effect of electron beam processing on structure of electroexplosion coating of ZnO-Ag system

D A Romanov[1], **S V Moskovskii**[1], **K V Sosnin**[1], **V E Gromov**[1], **A P Semin**[1], **A D Filyakov**[1]

[1] Department of natural science disciplines of Professor V. Finkel, Siberian State Industrial University, 42 Kirov street, Novokuznetsk, 654007, Russia

E-mail: romanov_da@physics.sibsiu.ru

Abstract. The structure, phase composition of the composite electroexplosion coating of ZnO-Ag system subjected to the irradiation by the high intense electron beam have been studied for the first time. The combined processing including the electroexplosion spraying of the coating of ZnO-Ag system and its subsequent electron beam processing results in the formation of multilayer structure of the coating. The electron beam melts and homogenizes the surface layer of the composite electroexplosion coating of ZnO-Ag system to the depth of 35-40 μm. The effect of the electron beam leads to the formation of the nanocrystalline phases of Ag, ZnO, Cu, CuZn, Ag_5Zn_8, Cu_2O, Ag_2O, $Cu_{0.67}Zn_{0.33}$, AgZn. The dependences of changes in crystal lattice parameters and coherent scattering regions of the phases being formed depending on the processing regime by high intense electron beam have been found.

1. Introduction

The touch of the electrical contacts takes place along not the entire surface but only in separate points due to the roughnesses on the surface of the contact [1-3]. In practice, the touch of their contact surfaces always occurs along small areas [4-6] independent of the electrical contact type. It is explained by the fact that the surface of the contact elements cannot be ideally even. Therefore, practically, on the approaching of the contact surfaces, at first, some projecting peaks (points) come into contact and then, as the pressure increases, the deformation of the contact materials takes place and these points transform into small areas [7-9]. The lines of the electric current, when passing from one contact to another, gather to these points of the touch [10, 11]. Therefore, the contact introduces some additional contact resistance into the circuit commutated by it. The transient resistance [12, 13] is one of the main parameters of the electrical contacts' quality because it characterizes the quantity of energy being absorbed in the contact connection which transforms into heat and heats up the contact. The method of processing of the contact surfaces and their state [14, 15] may strongly affect the transient resistance. Nowadays, it is proved by the scientists of different countries that the application of electron beam processing (EBP) enables to obtain a mirror-smooth surface of the materials [16], nanostructurize [17] and homogenize the material [18]. The application of EBP permits the increase in different service properties of the materials [19]. Knowing this information about EBP it seems to be effective to use the technology for the processing of the surface of electrical contacts [20]. In particular, it will make possible to minimize the transient resistance of the electrical contacts and to increase in the service life of their operation. In this connection, the purpose of the research was to study the EBP effect on the structure and phase composition of the electroexplosion coating of ZnO-Ag system.

IOP Conf. Series: Materials Science and Engineering **681** (2019) 012036 doi:10.1088/1757-899X/681/1/012036

2. Materials and methods

The investigation is concerned with the copper weak-current electrical contacts of the controller KKT-60 on the contact surfaces of which the electroexplosion coating of ZnO-Ag system was formed by the electroexplosion method [20]. In the electrical devises with low power the electrical arc on the contacts arises rarely but, however, the sparking, dangerous for the particularly sensitive devices, appears very often or the insulation gap breakdown is possible. It may arise in weak-current circuits in case of the fast opening of the contacts and frequently results in the false switching off and reduces considerably the service life of the contacts. Therefore, for the protection of the electrical contact surface of the controller KKT-60 it is enough to from the electroexplosion coating of ZnO-Ag system at most 40 μm thick on its surface. The silver foil 250 mg in mass was used as the current carrying material being exploded on the surface of which the weighed sample of ZnO powder 80 mg in mass was placed in the region of explosion. The time of plasma effect on the sample surface was ~100 μs, the power density being absorbed on the jet axis ~ 5.5 gW/m^2, the pressure in the shock-compression layer near the surface being irradiated ~ 12.5 MPa, the residual gas pressure in the working chamber ~ 100 Pa, the plasma temperature on the nozzle's cut ~ 10^4 K, the thickness of thermal effect zone ≈ 50 μm. After the electroexplosion spraying EBP was performed for the purpose of reducing the roughness of the coatings and the homogenization of their structure. The irradiation of the electroexplosion coating of ZnO-Ag system by the electron beam was done at the following values of the power density of electron beam: 0.3, 0.45, 0.5, 0.55 and 0.6 MW/cm^2. The structure and the morphology of the coating were analyzed by the methods of the scanning electron microscopy (device Carl Zeiss EVO50) and X-ray structural analysis (X-ray diffractometer ARLX'TRA).

3. The results and discussion

As a result of the combined processing including the electroexplosion spraying (EES) of the coating of ZnO-Ag system and its subsequent electron beam processing (EBP), the coating is formed in which four layers are distinguished by the structural state. The layers are the following: the surface layer obtained by two-stage processing EES+EBP (the layer 1 in Fig. 1); the boundary between the surface layer EES+EBP and the layer of EES unaffected by EBP (designated by the dotted line in Fig. 1); the layer of the electroexplosion coating unaffected by EBP (the layer 2 in Fig. 1) and the boundary of the electroexplosion coating with copper substrate (designated by the arrow in Fig. 1).

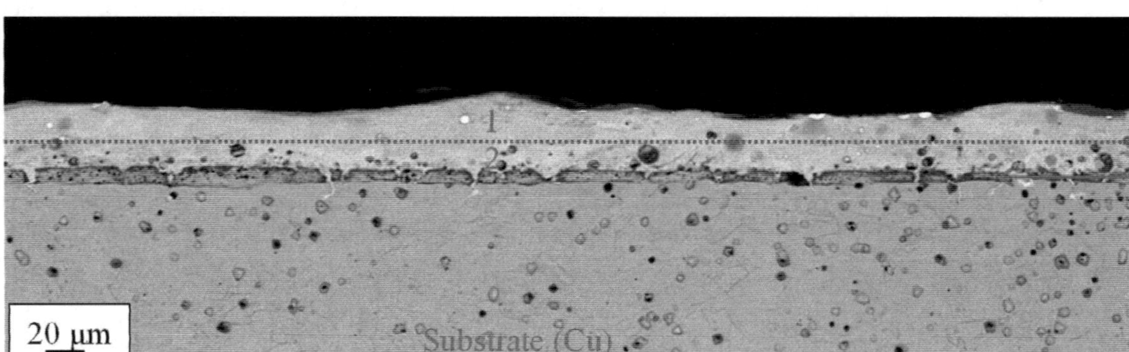

Figure 1. Structure of electroexplosion coating of ZnO-Ag system after electron beam processing. The straight metallographic section. Scanning electron microscopy. 1- the surface layer obtained by two – stage processing EES + EBP; 2 – the layer of electroexplosion coating unaffected by EBP; the dotted line designates the boundary between the layers 1 and 2; the arrow designates the boundary of electroexplosion coating with the copper substrate.

The parameters of the crystal lattice and the values of coherent scattering regions of the phases being formed in the surface layer obtained by two-stage processing EES + EBP (the layer in Fig. 1) were studied by the method of X-ray structural analysis. Let's analyze the dependence of the volume fraction of the electroexplosion coating phase of ZnO-Ag system on the power density of the electron beam (Fig. 4). The phase composition of the layer represents the following phases: Ag, ZnO, Cu, CuZn, Ag_5ZN_8, Cu2O, Ag_2O, $Cu_{0.67}Zn_{0.33}$, AgZn.

The initial value of Ag volume fraction (Fig. 2, blue marker) amounts to 19.16% and is the least one at 0.3 MW/cm^2. With the increase in the power density the volume fraction increases to the maximum value being equal to 85.97% at 0.45 MW/cm^2. After it Ag volume fraction decreases to the finite value 21.98% at 0.6 MW/cm^2.

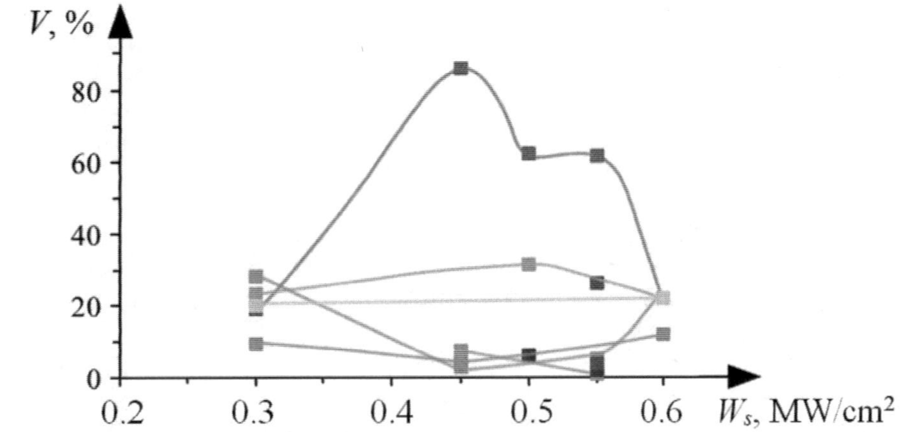

Figure 2. Dependence of the volume fraction of the electroexplosion coating (V) of ZnO-Ag system on the power density of the electron beam (Ws).

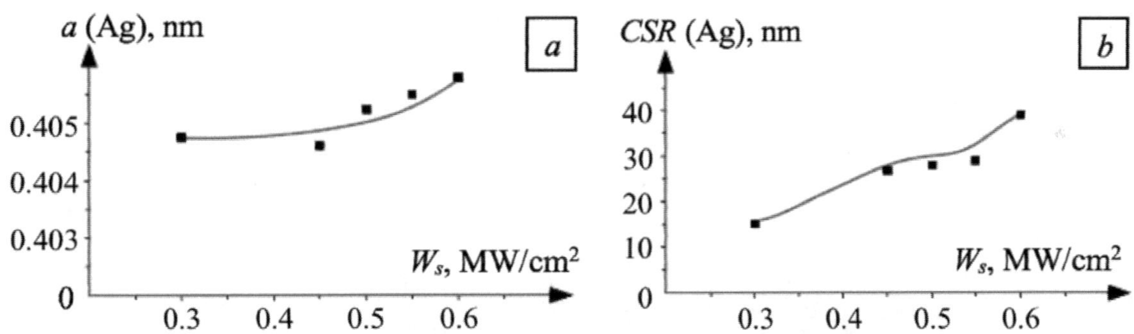

Figure 3. Dependence of parameter a of the crystal lattice (a) and the coherent scattering regions (b) of Ag (Fm3m) phase on the power density of electron beam.

The initial value of the crystal lattice parameter *a* equals to 0.40475 nm at 0.3 MW/cm^2 (Fig. 5, a) and decreases to its minimum of 0.40461 nm at 0.45 MW/cm^2. At 0.55 MW/cm^2 the parameter a equals to 0.40549 nm. At 0.6 MW/cm^2 the maximum value of the parameter *a* on Ws for Ag (Fm3m) phase takes the form of the power function. CSR of Ag (Fm3m) phase (Fig. 3, b) increases linearly, in practice, with Ws growth.

The volume fraction of Cu phase (Fig. 2, orange marker) amounts to 28.31% at 0.3 MW/cm^2 and decreases to the minimum of 2.93% at 0.45 MW/cm^2. At the power density being equal to 0.5 MW/cm^2 Cu phase fails to be observed. At the change in the power density from 0.55 MW/cm^2 to 0.6 MW/cm^2 the volume fraction varies within the limits from 21.98% to the finite value of 5.02%.

IOP Conf. Series: Materials Science and Engineering **681** (2019) 012036 doi:10.1088/1757-899X/681/1/012036

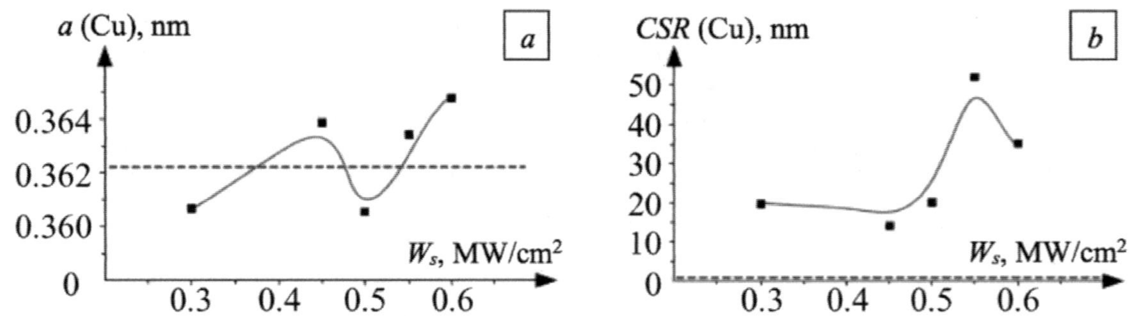

Figure 4. Dependence of crystal lattice parameter *a* (a) and coherent scattering regions (b) of Cu (Fm3m) phase on power density of electron beam. The red dotted line designates the values of these parameters for copper electric contact used as a substrate for the formation of electroexplosion coating.

In this case, the crystal lattice parameter *a* of copper (Fig. 4, a) at the power density being equal to 0.3 MW/cm^2 amounts to 0.360 nm. With the increase in the power density to 0.45 MW/cm^2 the value of the parameter *a* increases to 0.36387 nm after which at 0.5 MW/cm^2 the parameter *a* reaches its minimum being equal to 0.36953 nm. At the subsequent increase in the power density the parameter *a* increases to its maximum being the finite value and being equal to 0.36478. CSR of cupper (Fig. 4, b) at 0.3 MW/cm^2 equals to 19.57 nm then with the increase in the power density CSR decreases to its minimum being equal to 13.95nm. With the increase in the power density to 0.55 MW/cm^2 CSR increases to its maximum being equal to 51.95 nm. At 0.6 MW/cm^2 CSR decreases to 35.18 nm.

The volume fraction of CuZn phase at 0.3 MW/cm^2 amounts to 23.32% (Fig. 2). At 0.45 and 0.55 MW/cm^2 CuZn phase fails to be observed. The largest value of the phase equals to 31.42% at 0.5 MW/cm^2 and it decreases to the finite value of 5.02% at 0.6 MW/cm^2.

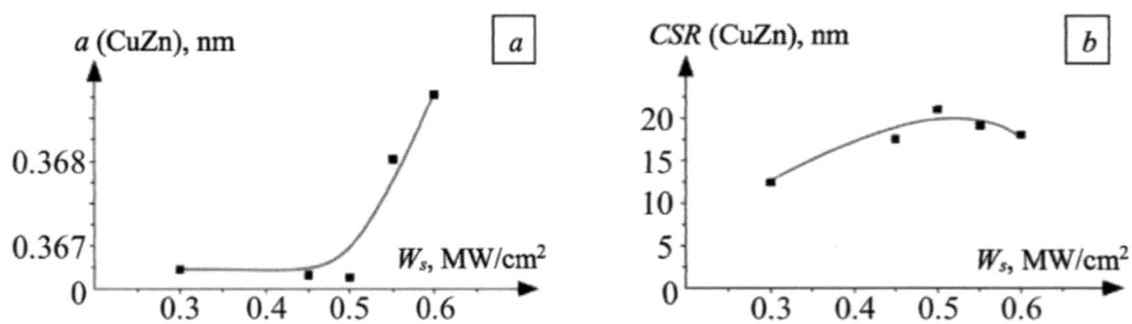

Figure 5. Dependence of crystal lattice parameter *a* (a) and coherent scattering regions (b) of CuZn phase on power density of electron beam.

The crystal lattice parameter *a* of CuZn phase (Fig. 5, a), when Ws changes from 0.3 to 0.5 MW/cm^2, remains constant and amounts to 0.36613 nm. After Ws increase to 0.55 and 0.6 WM/cm^2 the sharp increase in the crystal lattice parameter *a* of CuZn phase to the maximum value of 0.36828 nm takes place. CSR minimum value of CuZn phase (Fig. 5, b) amounts to 15.08 nm and is observed at 0.3 MW/cm^2. CSR minimum value of 38.97 nm is observed at 0.6 MW/cm^2.

For ZnO (Fm3m) phase the initial and the minimum value of the volume fraction amounts to 19.99% at 0.3 MW/cm^2 (Fig. 2), and the finite and the largest one – 21.93% at 0.6 MW/cm^2. The electron beam effect fails to result in the considerable change in the volume fraction of ZnO phase.

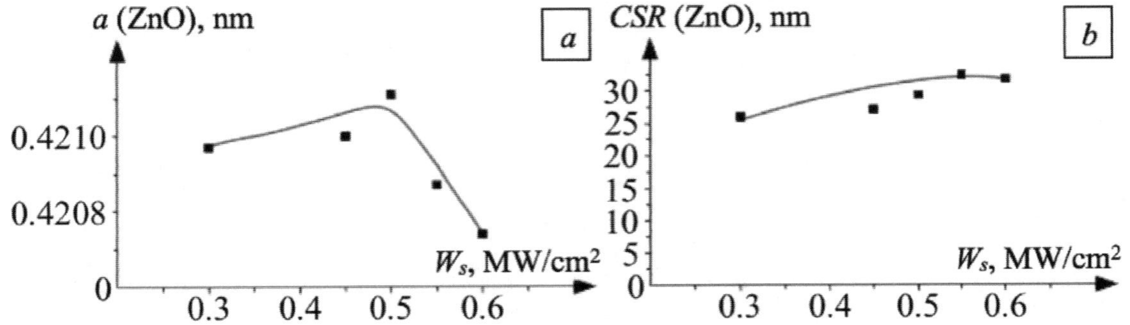

Figure 6. Dependences of crystal lattice parameter *a* (a) and coherent scattering regions (b) of ZnO (Fm 3m) phase on power density of electron beam.

The crystal lattice parameter *a* of ZnO (Fm3m) phase (Fig. 6, a) with the change in Ws from 0.3 to 0.5 MW/cm^2 increases from 0.42097 nm. With the further increase in Ws the sharp drop of the crystal lattice parameter *a* of ZnO (Fm 3m) phase to the minimum value of 0.42074 nm takes place. CSR minimum value of ZnO (Fm3m) phase (Fig. 6, b) being observed at 0.3 MW/cm^2 equals to 25.98 nm, and the maximum one – 31.82 nm at 0.6 MW/cm^2. The dependence of CSR change on *Ws* is approximated by the linear function.

For Ag_5Zn_8 phase the initial value of the volume fraction amounts to 9.22% at 0.3 MW/cm^2 (Fig. 4). With the increase in the power density the volume fraction falls to its minimum of 3.82 % at 0.45 MW/cm^2. At 0.6 MW/cm^2 the volume fraction of Ag_5Zn_8 phase is the maximum one and amounts to 12.13 %.

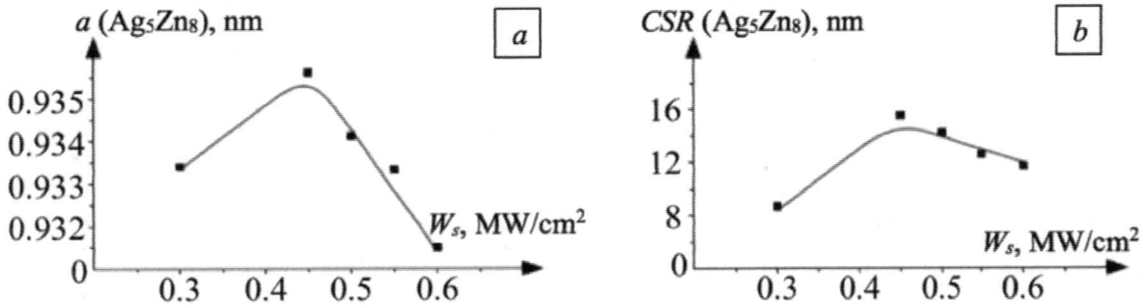

Figure 7. Dependences of crystal lattice parameter *a* (a) and coherent scattering regions (b) of Ag_5Zn_8 phase on power density of electron beam.

The initial value of the crystal lattice parameter *a* of Ag_5Zn_8 phase at 0.3 MW/cm^2 amounts to 0.9334 nm (Fig. 7, f). The subsequent growth of the parameter reaches its maximum of 0.93561 nm at 0.45 MW/cm^2. After the growth of the crystal lattice parameter *a* of Ag_5Zn_8 phase its descend occurs. The crystal lattice parameter of Ag_5Zn_8 phase reaches its minimum of 0.9315 nm at 0.6 MW/cm^2. CSR initial value of Ag_5Zn_8 phase (Fig. 7, *b*) is also the minimum one – 8.67 nm at 0.3 MW/cm^2. The gradual increase in the parameter reaches its maximum value of 15.51 nm at 0.45 MW/cm^2. CRS finite value of Ag_5Zn_8 phase amounts to 11.73 nm at 0.6 MW/cm^2.

The initial and the minimum value of the volume fraction of Cu_2O (Fm3m) phase amounts to 7.28% (Fig. 2) at 0.45 MW/cm^2, and the finite and the least one – 1.07 % at 0.55 MW/cm^2.

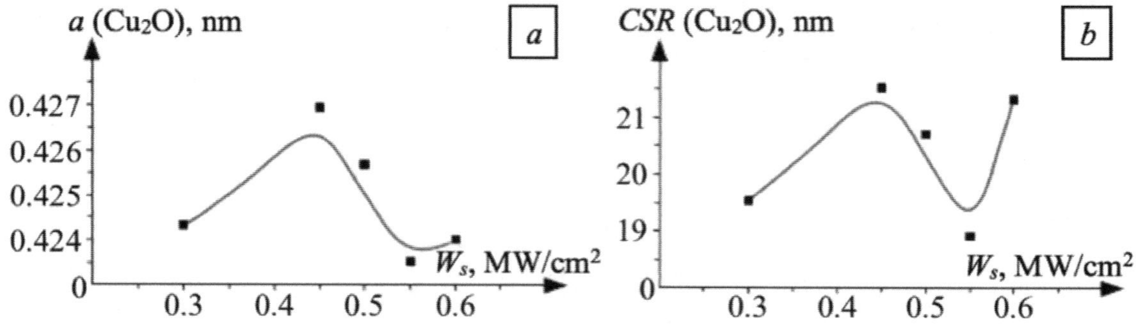

Figure 8. Dependences of crystal lattice parameter *a* (a) and coherent stuttering regions (b) of Cu_2O (Fm3m) phase on power density of election beam.

The maximum value of the crystal lattice parameter *a* of Cu_2O (Fm3m) phase amounts to 0.42694 nm and is observed at 0.45 MW/cm^2 (Fig. 8, a). The minimum value 0.42353 nm is observed at 0.55 MW/cm^2. CSR maximum value of Cu_2O phase being observed at 0.45 MW/cm^2 equals to 21.52 nm, and the minimum one – 18.92 nm at 0.55 MW/cm^2 (Fig. 8, b).

The initial and maximum value of Ag_2O phase (Fig. 9) amounts to 6.05 % at 0.5 MW/cm^2, and the finite and the largest one – 2.01 % at 0.55 MW/cm^2.

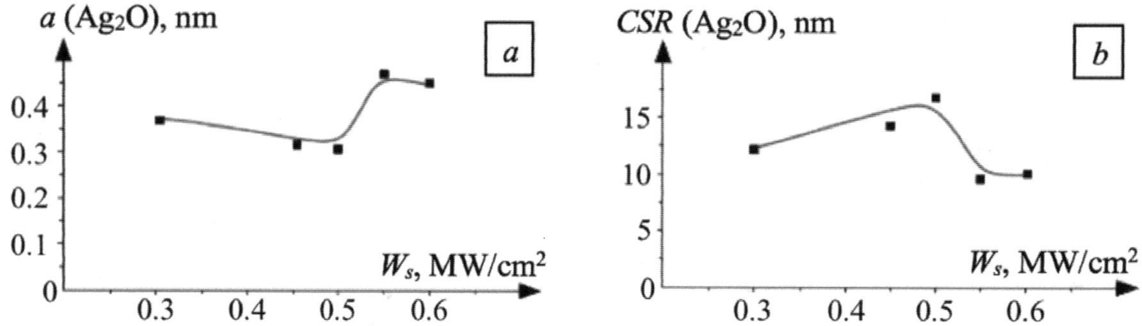

Figure 9. Dependences of crystal lattice parameter *a* (a) and coherent scattering regions (b) of Ag_2O (Pm3) phase on power density of electron beam.

The crystal lattice parameter *a* of Ag_2O (Pm) phase (Fig. 9, a) with the change in Ws from 0.3 to 0.5 MW/cm^2 remains constant and amounts to 0.30626 nm. After the increase in Ws to 0.55 and 0.6 MW/cm^2 the sharp increase in the crystal lattice parameter *a* of Ag_2O (Pm3) phase to the maximum value of 0.47017 nm takes place. CSR maximum value of Ag_2O (Pm) phase (Fig. 9, b) being observed at 0.5 MW/cm^2 equals to 16.78 nm, and the minimum one – 9.62 nm at 0.6 MW/cm^2.

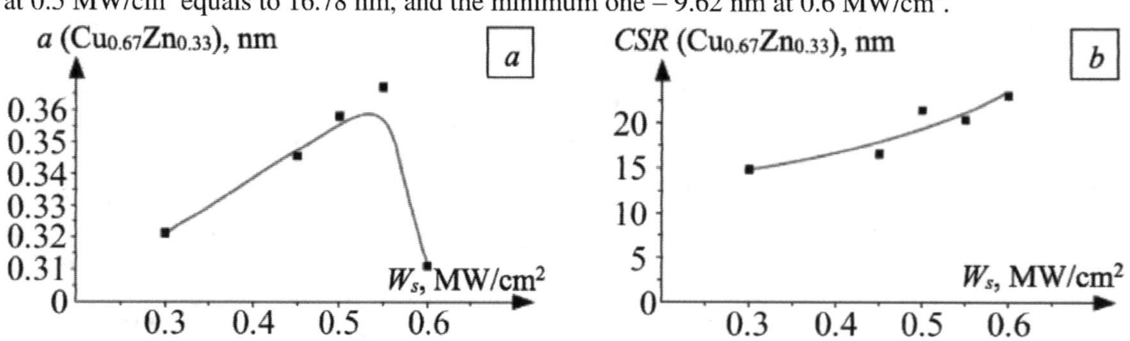

Figure 10. Dependences of crystal lattice parameter *a* (a) and coherent scattering regions (b) of $Cu_{0.67}Zn_{0.33}$ phase on power density of electron beam.

The maximum value of $Cu_{0.67}Zn_{0.33}$ volume fraction is 26.34% (Fig. 10) at 0.55 MW/cm^2, in this case, the crystal lattice parameter *a* is equal to 0.36695 nm, and CSR – 20.42 nm. The similar situation is observed with AgZn phase. Its volume fraction amounts to 3.92 % at 0.55 MW/cm^2.The crystal lattice parameter *a* equals to 0.31672 nm, and CSR – 15.96 % nm.

The similar changes in the crystal lattice parameters and CSR (Fig.s 3–10) of the detected phases are connected with the change in the dominating thermodynamic processes, the active phase transformations, the possible fragmentation of the separate phases, the formation of the new internal boundaries in them as a result of the thermal stress growth occurring under the processing of the electroexplosion coating by the electron beam. As a result of EBP effect the layers 1 and 2 in Fig. 1, as well as their boundary begin to behave themselves asynchronously, most likely, here the temperature lag of the structure transformation processes develop. The increase in CSR parameter of the separate phases is connected with the 'healing' of low angle boundaries due to the facilitating of the intergranular slippage and coming out of the dislocations to the surface. These processes are facilitated by the formation of the liquid – phase interlayers and the decrease in their viscosity as well as the decrease in the mechanical stresses in solid – phase carcass. CSR dimensions in the coating being formed depend substantially on the structural transformations under the temperature effects caused by EBP. The changes in CSR dimension and the enumerated parameters correlate. CSR dimension seem to be a rather sensitive signal for the control of the changes occurring in the structure of the coating [21-23]. These results are in the logical connection with EBP technological parameters as well. These issues require special investigations.

4. Conclusion

The electroerosion resistant coatings of ZnO-Ag system have been obtained for the first time on the surface of the electrical contact of the electromagnetic starter KKT-60. The formation of the coatings of ZnO-Ag system was produced due to the processing of the electrical contact surface by plasma formed at electrical explosion of silver foil with the weighed sample of ZnO powder. The electron beam melts and homogenizes the surface layer of the composite electroexplosion coating of ZnO-Ag system to the depth of 35-40 μm. The effect of the electron beam leads to the formation of the nanocrystalline phases of Ag, ZnO, Cu, CuZn, Ag_5Zn_8, Cu_2O, Ag_2O, $Cu_{0.67}Zn_{0.33}$, AgZn. The dependences of changes in crystal lattice parameters and coherent scattering regions of the phases being formed depending on the processing regime by high intense electron beam have been found. The changes in the crystal lattice parameters and CSR of the detected phases are connected with the change in the dominating thermodynamic processes, the active phase transformations, the possible fragmentation of the separate phases, the formation of the new internal boundaries in them as a result of the thermal stress growth occurring under the processing of the electroexplosion coating by the electron beam.

Acknowledgments
The present work was performed within Russian Science Foundation project No. 18-79-00013.

References

[1] Baek S J, Kim M S, An W J and Choi J H 2019 *Compos. Struct.* **220** 179-184
[2] Tayebani B and Mostofinejad D 2019 *Construction and Building Mater.* **208** 75-86
[3] Roh H D, Lee S Y, Jo E, Kim H, Ji W and Park Y B 2019 *Compos. Struct.* **216** 142-150
[4] Zhang Y, Hang T, Dong M, Wu Y, Ling H, Hu A and Li M 2019 *Thin Solid Films* **677** 39-44
[5] Zhu Y, Xu Y, Li K, Wang X, Liu G and Huang Y 2019 *Meas.: J. of the Int. Meas. Confederation* **138** 8-24
[6] Kim Y I, An S, Kim M W, Jo H S, Kim T G, Swihart M T, Yarin A L and Yoon S S 2019 *J. Alloys Compd.* **790** 1127-1136
[7] Das A, Ashwin T R and Barai A 2019 J. of Energy Storage **22** 239-248
[8] Mohammed A G, Ozgur G and Sevkat E 2019 *Cold Regions Sci. and Technol.* **160** 128-138

IOP Conf. Series: Materials Science and Engineering **681** (2019) 012036 doi:10.1088/1757-899X/681/1/012036

[9] Astacio R, Gallardo J M, Cintas J, Montes J M, Cuevas F G, Prakash L and Torres Y 2019 *Int. J. Refract. Met. Hard Mater.* **80** 259-269
[10] Zhang D, Yi P, Peng L, Lai X and Pu J 2019 *Carbon* **145** 333-344
[11] Mombrú D, Romero M, Faccio R and Mombrú A W 2019 J. of Mater. Sci.: *Mater. in Electron.* **30** 5959-5964
[12] Kazemzadeh A, Ein-Mozaffari F and Lohi A 2019 *Chem. Eng. Res. Des.* 226-240
[13] Zhao T, Iso Y, Ikeda R, Okawa K and Takei M 2019 *Flow Meas. and Instrum.* **65** 90-97
[14] Tokarska M and Orpel M 2019 *Textile Res.* J. **89** 1073-1083
[15] Singh B K, Jain E and Buwa V V 2019 *Chem. Eng. J.* **358** 564-579
[16] Park S, Kil H S, Choi D, Song S K and Lee S 2019 *J. of Industrial and Engineering Chemistry* **69** 449-454
[17] Batrakov A V, Onischenko S A, Kurkan I K, Rostov V V, Yakovlev E V, Nefedtsev E V and Tsygankov R V 2018 *Proceedings - International Symposium on Discharges and Electrical Insulation in Vacuum, ISDEIV* **1** 8537014 77-80
[18] Batrakov A V, Onischenko S A, Kurkan I K, Rostov V V, Yakovlev E V, Nefedtsev E V and Tsygankov R V 2018 *Proceedings - 2018 20th International Symposium on High-Current Electronics, ISHCE 2018* **8521188** 126-129
[19] Sidorov A I, Yurina U V, Rakhmanova G R, Shinkarenko M N, Podsvirov O A, Fedorov Y K and Nashchekin A V 2018 *J. Non-Cryst. Solids* **499** 278-282
[20] Romanov D A, Moskovskii S V, Sosnin K V, Gromov V E and Bataev V A 2019 *Mater. Res. Express* **6** 5 055042
[21] Bugayev A and Elsayed-Ali H E 2019 *J. Phys. Chem. Solids* **129** 312–316
[22] Qin Y, Dong C, Song Z, Hao S, Me X, Li J, Wang X, Zou J and Grosdidier T J. 2009 *Vac. Sci. Technol. A* **27** 3 430–435
[23] Zinenko V I and Pavlovskii M S 2008 *JETP Letters* **87** 6 288–291

IOP Conf. Series: Materials Science and Engineering **681** (2019) 012037 doi:10.1088/1757-899X/681/1/012037

Assessment of welding working properties of power supplies

Yu N Saraev [1], D P Ilyashchenko[2], M A Krampit[2], V M Semenchuk[1], A S Nepomnyashchiy[1]

[1] Institute of Strength Physics and Materials Science SB RAS, Tomsk
[2]National research Tomsk Polytechnic University, Tomsk;

E-mail: mita8@rambler.ru

Abstract. The welding working properties of power supplies are the basic indicators of efficiency of their application when forming permnanent joints of critical duty constructions. Growth of the share of welded structures made of difficult-to-weld steels and alloys requires more advanced welding equipment characterized by better static and dynamic characteristics and ensuring optimal thermal welding cycles. In the paper the authors present their own view of the necessity of power supplies welding working properties assessment from the perspective of quantitative assessment of welding conditions indicators. It allows most complete characterizing of the process flow parameters.

1. Introduction

Today fusion welding is one of the key technologies in all industries. Meanwhile the quality factors of the weld joints and deposited surfaces depend not only on the technological capacity of the mechanical equipment but also on the versatility of the process implemented. The versatility of the process means the ability to keep the electrical and thermal characteristics unchanged or changing periodically according to a specified program at the level of instantaneous values [1]. Solution of the given problem is becoming more and more relevant as the technological development sets ever increasing requirements to the quality of the welding constructions and efficiency of the welding production.

In Russia the share of welded structures made of difficult-to-weld steels and alloys grows every year. Welding this type of steels and alloys requires equipment ensuring optimal thermal cycles [2] which demands application of new generation power supplies with improved static and dynamic performance [3]. The given characteristics determine the welding working properties of power systems and their quantitative assessment provides complete characteristics of the quality factors of the technical process. Up to now the welding working properties of power supplies have been registered in strict accordance with State Standard 25616-83 and State Standard 9466-75. The given standards allow getting a quality assessment by counting the number of points – good, satisfactory, bad. This way one cannot get objective information about the advantages of this or that power supply or applied welding materials.

Goal of research: Estimating the relationships between the quantitative indicators of the welding working properties of power supplies for manual metal-arc welding and the process performance specified by current standards in their assessment.

2. Experimental method and procedure.

To eliminate the shortcoming described above we developed a method of quantitative assessment of the welding working properties [4]. We chose the following power supplies as the basic ones: power supply made in accordance to the classic circuit of welding rectifier – VD-306E type (diode rectifier) and inverter power supplies: "ARC 250" (WS Weldeship) (inverter 1); "Resanta: SAI 250 PROF" (inverter 2); "Fubag INTIG 200 SYN PLUS" (inverter 3). The power supplies listed above have different static and dynamic performance determined according to the external characteristics presented in Figure 1. The dynamic properties of the power supplies can be estimated by analyzing the oscillograph charts of current and arc voltage presented in Figure 2.

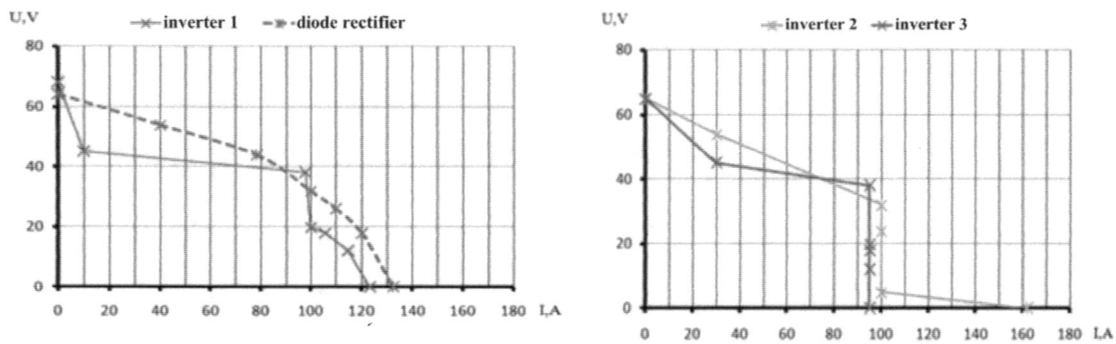

Figure 1. External characteristics of power supplies.

According to the method mentioned above we studied how the alteration rate of the basic energy parameters within the separate welding microcycles completed with power supply from the sources with the different types of energy conversion upon affects the stability of the surfacing process according to the conditions described in Table 1.

Table 1. Parameters of experimental surfacing

Power supply	Electrodes	Mean values of parameters recorded with oscillograph AKIP-4122/10.
Diode rectifier	OZCh-4	I_{arc}= (84.4 ± 2.7) A; U_{arc}= (21.9 ± 0.6) B; V_{ws}= 0.25 m/min
Inverter 1		I_{arc}= (84.5 ± 2.7) A; U_{arc}= (22.5 ± 0.7) B; V_{ws}= 0.25 m/min
Inverter 2		I_{arc}= (87.5 ± 3.1) A; U_{arc}= (21.3 ± 0.5) B; V_{ws}= 0.25 m/min
Inverter 3		I_{arc}= (85.6 ± 2.4) A; U_{arc}= (21.2 ± 0.4) B; V_{ws}= 0.25 m/min.

Here, I_{arc} – mean value of arc current; U_{arc} – mean value of arc voltage; V_{ws} – estimated value of welding speed.

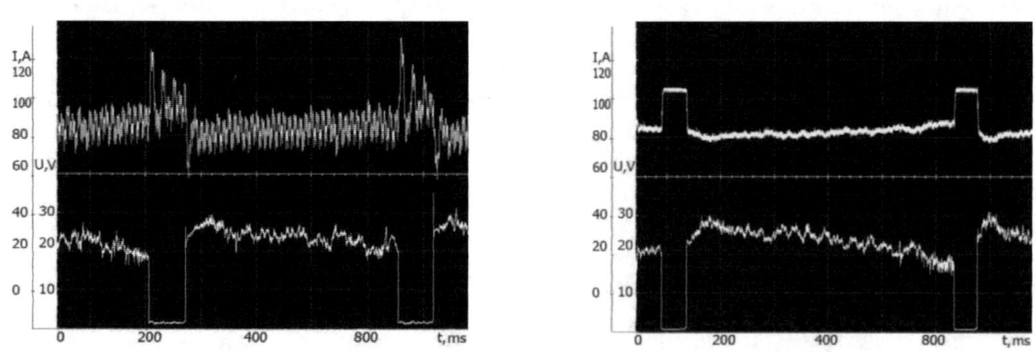

a b

IOP Conf. Series: Materials Science and Engineering **681** (2019) 012037 doi:10.1088/1757-899X/681/1/012037

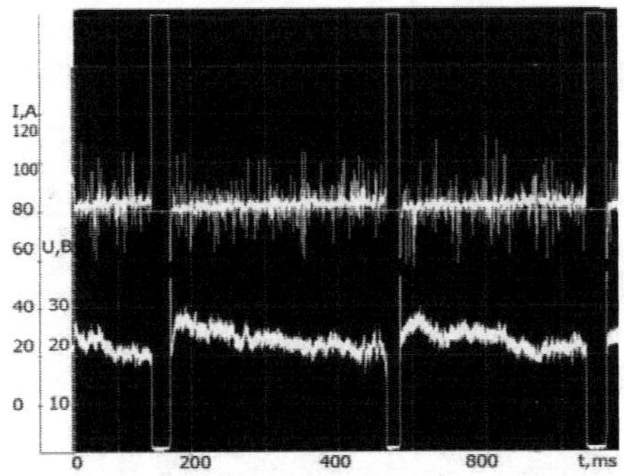

c

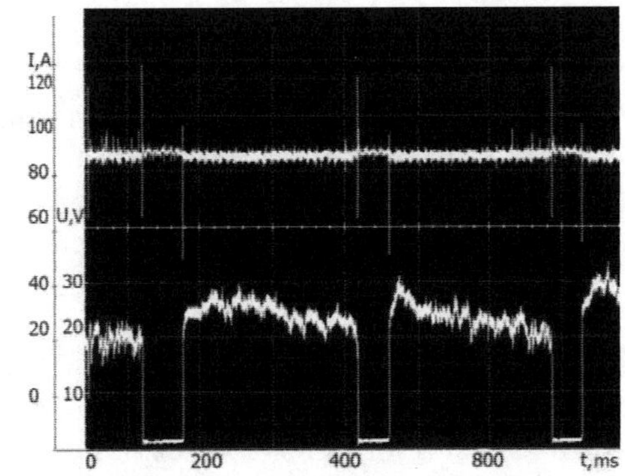

d

Figure 2. Oscillograph charts of the current in the welding circuit and voltages between the electrode and the workpiece (electrodes OZCh-4, 3 mm in diameter): a – diode rectifier; b – inverter 1; c – inverter 2; d – inverter 3.

3. Discussion of obtained results.

Initial arc striking (the first property of the welding working parameters according to State Standard 25616-83) does not cause any problems to inverter power supplies as they all have HOT START function while the diode rectifier does not.

The term "Arc stability" is the second property according to State Standard 25616-83. Its physical meaning was most completely formulated by Yu. N. Lankin in his paper [5]: "The process of welding with parameter deviation from mean values not exceeding the specified level is called stable. The material measure of stability is parameter deviation from its mean value. Parameter deviation from its mean value is understood as its dispersion, standard deviation or coefficient of variability". Analysis of obtained results (Table 1) shows growth of short circuits of the arc space under the identical values of the surfacing conditions parameters indicating the advantages of inverter rectifiers over the diode ones.

Table 2. Results of statistical analysis of current and voltage oscillograph charts recorded when completing the technical studies

Type of power supply	Parameter				
	τ_{sc}, ms ± $\sigma_{\tau sc}$, ms	T_{st}, ms ± σ_{Tst}, ms	I_{min}, A	$I_{k.z.}$, A	$\frac{I_{arc}}{I_{k.z.}} \to 1$
Diode rectifier	61.3±9.4	750±295	58.28±1.8	127±6.5	0.66
Inverter 1	45.9±9.7	658.1±118.8	76.9±1.5	105±3	0.8
Inverter 2	22.6±2.7	295.5±45.5	14.1±0.8	162±1.4	0.54
Inverter 3	37.1±11.8	310.2±34.9	83±1.1	88.1±0.9	0.97

Here τ_{sc} is the short circuit duration of the arc space; ± $\sigma_{\tau sc}$ – the standard deviation of short circuit duration; $T_{st} = (\tau_{sc} + \tau_{ad})$ – cycle time; τ_{ad} is arc duration; σ_{Tst} – standard deviation of cycle time; I_{arc} – mean value of arc burning current; I_{sc} - mean value of short circuit current.

We analyzed the obtained oscillograph charts for the current in the welding circuit and for the voltage between the electrode and the work piece (Figure 1, Table 2) when applying power supplies with different type of energy conversion with consideration to:

- parameters of electrode metal drop transfer from the electrode surface: the longest time of bridging is ensured by diode rectifier ($\tau_{К.З}$=61.3±9.4 msec) and the shortest – by inverter 2 ($\tau_{К.З}$=22.6±2.7 msec). Reduction of drop transfer time and increase of short circuits number when applying inverters proves that electrode metal is transferred by smaller electrode metal drops [6,7].

- maximal and minimal values of the welding current: the smallest value is ensured by inverter 3 (5 A) and the largest – by inverter 2 (148 A). High peak values increase the rate of the power supply response to the natural disturbing influences of the short circuits of the arc space, which improves the stability of the welding process.

With consideration to the facts mentioned above we can conclude that inverter power supplied ensure more stable process of arc burning as compared to a diode rectifier [8-11].

As for electrode metal spatter (third property according to State Standard 25616-83) we can confirm that it fully corresponds to the previously observed effect [12-14] and is connected to the welding process stability characteristics to a full extent. The more unstable the processes of arc burning and electrode metal transfer to the weld pool are the larger is spatter. To eliminate this negative effect inverter power supplies 1 and 3 use systems for limiting the explosion energy of the jumper between the electrode metal drop and the weld pool at the last moment of its life. Relation $I_{arc}/I_{K.z.} \rightarrow 1$, on this interval, is going to one which allows efficient reduction of electrode metal spattering.

4. Conclusion.

Objective assessment of arc stability, quality of electrode metal transfer and weld formation as basic indicators of welding working properties of power supplies ensuring high frequency energy conversion requires reconsidering the assessment criteria for these parameters according to State Standard 25616-83 as well as the assessment criteria for these parameters according to State Standard 25616-83 as they do not allow objective judgment about the advantages and disadvantages of the power supplies used. Further application of current standards makes it impossible to forecast the operational characteristics of the weld joints due to the unpredictability of the working welding properties of power supplies ensuring high frequency energy conversion and, as a result, it makes impossible obtaining the required properties of permanent joints in critical duty constructions [15,16].

Acknowledgments
The paper is completed under the agreement on scientific and technological cooperation between the National research Tomsk Polytechnic University and Institute of Strength Physics and Material Science SB RAS as well as the programs of fundamental scientific research of state academies of sciences for 2018-2019, project III.23.2.1.

References

[1] Exploratory studies to increase the reliability of critical metal structures operating under extreme loads and low climatic temperatures / Chapter 5 in the book: High technology in the projects of the RSF. Siberia / Ed. by S.G. Psahye and Yu.P. Sharkeev. – Tromsk: NTL publishers, 2017. – 428 p. ISBN 978-5-89503-607-5. P. 134-202.
[2] Efimenko L.A., Ramus A.A., Merkulova A.O. Features of the decomposition of austenite in the heat-affected zone during welding of high-strength steels // Physics of metals and metal science. – 2015, vol. 116. – p. 520 – 529.
[3] Loos A.V. Lukutin A.V., Saraev Yu.N. Power supplies for pulsed electrical processes. - Publishing Printing Company of TPU. – Tomsk. – 1998. – 159 p.
[4] Saraev, Y.N., Chinakhov, D.A., Ilyashchenko, D.I., Kiselev, A.S., Gordynets, A.S. Investigation of the stability of melting and electrode metal transfer in consumable electrode arc welding using power sources with different dynamic characteristics (2017) Welding International, 31 (10), pp. 784-790.
[5] Lankin, Yu.N. Stability indices of the process of arc welding with a consumable electrode / Yu.N. Lankin // Automatic welding. – 2011. – № 1. – P. 7–15.
[6] Mazel A.G. Technological properties of the electric arc. "Mashinostroyeniye", 1969, 178 p.
[7] Erokhin A.A. Basics of fusion welding. Physico - chemical laws. "Mashinostroyeniye". 1973, - 448p.
[8] Saraev Yu.N., Sleptsov O.I, Bezborodov V.P., Nikonova I.V., Tyutev A.V. Fatigue fracture analysis, structure and properties of welded joints used in Siberia and the Far North //

IOP Conf. Series: Materials Science and Engineering **681** (2019) 012037 doi:10.1088/1757-899X/681/1/012037

Installation and special works in construction. – 2005. - № 8. – p. 18 - 22.

[9] Saraev Yu.N. Experience in the development and practical application of adaptive pulsed-arc welding methods for the construction and repair of trunk pipelines // Problems of Materials Science. – 2015. – № 1 (81). – p. 210 - 218.

[10] Evaluation of the welding properties of equipment with inverter power supplies according to energy characteristics Gladkov E.A, Yushin A.A, Perkovsky R.A., Mymrikov S.A., Brodyagin V.N. Welding and diagnostics. 2011. № 1. P. 31-35.

[11] Development of criteria for an operational assessment of the welding properties of modern welding equipment Yushin A.A., Gladkov E.A. Welding and diagnostics. 2010. № 4. P. 52-54.

[12] Fedko V.T. Theory, technology and means of reducing spatter and labor intensity when welding in carbon dioxide. Tomsk: Tomsk State University, 1998. – 432 p.

[13] Fed'ko, V.T., Zernin, E.A. Main spattering reasons and protection methods of parts to be welded from welded metal spatters at CO2-welding (2005) *Svarochnoe Proizvodstvo*, (1), pp.37-3

[14] Il'yashchenko, D.P., Zernin, E.A. Determination of metal losses through splashing using different power sources (2011) *Welding International*, 25 (1), pp. 69-72.

[15] Saraev Yu.N., Bezborodov V.P., Grigoryeva A.A., Lebedev V.A., Maksimov S.Yu., Golikov N.I. Management of the structure and properties of welded joints of critical systems of engineering for adaptive pulsed-arc welding methods // Problems of Materials Science. – 2015. - № 1 (81). – p. 127 -132.

[16] Saraev Yu.N., Shpigunova O.I. A mathematical model of melting and transfer of electrode metal with systematic short circuits of the arc gap // Welding production. – 1992. - № 6. – p. 28 - 31.

IOP Conf. Series: Materials Science and Engineering **681** (2019) 012038 doi:10.1088/1757-899X/681/1/012038

Enhancing an arc welding technology by the methods of adaptive pulsed control of energetic parameters

Y N Saraev[1], A G Lunev[1], V M Semenchuk[1], A S Nepomnyashchiy[1], P A Strelnikov[2]

[1]Institute of Strength Physics and Materials Science SB RAS, Academicheskiy ave., 2/4, Tomsk, 634055, Russia.
[2]Tomsk State University of Control Systems and Radioelectronics Tomsk, 634050, Russia Lenin Ave, 40

E-mail: litsin@ispms.tsc.

Abstract. The work scrutinizes the main directions of improving conventional methods of arc, plasma and electroslag welding and cladding that are based on the algorithms of adaptive pulsed control of energetic parameters of the process. The deliverables demonstrate that the adaptive pulsed technologies feature a range of undeniable advantages against conventional methods that implement a strictly specified program for changing main energetic parameters of the process. The treatise justifies the main principles of adaptive technologies that ensure stability of heat and mass transfer and production of permanent joints of demanded quality under various disturbances affecting the system: power source, arc, weld pool, product.

1. Introduction

The emergence of pulsed welding and cladding technologies in world's industry has conceptually changed the perception of what conventional arc, plasma and electroslag technologies are capable of in formation of critical permanent joints [1]. The main strong points of pulsed technologies are controllable melting and transfer of each electrode metal droplet, controllable seam metal formation in different spatial positions, shrinkable structurally nonuniform regions, diminishable after-welding stresses and strains. However, the practice shows that fusion welding is accompanied by a plethora of disturbances that impose negative effects on the reliability of permanent joints in manufactured critical structures. This often reduces their performance, especially in extreme conditions and low climatic temperatures [2]. The disturbances are frequently caused by the characteristics of heat and mass transfer accompanying the formation and transfer of electrode metal into a weld pool [3]. Liquid metal formation itself on the edge of continuously fed electrode is stochastic, which finally causes instability that increases sputter during welding and general labor intensity of manufacturing, raises the number of faulty products and total cost of installation and welding [4].

The experience in elaboration of the theory of melting and formation of metal during consumable electrode welding shows that the most effective methods that reduce the impact of disturbance factors on the heat and mass transfer stability are the methods of adaptive pulsed control. As of today, these methods are widely incorporated in modern installation, welding and repair enterprises [5–6]. The developed method is based on strict dosage of energy for melting and transfer of each electrode metal droplet [7]. The whole technological complex (power source, electrode wire feeder, arc, weld pool, welded part) is considered as an integrated closed automatic control system [8]. The system during the whole welding cycle performs continuous control of heat and mass transfer stability, changing instantaneous values of the majority of technological parameters responsible for the operational indicators of welded connections. The specificities above require additional comprehensive theoretical and experimental studies covering the processes of heat and mass transfer of electrode metal during subprocesses of melting, transfer and crystallization of the seam metal from the melt. Such studies should deliver formulated principles for improving electric arc welding and cladding that ensure long life and high durability of critical engineering systems working in extreme operating conditions and low climatic temperatures [9].

Research goal: Find ways to improve conventional technologies of arc welding and cladding based on the methods of adaptive pulsed control of processing energetic parameters in installation and welding works, including those performed at low climatic temperatures.

2. Main operation principles of closed and open systems for automatic control of arc fuse welding.

The developed direction of works is performed with due account for the main statements of the automatic control theory developed for arc welding and cladding as the most widely spread industrial method for manufacturing of various metal structures and articles. There is a huge variety of automatic systems featuring different functions for controlling physical processes that can be implemented both directly in welding instruments and in technologies involving them. Such systems combine various designs of mechanical and electrical devices that integrate into a sophisticated complex of interacting constituents.

The examples of automatic systems of controlling arc welding processes are:

- thermal protection of welding power source;
- engagement and disengagement of the power source and electrode wire feeder in the beginning and in the end of the process;
- system for monitoring and aligning the welding head along the joint of a welded part;
- adjustment of the electrode wire feeding depending on the electrode stick-out both before and during welding;
- programmable displacement of the welding head along the welded joint;
- automatic arc length preservation regardless of changing position of weld pool and other parameters.

Almost all automatic control systems can be divided into two categories:

1) systems for single or multiple operations: switching welding equipment on and off, setting electrode wire feeding rate, setting parameters of a power source, etc.;
2) systems that during welding change (or retain) physical values, for instance, coordinates of a moving object, welding rate, arc voltage, stabilize supply voltage, dose energy necessary for melting and transfer of each droplet of electrode metal, control thermal cycles and control seam metal formation in different spatial positions, etc.;

The second type of automatic control systems is the most sophisticated, though more versatile. Such systems can be closed or open. The first principle is used almost in all conventional supply systems (Fig. 1).

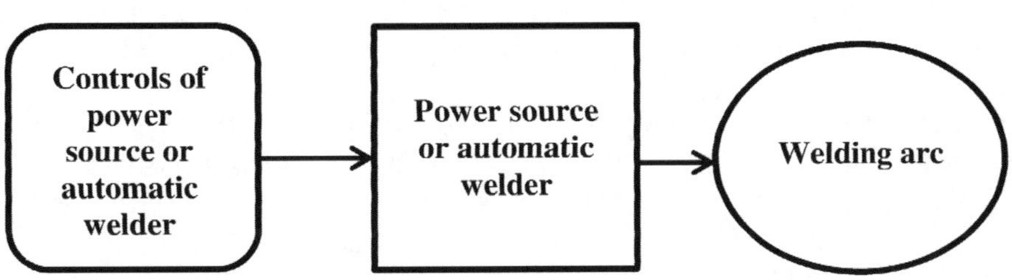

Figure 1. Open automatic control system

IOP Conf. Series: Materials Science and Engineering **681** (2019) 012038 doi:10.1088/1757-899X/681/1/012038

The scheme in Fig. 1 is an elementary automatic control system when the power source is controlled by a man. A characteristic feature of open automatic control systems is that its operation has no effect on the result of object control. Therefore, a natural improvement of an automatic system is closing its output with the input of a controllable power source via a feedback unit, which allows adjusting its operation during welding. Besides, the inputs of the controllable power source are connected with the channels that conduct signals proportional to the disturbances that tend to destabilize the system. The structural scheme of an open automatic control system is depicted in Fig. 2.

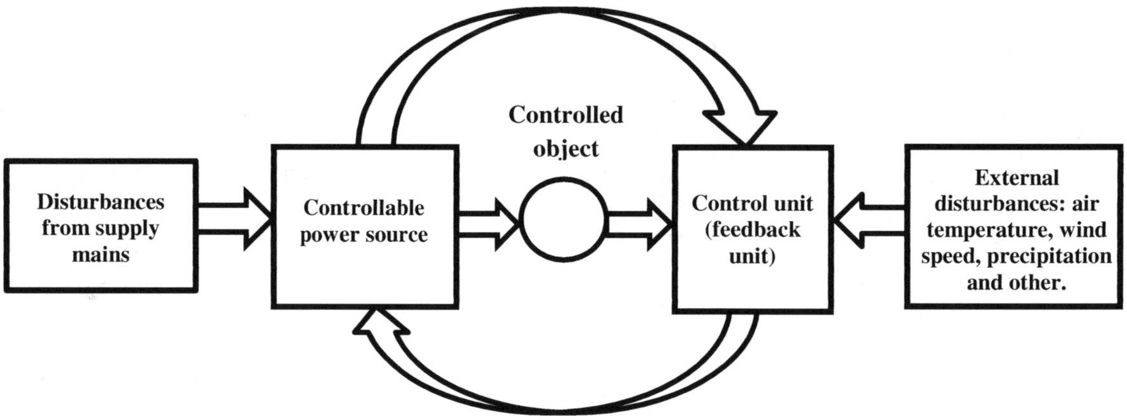

Figure 2. Closed automatic control system

Evidently, a closed automatic control system features a complete interdependence of all constituents. Such design of closed systems is the principle discrepancy from open ones and makes them irreplaceable for creating welding instruments.

3. Directions and methods for designing closed automatic control systems.

The practical implementation of such approach requires solving a number of elaborated technological and electromechanical problems [10].

First group of problems: droplet formation control, transfer of electrode metal, crystallization of weld pool, controllable main welding parameters (arc voltage, welding current, instantaneous arc power) by welding equipment [11].

The second group of problems (electromechanical) requires introducing special units (choppers) into the welding equipment that are reliable and capable of switching maximum pulse power, having low inertia and adequate dimensions and weight [12]. Such requirements are hard to meet because they involve switching of huge pulse power (more than 50 kW) over microsecond durations.

Indeed, this approach was put into practice in 1970s-1980s in the Soviet Union as methods of manual shielded metal-arc welding, shielded machine welding in CO_2, Ar and their gas mixtures featuring controllable and uncontrollable transfer of electrode metal by solid and powder wires [13–17]. However, despite the efficiency of existing technologies and instruments, they are not suitable for all the cases of increasing the life, durability and reliability of critical systems working under low climatic temperatures.

From our perspective, this goal can be achieved by solving three interconnected problems:

- Development and practical implementation of new-generation welding and cladding materials with submicrocrystalline and nanosized components modifying the zone of permanent joint [18].
- Development and practical implementation of novel methods for diagnosing rapid heat and mass transfer processes accompanying melting, transfer and crystallization of seam metal from melt that have dominating impact on the stress-strain state of critical structures.

- Development of practical implementation of new-generation power sources with high dynamic properties and algorithms of adaptive pulsed control [19].

The first problem of developing and implementing new-generation welding and cladding materials is to be solved by using submicrocrystalline and nanosized powders as modifiers.

In late XX-early XXI centuries, the USA, Canada, European Union, Russia, China and CIS countries have started working on this conceptual problem by applying alloying chemical materials with hard alloys, solid solutions, including those of aluminium, silicon, zirconium and powders of refractory compositions of chrome, titanium, nickel and others. Under pulsed highly concentrated energy fluxes with low pulse duration, they can activate structure, phase and thermochemical transformations, which provide increased physicomechanical and operational properties of coatings [20].

The second problem of developing and implementing novel methods for diagnostics of high-rate processes of heat and mass transfer accompanying melting, transfer and crystallization of seam metal from melt completely depends on the new approaches to diagnosing the micro-metallurgical processes during melting, electrode metal transfer and crystallization of weld pool during formation of a permanent joint.

The most complicated problem is development and practical implementation of new-generation power sources. In principle, it requires creating new sources of pulsed energy for pulsed-arc, pulsed-plasma, pulsed laser, high-rate gas-flame, electro-contact and detonation instruments that would be able to create directed highly concentrated energy fluxes with low pulse duration and using systems of adaptive control to stabilize operational characteristics [21].

Fig. 3 presents a scheme of adaptive automatic control system for shielded welding.

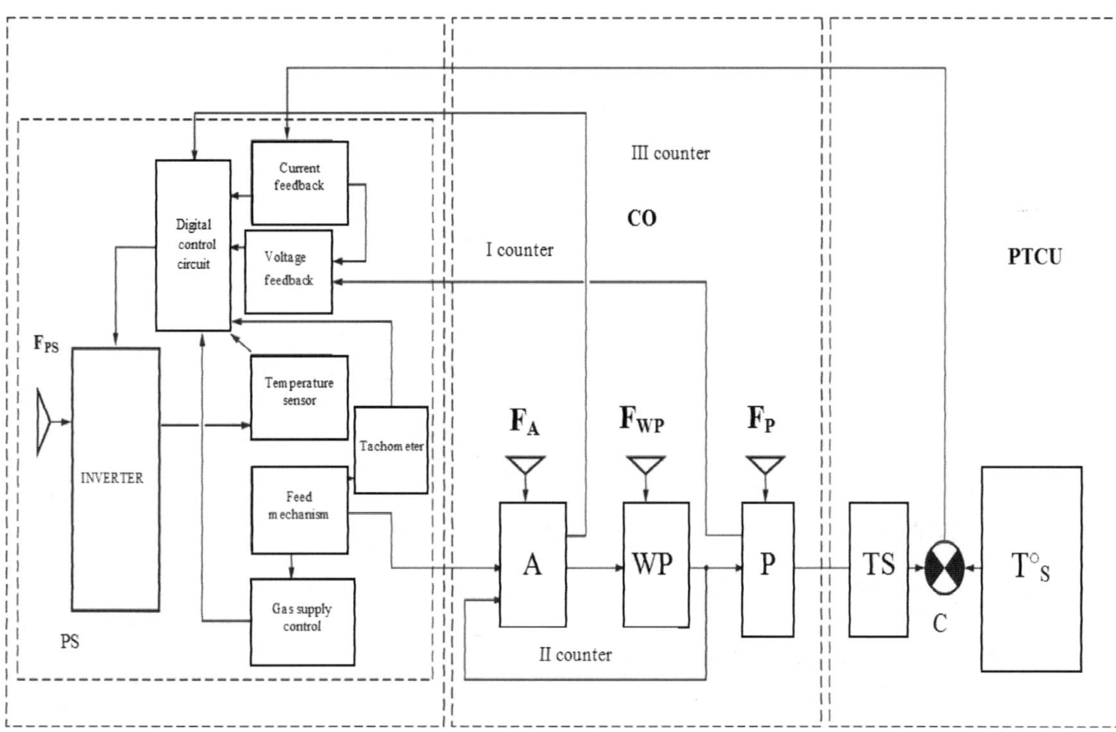

Figure 3. Scheme of adaptive automatic control system for welding process: PS – power supply, A – arc, P – product, WP – weld pool, CO – controlled object, PTCU– product temperature control unit, TS – temperature sensor, T°S – product temperature setting unit, C – comparator, F_{PS}, F_A, F_{WP}, F_P – disturbances.

The technical essence of the power source scheme is in the presence of different intercorrelated units in its structure. The units control the state of the controlled object and play the role of actuating elements in the supply system retaining a preset wire feeding rate, adjusting the operating temperature of the power circuit elements by automatic heating, and changing surrounding air temperature. The most important characteristic of the created power supply systems is the control reaction speed, because modern technological processes frequently require the reaction speed of power supply system to the correction of control algorithm from 0.1 to 0.3 milliseconds.

An important element of the automatic control system is the presence of feedback channels that continuously control the disturbances affecting the object of automatic control. They are processed using the changes in instantaneous values of the main energetic parameters of the process. One of the main elements of presented system in the controllable power source (WPS), which power circuit scheme is presented in Fig. 4.

Based on the experience of creating adaptive pulsed supply systems, we may note that their design now allows eliminating the necessity to control the output voltage of the power source. The necessary arc voltage can be retained during changing electrode feeding rate by using the current pulse duration (arc duration).

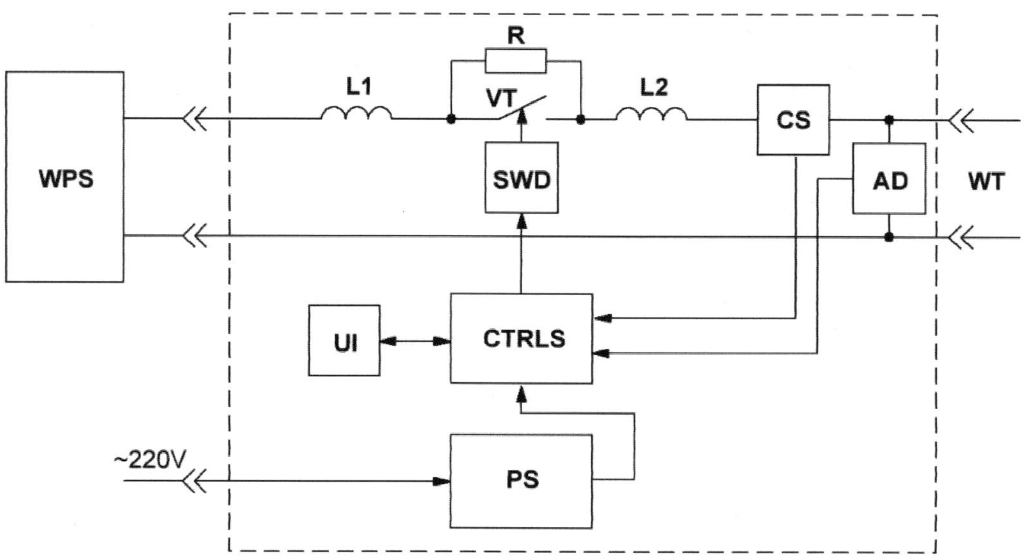

Figure 4. Scheme of the power source power circuit: WPS – Welding Power Source, L1 – Input Series Inductor, L2 – Welding cables self-inductance, R – Current limiting resistor, VT – Transistor Switch, SWD – Switch Driver, CS – Current Sense, AD – Arc Detector, CTRLS – Control System, PS – input terminal for 220 VAC power supply, UI – User Interface, WT – Welding Torch.

Such design allows eliminating the need in conventional smoothing coils on the stage of short-circuit. This allows for 30-40% reduction of lasting short-circuits, which appreciably increases the efficacy of the process and makes it less sensitive to the disturbances at the stage of electrode metal transfer into the weld pool.

4. Conclusion.

1. The method of adaptive pulsed technologies (APT), ensures stable energy parameters of the processes during welding in different spatial positions, capability of programmable introduction of heat into the welding zone, and controllable melting and transfer of each droplet of electrode metal.

Such factors ensure the most favorable conditions for forming highly dispersed structure of seam metal, reducing the heat-affected zone and residual deformations of the welded joints.

2. Considering the efficacy of APT methods in producing critical systems under low climatic temperatures, their application will facilitate the growth of the performance of critical structures. This opens new horizons for achieving the main goals of fundamental investigations: development of mechanical engineering, power engineering, chemical, extraction and processing industries in the Far North and Arctic.

3. Further fundamental and targeted studies in this direction are high on the scientific and engineering agenda. Their success will affect future development of the world's industry.

Acknowledgments
The work was carried out within the program of fundamental scientific research of state academies of sciences for 2013–2020, project No. III.23.2.1.

References

[1] Saraev Yu.N. Pulse technological processes of welding and cladding (in Russian) / Saraev Yu.N. / Nauka Press. Executive Editor A.S. Zubchenko. Novosibirsk, 1994. – 107 p.

[2] Larionov V.P., Sleptsov O.I., Saraev Yu.N., Bezborodov V.P. New approaches to the development of modern welding and coating technologies to ensure the reliability of metal structures and products operated in Siberia and the Far North (in Russian) // Bulletin of the Academy of Military Sciences. – 2008. – No. 3. – P. 67-69.

[3] Saraev Yu.N. Investigation of the properties of welded joints of manganese steel obtained by low-frequency pulsed-arc welding (in Russian) / Yu.N. Saraev, V.P. Bezborodov, S.V. Gladkovsky, N.A. Golikov // Deformation and destruction. – 2016. - No. 1. – p. 36–41.

[4] Saraev Yu.N. et al. Exploratory studies to improve the reliability of critical metal structures operating under extreme loads and low climatic temperatures (in Russian) / Chapter 5 in book: High-tech in RSF projects. Siberia. / Edited by S.G. Psakhie and Yu.P. Sharkeev. – Tomsk: NTL Publishing House, 2017 – 428 p. ISBN 978-5-89503-607-5. - p. 134-202.

[5] Saraev Yu.N. Adaptive pulse-arc methods of mechanized welding in the construction of trunk pipelines (in Russian) / Saraev Yu.N. // Welding production. 2002. No. 1. P. 4-7.

[6] Saraev Yu.N. Control of electrode metal transfer during CO_2-shielded welding with short circuits of the arc gap (Review) (in Russian) / Automatic welding. 1988. No. 12. P. 16-23 .

[7] Saraev Yu.N. Justification of the concept of improving the safety and survivability of technical systems operating in the regions of Siberia and the Far North, based on the use of adaptive pulsed welding technologies (in Russian) // Heavy Machinery Engineering. 2010. No. 8. P. 14-19.

[8] Paton B.E. Improving the technological processes of welding and cladding based on the methods of controlled high-energy impact on the characteristics of the melting and transfer of electrode metal (in Russian) / Paton B.E., Saraev Yu.N., Lebedev V.A. / Proceedings of the International scientific-practical conference with elements of a scientific school for young scientists "Innovative technologies and economics in mechanical engineering." May 20-21, 2010 Yurga. – p. 15 – 22.

[9] Saraev Yu.N. Improving the reliability of metal structures during operation in low climatic temperatures through the integrated application of modern methods of modifying the weld zone (in Russian) / Yu.N. Saraev, V.P. Bezborodov, S.V. Gladkovsky, N.I. Golikov // Welding production. – 2016. - No. 9. – p. 3 – 9.

[10] Saraev Yu.N. Ensuring the operational reliability of critical large-sized metal structures at the stage of their manufacture and repair using adaptive pulsed welding technologies (in Russian) / Saraev Yu.N., Bezborodov V.P., Demyanchenko A.A. // Metal processing. – 2011. - No. 2. – p. 18 – 22.

[11] Saraev, Yury The development and application of adaptive pulse-arc welding methods for construction and repair of pipelines. IEEE: International Conference on Mechanical Engineering, Automation and Control Systems (MEACS). Tomsk, RUSSIA.: Oct. 16-18, 2014. Published 2015.

[12] Saraev Yu N Improving Processes of Mechanized Pulsed Arc Welding of Low-Frequency Range Variation of Mode Parameters / Saraev Yu N, Solodskiy S A, Ulyanova O V // YIT-UPMME 2015 IOP Publishing. IOP Conf. Series: Materials Science and Engineering 127 (2016) 012019 doi:10.1088/1757-899X/127/1/012019.

[13] Saraev Yu.N. Influence of energy parameters of the welding process on the structure and properties of welded joints of low alloy steels (in Russian) / Saraev Yu.N., Bezborodov V.P. // Welding production. 2012. No. 8. P. 3-5.

[14] Saraev Yu.N. Experience in the development and practical application of adaptive pulsed-arc welding methods for the construction and repair of pipelines (in Russian) // Problems of Materials Science. – 2015. – No. 1. – P. 210-219.

[15] Saraev Yu.N., Makarova L.I., Ekimov V.S. et al. Ways to increase the efficiency of construction, operation and repair of oil and gas field equipment and trunk pipelines based on adaptive pulsed welding and cladding technologies (in Russian) // Welding production. 2001. No. 5. P. 31-37.

[16] Saraev Yu.N. Features of the formation of protective corrosion-resistant coatings during pulse-arc cladding of austenitic-grade steels (in Russian) / Saraev Yu.N., Bezborodov V.P., Selivanov Yu.V. // Welding production. 2009. No. 4. P. 20-25.

[17] Saraev Yu.N. Improving the structure and properties of welded joints of large diameter pipes of low alloy steel during pulse-arc welding (in Russian) // Saraev Yu.N., Bezborodov V.P., Poletika I.M., Tyutev A.V., Nikonova I.V., Kirilova N.V., Sevastyanov S.P. // Automatic welding. 2004. No. 12. P. 34-38.

[18] Saraev YN The Ways of Reliability Enhancement of Welded Metal Structures for Critical Applications in the Conditions of Low Climatic Temperatures / YN Saraev, VP Bezborodov, SV Gladkovsky, and NI Golikov // Advanced Materials with Hierarchical Structure for New Technologies and Reliable Structures 2016. AIP Conf. Proc. 1783, 020195-1–020195-4; doi: 10.1063/1.4966489. Published by AIP Publishing. 978-0-7354-1445-7. 020195-1.

[19] Saraev Yu.N. Improvement of welding and surfacing technological processes on the basis of adaptive high-energy pulse methods on the characteristics of electrode metal transfer and crystallization of the weld pool (in Russian) / Saraev Yu.N., Bezborodov VP, Grigoryeva A.A. // Welding and diagnostics. – 2013. - No. 5. – p. 40 – 47.

[20] Saraev, Y. N. Effect of Heat Input Pulse on the Structure and Properties of Welded Joints of Steels Ferritic-Pearlitic Class, Operating Under Low-Frequency Temperature-Force Loading / Saraev, Y. N.; Bezborodov, V. P.; Putilova, E. A. / IOP: 6th International Scientific Practical Conference on Innovative Technologies and Economics in Engineering. Natl Res Tomsk Polytechnic Univ., Yurga Inst. Technol., Yurga, RUSSIA: MAY 21-23, 2015.

[21] Saraev Yu.N. The influence of the energy parameters of welding conditions on the distribution of residual stresses in the zone of permanent connection (in Russian) / Saraev Yu.N., Golikov N.I., Sidorov M.M., Lunev A.G., Perovskaya M.V. // Urgent problems in mechanical engineering. Vol. 5. No. 3-4. 2018. – p. 30-35.

IOP Conf. Series: Materials Science and Engineering **681** (2019) 012039 doi:10.1088/1757-899X/681/1/012039

Effect of ultrasonic impact treatment on redistribution of residual stresses in dissimilar thickness girth welds

M M Sidorov[1] and N I Golikov[1]

[1] Larionov Institute of Physical and Technological Problems of the North, the Siberian Branch of the Russian Academy of Science, Federal Research Center 'Yakut Scientific Center of the Siberian Branch of the Russian Academy of Sciences', Yakutsk, Russia, 677980

E-mail: sidorovmm@bk.ru

Abstract. Effect of repeated ultrasonic impact treatment on redistribution of residual stresses in a dissimilar thickness girth butt weld of pipes with a diameter of 530 mm of low alloyed (0.9%C-2%Mn-1%Si) steel was studied. Residual stresses were measured by the X-ray method based on measurement of microdeformation of a crystal lattice of a material under its influence. The research results showed that tensile residual stresses in the girth weld were formed at the inner side of the pipe after welding. Ultrasonic impact treatment of the weld at the inner side caused complete transformation of residual stresses from tensile to compressive ones when speed was in the range from 0.06 to 0.12 m/min. However, reduce in speed of treatment down to 0.04 m/min gave a negative effect. The values of compressive residual stresses in heat affected zone on the side of the thinner pipe were decreased.

1. Introduction

As a rule, tensile residual stresses are present in welded joints and adversely affect their fatigue strength [1, 2]. This negative factor is not always possible to minimize using technological processes before or during welding. Therefore, post-weld treatment of welds is required in many cases. Sometimes, post-weld treatment is necessary to reduce tensile residual stresses. Various methods and devices are used for this purpose [3–6]. Ultrasonic impact treatment (UIT) is one of the most effective and widely spread among post-weld treatment methods.

A principle of the UIT operation is based on impact of low-frequency ultrasonic vibrations of a vibro-impact tool on the surface of a welded joint. As a result, redistribution of hazardous tensile residual stresses to compressive ones occurs on the treated surface [7–9]. It is known that compressive residual stresses improve fatigue strength of welded joints.

Investigation of redistribution of residual stresses in welded joints before and after treatment is a rather laborious process. It is not always possible to predict how residual stress fields will be formed after treatment. The result depends on joint dimensions, residual stresses, energy input, duration and a method of treatment. Therefore, research in this direction is an urgent task. This may help correctly apply promising processing technologies in practice and extend lifecycle of critical welded structures.

The purpose of this paper is to study the effect of repeated UIT on redistribution of residual stresses in dissimilar thickness girth welds.

2. Materials and methods

A girth butt weld of pipes with a diameter of 530 mm was investigated. The wall thickness of the pipe No.1 was 10 mm, the pipe No.2 was 8 mm (Figure 1). Both pipes were of low-alloyed (0.9%C-2%Mn-1%Si) steel. The pipes were joined by shielded metal arc welding. The number of passes was three (root, intermediate and cover). Mean values of mechanical properties of the base metal and the weld metal are given in Table 1. Chemical composition the pipe metals based on the results of spectral analysis as well as Russian State Standard GOST 19281-2014 are presented in Table 2. The results of mechanical tests and spectral analysis were obtained using facilities of the Center for Collective Use 'Station of low-temperature tests' (Larionov Institute of Physical and Technological Problems of the North, the Siberian Branch of the Russian Academy of Sciences).

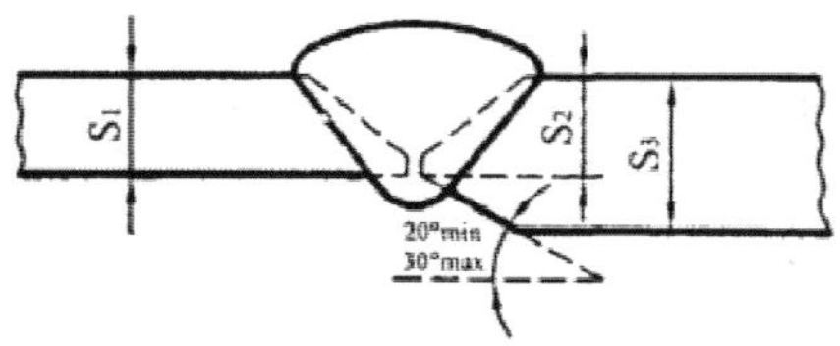

Figure 1. Scheme of the edge shapes and the dissimilar thickness girth weld.

Table 1. Mechanical properties of the base metal and the weld metal

Material	Mean values				
	Yield strength σ_T, MPa	Ultimate tensile strength σ_B, MPa	Percent elongation $\delta_{0,5}$, %	Impact strength *a*, J/cm^2	
				KCV^{-20}	KCV^{-60}
Base metal of the pipe No.1 (wall thickness of 10 mm)	560	668	27	109	41
Base metal of the pipe No.2 (wall thickness of 8 mm)	463	585	32	152	137
Weld metal	486	609	24	89	35
Fusion zone		-		55	22

Table 2. Chemical composition the pipe metals based on the results of spectral analysis as well as Russian State Standard GOST 19281-2014

Material	Chemical composition, %										
	C	Si	Mn	Cr	Ni	Al	Mo	P	S	Cu	Fe
Base metal of the pipe No.1	0.10	0.57	1.31	0.12	0.12	0.03	0.02	-	-	0.20	Bal.
Base metal of the pipe No.2	0.12	0.50	1.41	0.03	0.03	0.03	0.02	-	-	0.03	Bal.
Steel 0.9%C-2%Mn-1%Si according to Russian State Standard GOST 19281-2014	≤0.12	0.5-0.8	1.30-1.70	≤0.30	≤0.30	-	-	≤0.03	≤0.04	≤0.30	Bal.

UIT was done using a special technological setup. It consisted of an UZGT 0.5/27 ultrasonic generator and a 'Bumblebee' unit at the inner side of the pipe wall. The method was described in detail in [10]. Each surface area was treated three times at different speeds and, as a result, with different performance (Table 3). Residual stresses at the outer and inner sides of the girth weld were measured after each treatment.

Table 3. Parameters of the UIT process

Treated surface area	Treatment time and speed		
	UIT-1	UIT-2	UIT-3
Section (60 mm long and 40 mm wide) symmetrically to the weld axis	30 s	30 s	30 s
	0.12 m/min	0.06 m/min	0.04 m/min

Residual stress fields in the girth weld were determined by X-ray method [11]. Stresses were measured in the axial (σ_z) and circumferential (σ_θ) direction relatively to the pipe axis at points located at different distances from the weld center (the maximum distance was 20 mm). The measurement scheme is shown in Figure 2. Before measurement, the tested surface was cleaned, manual grinded, and etched to a depth of 200 μm with a mixture of concentrated nitric and hydrochloric acid in a ratio of 1:3 to remove the deformed layer.

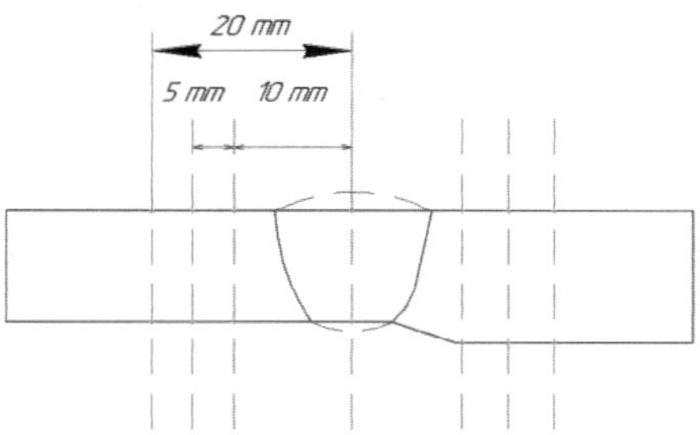

Figure 2. Scheme of measurement of residual stresses at the outer and inner sides of the weld.

3. Results and discussion

The results of measurement of residual stresses at the outer and inner sides of the weld before treatment are shown in Figures 3 and 4. Figure 3 shows compressive residual stresses at the outer side of the girth weld in the circumferential and axial directions which were formed during welding. Compressive residual stresses in heat affected zone (HAZ) on the side of the pipe No.2 were higher than on the side of the pipe No.1.

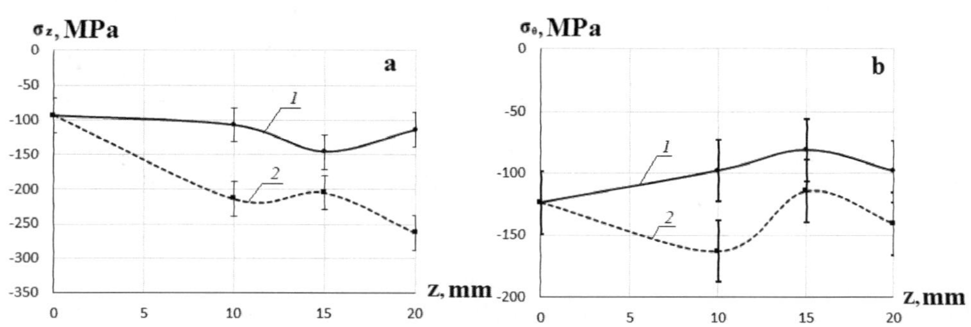

Figure 3. Distribution of axial (a) and circumferential (b) stresses in the girth weld at the outer side: *Z* is distance from the weld axis, *1* is the pipe with a wall thickness of 10 mm; *2* is the pipe with a wall thickness of 8 mm.

Tensile residual stresses were observed at the inner side of the weld. Their values in HAZ on the side of the pipe No.2 were also higher than on the side of the pipe No.1 (Figure 4). The obtained results show that wall thickness of the pipes played an important role in formation of axial and circumferential stresses in the weld. The highest compressive and tensile residual stresses were formed in the pipe with a thinner wall as it had been more heated and deformed during welding.

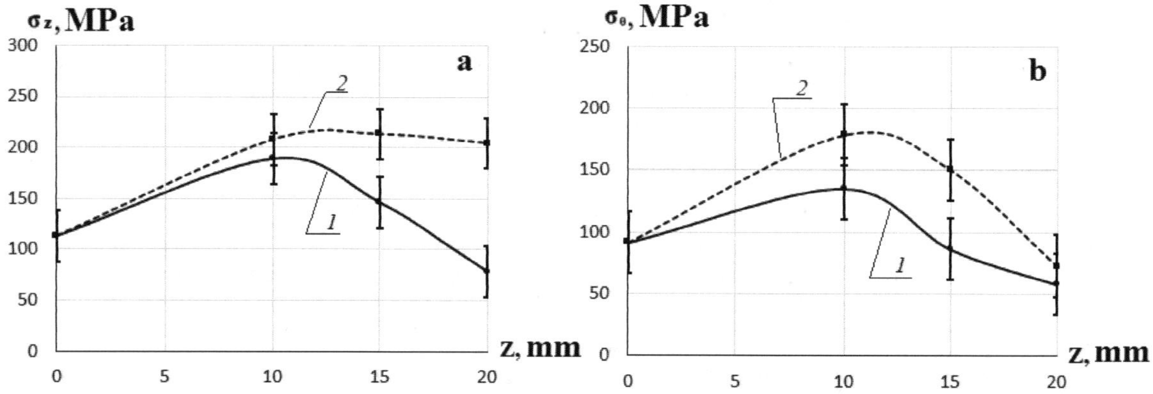

Figure 4. Distribution of axial (a) and circumferential (b) stresses in the girth weld at the inner side: Z is distance from the weld axis, *1* is the pipe with a wall thickness of 10 mm; *2* is the pipe with a wall thickness of 8 mm.

Figure 5 shows redistribution of residual stresses at the inner side of the weld after the first UIT. Axial and circumferential stresses partially transformed into compressive ones. Tensile residual stresses remained at the level of 20-50 MPa on the side of the pipe No.1 at a distance of 20 mm from the weld axis.

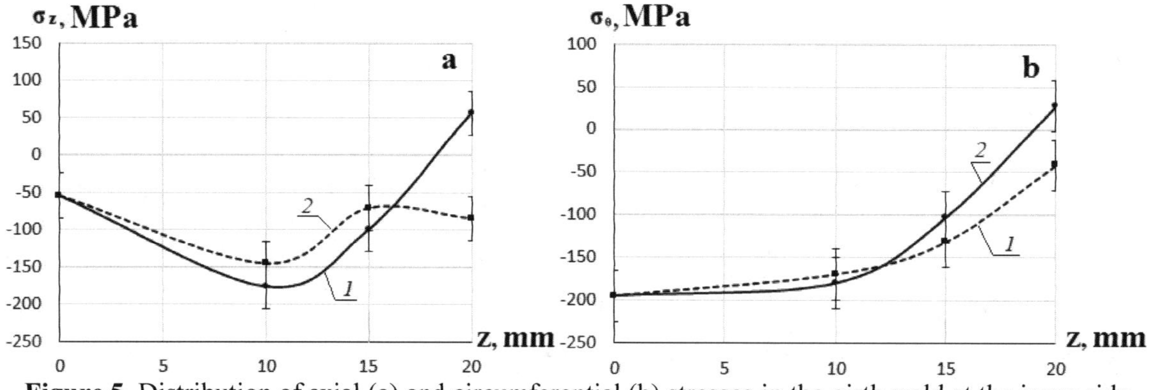

Figure 5. Distribution of axial (a) and circumferential (b) stresses in the girth weld at the inner side after UIT-1: Z is distance from the weld axis, *1* is the pipe with a wall thickness of 10 mm; *2* is the pipe with a wall thickness of 8 mm.

Complete transformation of tensile residual stresses to compressive ones, as well as an increase in the level of compressive residual stresses obtained after the first treatment, occurred after the second UIT at a speed of 0.06 m/min (Figure 6).

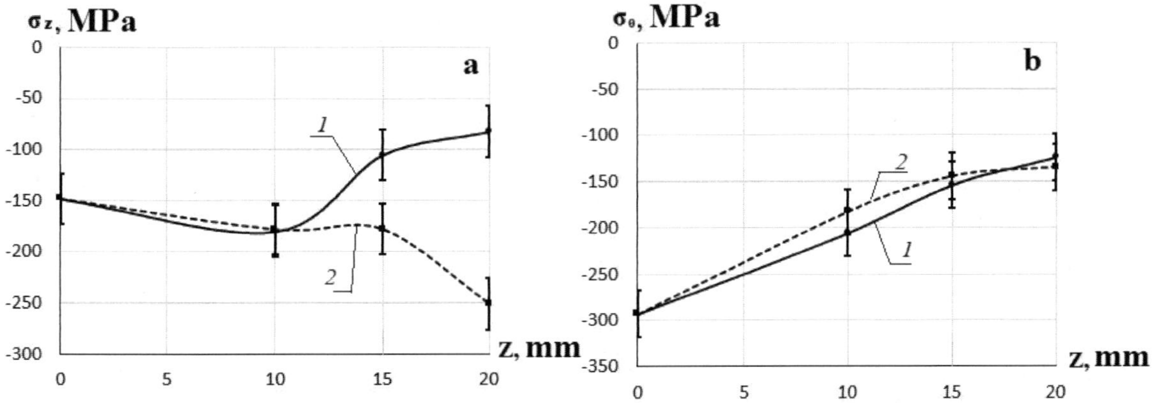

Figure 6. Distribution of axial (a) and circumferential (b) stresses in the girth weld at the inner side after UIT-2: *Z* is distance from the weld axis, *1* is the pipe with a wall thickness of 10 mm; *2* is the pipe with a wall thickness of 8 mm.

Redistribution of residual stresses at the inner side of the weld after the third UIT is shown in Figure 7. The values of compressive residual stresses increased over 300 MPa in the weld and HAZ on the side of pipe No.1. This caused a decrease in compressive residual stresses in HAZ on the side of the pipe No.2. Based on these results, it can be concluded that the UIT of the dissimilar thickness girth weld with a speed of 0.04 m/min gave a negative effect. Therefore, wall thickness of the pipes and distribution of residual welding stresses should be taking into account when treating this type of welds. Also, the results of the study showed that the level of residual stresses at the outer side of the weld were almost unchanged after UIT at the inner side.

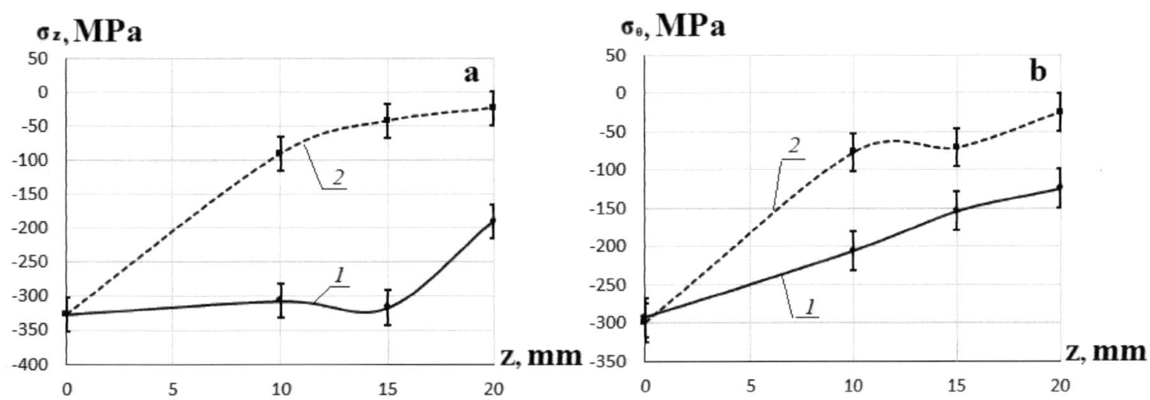

Figure 7. Distribution of axial (a) and circumferential (b) stresses in the girth weld at the inner side after UIT-3: *Z* is distance from the weld axis, *1* is the pipe with a wall thickness of 10 mm; *2* is the pipe with a wall thickness of 8 mm.

4. Conclusions

The following conclusions can be drawn from the obtained results:

1. Compressive residual welding stresses were found in the dissimilar thickness girth weld of low alloyed (0.9%C-2%Mn-1%Si) steel at the outer side but tensile stresses were on the inside. The values of tensile residual stresses were higher by 40-50% in the thinner wall pipe than in the thicker wall one.
2. Ultrasonic impact treatment of the girth weld at the inner side of the pipe caused transformation of residual welding stresses from tensile to compressive ones. As a result of such treatment, the maximum compressive residual stresses reached values of -200…-300 MPa.

IOP Conf. Series: Materials Science and Engineering **681** (2019) 012039 doi:10.1088/1757-899X/681/1/012039

3. Reduce in speed of ultrasonic impact treatment of the dissimilar thickness girth weld to 0.04 m /min gave a negative effect. The values of compressive residual stresses in heat affected zone on the side of the thinner pipe were reduced.

References

[1] Hensela J, Nitschke-Pagela T, Tchoffo Ngoula D, Beier H-Th, Tchuindjang D and Zerbstd U 2008 *Eng. Fract. Mech.* **198** 123-41

[2] Shena F, Zhaoa B, Lib L, Chuaa Ch K and Zhou K 2017 *Int. J. of Fat.* **103** 272-9

[3] Du Ch., Wang X and Luo Ch 2019 *J. of Mat. Proc. Tech.* **263** 138-50

[4] Wei X L and Ling X 2015 *Procedia Eng.* **130** 948-55

[5] Khajeiana A, Mahmoudi AH and Mehmanparast A 2019 *Marine Struct.* **66** 106-20

[6] Chaobo G, Zhijiang W, Dongpo W and Shengsun H 2015 *Appl. Surf. Sc.* 347 596-601

[7] Ganiev I, Gafurov I and Vagapov I 2016 *Appl. Acoust.* 103 195-201

[8] Statnikov E S, Korolkov O V and Vityazev V N 2006 Ultras. 44 e533-8

[9] Golikov NI and Sidorov MM 2014 *Weld. Intern.* **28** **12** 970-2.

[10] Patent RUS 2444423 2010 [Sidorov M M, Golikov N I and Ammosov A P] *Method of relieving residual weld stress in pipe wielded butt joints*

[11] Golikov N I and Ammosov A P *Strength of welded joints of tanks and pipelines operating in the conditions of the North* (Yakutsk: North-Eastern Federal University press) p 323

Experiments on the effect of thermoelectric current magnetic field on welding of 60-mm thick dissimilar steels

A P Sliva[1], A L Goncharov[1], E V Terentyev[1], I A Kharitonov[1], V K Dragunov[1]

National Research University Moscow Power Engineering Institute,
ul. Krasnokazarmennaya, 14, Moscow, 111250

E-mail: SlivaAP@mpei.ru

Abstract. The work describes a method for measuring the magnetic field induced by thermoelectric currents in control points above welded specimens and presents the measurement results. The dissimilar 60-mm thick steel samples are welded by electron-beam welding (EBW). The grades of the pairs of steels are 09Mn2Si + 12Cr18Ni10Ti, 09Mn2Si + 20Cr13 and 20Cr13 + 12Cr18Ni10Ti. The study shows that the electron beam in the weld root deflects towards the material with larger positive potential of integral thermal EMF. The deflection is uneven along the specimens, which necessitates taking into account the geometry of welding joints when elaborating the mathematical models of the process. Maximum deflection for the studied pairs of materials correspondingly amounts to 0.6, 2,8 and 3.8 mm, which is critical for producing a high-quality seam. In addition, the alteration of magnetic induction in control points during EBW was obtained.

1. Introduction

The electron-beam welding of dissimilar steels leads to the generation of thermoelectric currents inducing magnetic field in the processing zone due to different thermal EMF of the materials. The interaction of the beam electrons with the magnetic field leads to the deflection of the beam electrons from a set trajectory, occurrence of weld-seam defects (such as lack of penetration [1–7]) and alters the material penetration degree. This changes the chemical composition, structure and properties of the seam metal. The deflection from the weld joint can reach up to several millimeters and increases with both the thickness of welded parts and difference in absolute coefficient of thermal EMF of the materials. An appreciable impact is made by relative magnetic permeability and electric resistance of the materials [2, 3, 7]. According to the experimental background, a deflection of the weld root from the joint of 0.3–0.5 mm can lead to lack of penetration, while that of 0.1–0.3 mm can change the fraction of each of the welded metals in common molten pool by dozens of percent.

To compensate the deflection of the electron beam, several methods can be considered: preliminary inclination and displacement of the electron beam [2, 8], passing compensating electric current through the welding joint [9], induction of local magnetic fields [10].

The choice of electron beam correction parameters is often based on numerical models connecting the regimes and geometry of welded parts with a deflection of the electron beam. The development of precise numerical models is impeded, because the EBW of dissimilar materials causes interdependent factors to occur. Thermal field from the penetration channel conditions the generation of thermoelectric currents inducing the magnetic field also depending on the magnetic properties of the materials. The magnetic field affects the electron beam by deflecting it, changing the penetration

IOP Conf. Series: Materials Science and Engineering **681** (2019) 012040 doi:10.1088/1757-899X/681/1/012040

channel, altering the thermal field, thermoelectric current spreading and magnetic field distribution [1, 2].

Current study is aimed at developing the methods of experimental verification of a numerical model of electron beam deflection during EBW of dissimilar materials and studying the effects of the magnetic field induced by thermoelectric currents on the formation of welding seams of 60-mm thick dissimilar steels.

2. Experimental methods

For increased efficacy of the verification of numerical models, the most promising and feasible technique is the verification of the electron beam trajectory by both macrosections of the welding seams and magnetic field induction. The control points for magnetic field measurement should not introduce appreciable disturbance into the magnetic field induced by EBW. Therefore, they should reside away from the welded specimens. Fig. 1 presents the suggested scheme of magnetic field measurement above the welded specimens.

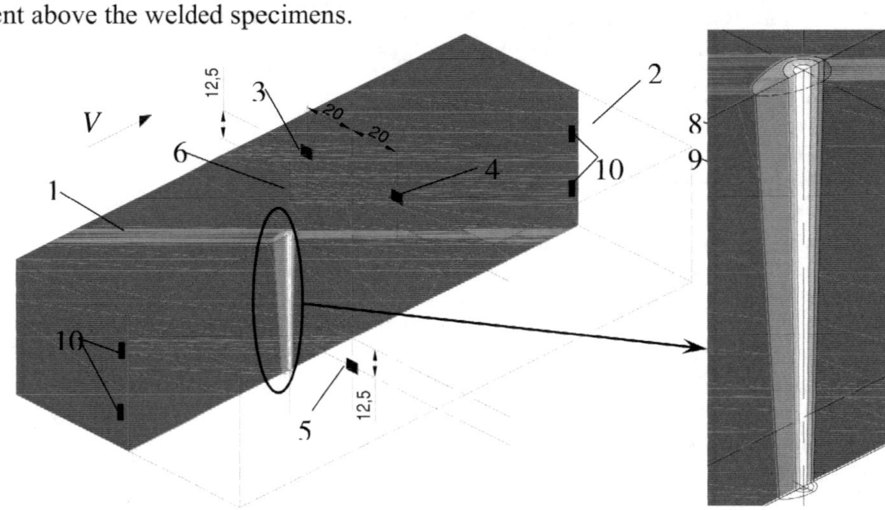

Figure 1. Scheme of magnetic field induction measurement during EBW of dissimilar materials: 1) and 2) welded specimens; 3) and 4) magnetic field induction sensors above the specimens; 5) magnetic field induction sensor in the weld root; 6) electron beam axis; 7) electron beam in the penetration channel; 8) gas-vapor channel; 9) molten pool; 10) tack welds.

Three pairs of structural steel specimens were studied: 12Cr18Ni10Ti + 20Cr13, 12Cr18Ni10Ti + 09Mn2Si and 09Mn2Si + 20Cr13. The dimensions were 60×50×200 mm, the former being the thickness. The specimens were welded on an AELTK-344-12 automatic EBW setup with an accelerating voltage of 60 kW. In the first place, the following EBW parameters were defined for burn-through welding: electron beam current of 250 mA, welding rate 15 m/h.

To measure the magnetic field induction during EBW, an experimental setup was elaborated based on the ACTest Pro automatic measuring complex (Laboratory of Measuring Systems (AC), Russia). The package includes a working station with sockets for electrically isolated LTR-modules, a ADC LTR 11 universal module for multi-channel data collection and a signal synchronization unit with DSCA40-03 DataForth signal normalizers with electrical isolation. The complex allows recording measured values, processing results and compiling measurement reports.

To measure the magnetic field parameters, the following sensors can be used: Hall sensors, ferroprobes, magnetoresistive sensors and magnetic-induction sensors. The anticipated values of magnetic field induction in the control points during EBW of dissimilar materials stay within a range of 0.1–10 mT. The magnetic field is almost unchanged in time. Therefore, the testbed was equipped with Hall sensors, because they cover the necessary measurement range, can be used for constant magnetic fields and have wide operating temperature range. The testbed was integrated with the automatic measurement complex using an AD22151 magnetic field sensor (Analog Devices). The sensor outputs a voltage proportional to the magnetic field applied in perpendicular to the top surface of its body. The sensors have high magnetic sensitivity and thermal stability of the main parameters in a wide temperature range. It can be adjusted for the necessary measurement range.

Sensors 1 and 2 (Fig. 1) should be positioned as closely as possible to the weld joint, though stay out of the action of the electron beam and liquid metal of the molten pool during EBW. Thus, the distance from the joint was 15–20 mm at the surface of the welded specimens. According to the preliminary estimates, the magnetic field induction in this region should not exceed 1.5 mT [4, 7]. Sensor 3 is located in the weld root and in the case of burn-through can be damaged by the electron beam, and the magnetic field induction in this region can be appreciably higher. The sensors were mounted on a breadboard and sealed by an epoxy-resin compound. The actual values of amplification coefficient and magnetic field induction measurement limit for sensors 1 and 2 were $K_a = 251$ and $B_{max}=\pm 2.49$ mT, correspondingly; for sensor 3 they were $K_a = 46.46$ and $B_{max} = \pm 13.45$ mT. The sensors were calibrated using a verified MX-10 magnetometer.

The weld specimens were mounted in a vacuum chamber as per the elaborated scheme (Fig. 1). The sensors were mounted directly on the welded specimens using special paramagnetic clamps (Fig. 2). The arrangement of the sensors was recorded for each of the experiments. The electron beam deflection was assessed by burn-throughs and macroscale polished sections of the specimens.

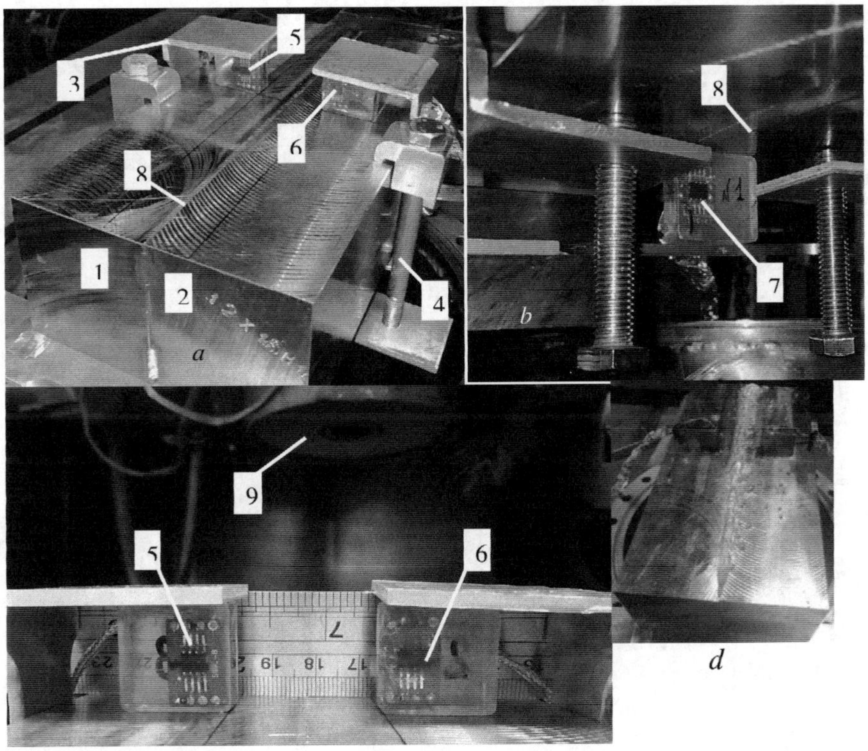

Figure 2. Arrangement of weld specimens and Hall sensors:
a) general view; b) weld root sensor; c) specimen surface sensors; d) specimens after EBW
1) and 2) welded specimens; 3) and 4) clamps for Hall sensors; 5) and 6) Hall sensors on the specimen surface; 7) weld root Hall sensor; 8) weld joint; 9) electron-beam gun.

3. Investigation Results

Measured magnetic field induction for 12Cr18Ni10Ti + 20Cr13 pair is presented in Fig. 3. After enabling the electron beam, the sensor above the 12Cr18Ni10Ti specimen failed, so its readings are not presented. The maximum magnetic field induction reached 0.89 mT in the weld root and 0.77 mT above 20Cr13 steel. The electron beam deflects in the weld root to 20Cr13 steel and demonstrates unsteady behavior: increases from 1.8 mm in the beginning of the seam up to 3.8 in the middle and decreases down to 1.6 mm in the end (Fig. 3). In the experiment, the electron beam crossed the location of the weld root sensor and damaged it, so in the rest of the experiments we decided to work without them.

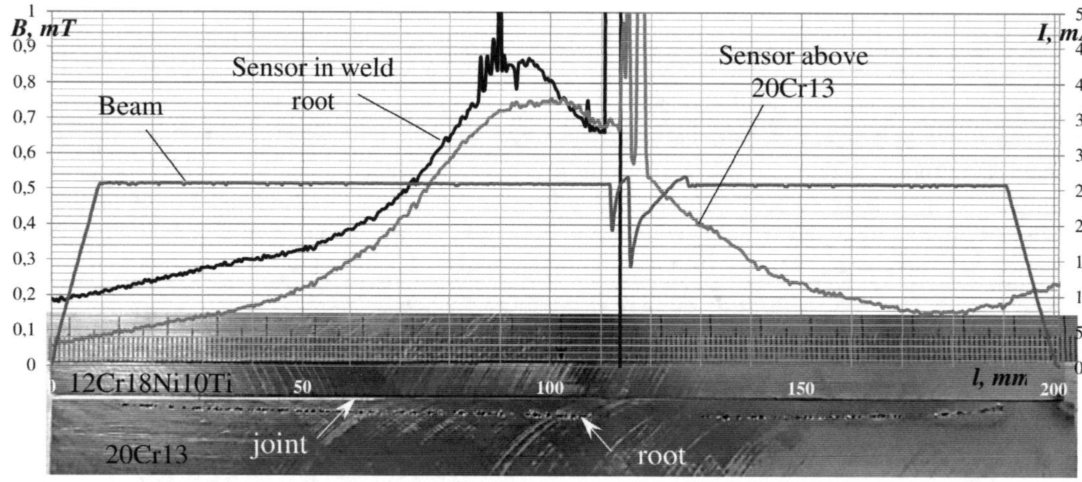

Figure 3. Magnetic field induction from thermoelectric currents during EBW and electron beam deflection in the weld root of 12Cr18Ni10Ti and 20Cr13 steels

During EBW of 12Cr18Ni10Ti and 09Mn2Si steels, the distance from the sensor sealing edge to the weld joint was increased up to 20 mm. The noise in Hall sensors caused by electron beam lead to inadequate values of the magnetic field induction, so this data is not presented. The electron beam in the weld root deflects towards 09Mn2Si (Fig. 4) and again has nonuniform character: in the beginning of the seam it is 0.64 mm, then reduces and almost disappears in the end.

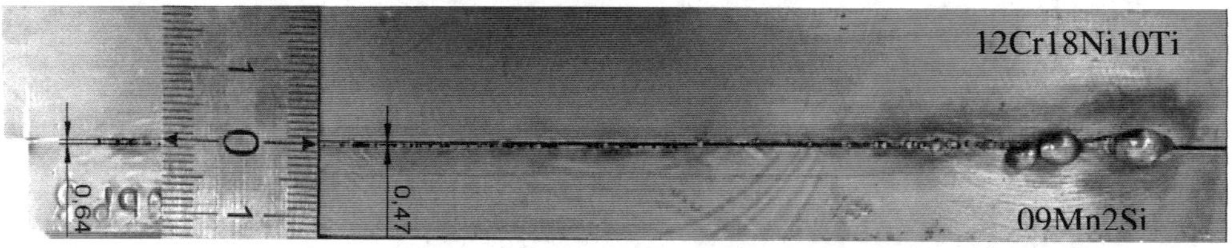

Figure 4. Electron beam deflection in the weld root of 12Cr18Ni10Ti and 09Mn2Si steels

The EBW of specimens from 20Cr13 and 09Mn2Si steels, the sensors were arranged at 20 mm from the weld joint and shielded by several layers of aluminium foil. Measured magnetic field induction values are presented in Fig. 5. The maximum magnetic field induction above 20Cr13 steel amounted to 0.54 mT and to 0.47 mT above 09Mn2Si steel. Interestingly, the maximums are observed simultaneously. The deflection of the electron beam in the weld root (Fig. 5) is nonuniform: increases from 2.1 mm in the weld beginning up to maximum of 2.8 in the middle and decreases in the end.

4. Discussion

For all welded specimens, the electron beam deflection along the joint length is nonuniform. This is caused by the changed resistance between welded specimens and altered conditions of thermoelectric current spreading in the course of welding seam formation: going farther from and closer to the boundary of the welded specimens. Besides, during EBW the temperature distribution changes: the high-temperature region increases, which changes the thermal EMF source power, hence, increases the thermoelectric currents, magnetic field induction and electron beam deflection from the weld joint. Taking into account the above, the deflection character during EBW of closed and open weld seams should be appreciably different. Thus, the numerical models that do not account for the geometry of welded parts, the change of thermal regime and contact resistance between the welded specimens during EBW yield incorrect results.

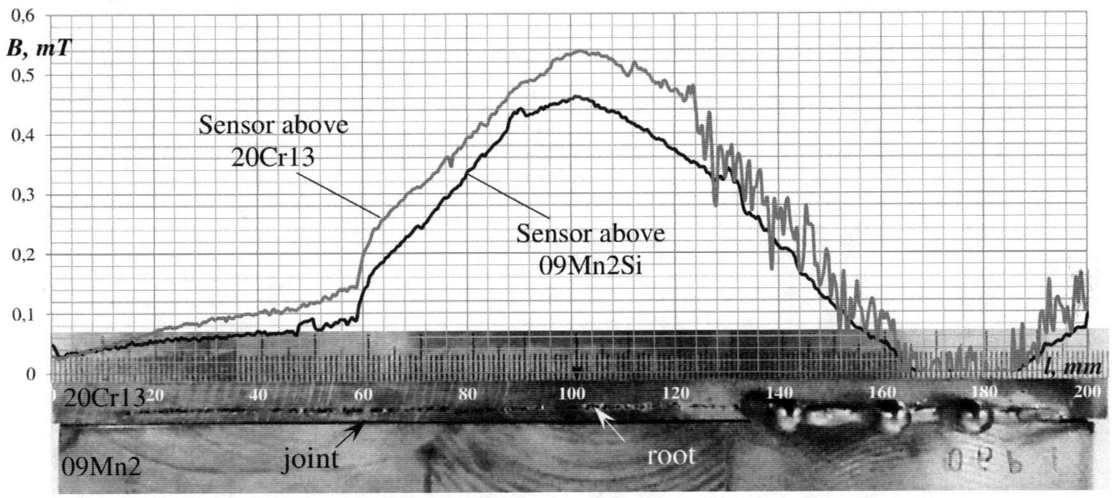

Figure 5. Magnetic field induction from thermoelectric currents during EBW and electron beam deflection in the weld root of 09Mn2Si and 20Cr13 steels

Absolute thermal EMF (E800) of the steels under study are as follows: minus 2.2 mV for 12Cr18Ni10Ti, plus 7.6 mV for 20Cr13 and minus 0.4 mV for 09Mn2Si. Thus, the deflection occurs towards the material with more positive thermal EMF potential, which agrees with [2, 3].

Fig. 6 presents the lateral macrosections and trajectories of the electron beam. The trajectories of the beam were determined by the coordinates of the weld seam middle measured on macrosections. The difference in thermal EMF (E800) of 20Cr13 + 12Cr18Ni10Ti and 20Cr13 + 09Mn2Si pairs equals 9.8 and 8.0 mV, correspondingly, which during the welding of 60-mm thick parts leads to lack of penetration in the weld root with a magnitude of more than a half of the joint thickness (Table 1).

Even with small difference in the thermal EMF (E800) for 12Cr18Ni10Ti + 09Mn2Si (only 1.8 mV), the lack of penetration in the weld root amounts to about 10 mm. Thus, welding of thick dissimilar materials should take into account the deflections of the electron beam from the joint even at negligible difference in thermal EMF.

Table 1. Deflection of the electron beam during EBW of dissimilar steels

Position of the section from the seam beginning		**Deflection of beam from joint** / lack of penetration in weld root [mm]		
		20Cr13 + 12Cr18Ni10Ti	12Cr18Ni10Ti + 09Mn2Si	20Cr13 + 09Mn2Si
25 mm	On surface	**0.51** / -	**0.12** / -	**-1.98** / -
(Fig. 6a)	In weld root	**-2.2** / 27	**0.47** / 9	**-2.4** / 42
95 mm	On surface	**-0.19** / -	**0.24** / -	**-1.33** / -
(Fig. 6b)	In weld root	**-3.42** / 35	**0.50** / 10	**-2.84** / 39

IOP Conf. Series: Materials Science and Engineering **681** (2019) 012040 doi:10.1088/1757-899X/681/1/012040

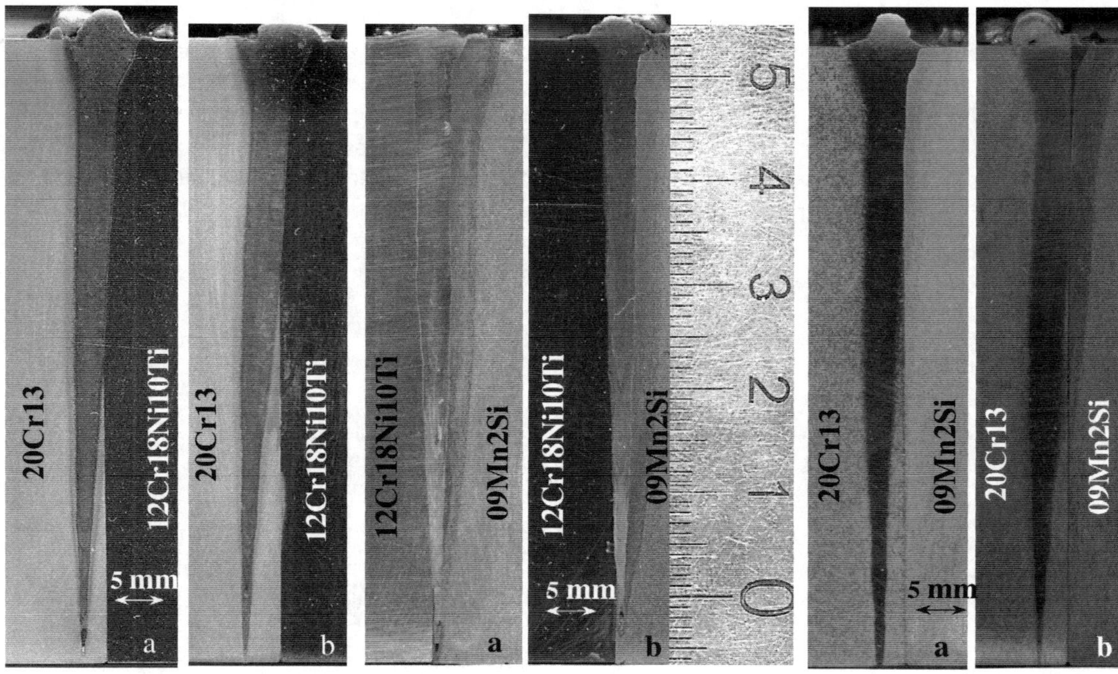

Figure 6. Macrosections and shape of burn-throughs in weld seams: a) at 25 mm from the edge of welded specimens; b) at 95 mm from the edge of welded specimens

A substantial difference of the electron beam trajectory during EBW for 20Cr13 + 12Cr18Ni10Ti and 20Cr13 + 09Mn2Si should be noted. Despite the lower difference in thermal EMF (E800) and deflection in the weld root of 20Cr13 + 09Mn2Si pair, the dimensions of lack of penetration in the weld root are significantly larger: 42 and 39 mm versus 27 and 35 mm. This is caused by appreciable deflection of the electron beam during welding of 20Cr13 + 09Mn2Si pair above the welded parts. It is also uneven, similarly to the deflection in the weld root, while the character of changes is contrary: minus 1.98 mm at 25 mm from the seam beginning and plus 1.33 mm at 95 mm (Table 1).

It should be regarded that above the pair of ferromagnetic 20Cr13 and 09Mn2Si, the field is much larger than above the pair of ferromagnetic 20X13 and paramagnetic 12X18H10T. This estimation cannot be unambiguous using the magnetic field strength, because of different distance of magnetic induction sensor mounting from the joint in the experiments.

5. Conclusions

During EBW of dissimilar materials, the deflection of the electron beam from the joint plane alters. This is due to the geometry of the weld joints, change in the thermal field, field of electric potentials and thermoelectric current spreading, change in the electric contact between welded edges under welding stresses and strains. For instance, during EBW of 60-mm thick 20Cr13 + 12Cr18Ni10Ti pair, the deflections changed from 1.8 mm in the weld seam beginning up to maximum value of 3.8 mm in the middle. This should be taken into account when elaborating numerical models of the process and methods for preventing the beam deflection from the weld joint.

During EBW of 60-mm thick dissimilar steels, the deflection of the electron beam from the weld joint was found to cause lack of penetration in the weld root. Even with the difference in thermal EMF (E800) of 1.8 mT for 12X18H10T + 09Mn2Si pair, the detected lack of penetrations amounted up to 10 mm.

The electron beam deflection from the joint plane in the weld root of considered sections of ferromagnetic + paramagnetic pair (20Cr13 + 12Cr18Ni10Ti) with the difference in thermal EMF (E800) of 9.8 mV amounts to 2.2 and 3.4 mm. For the pair of ferromagnetic materials

(20Cr13 + 09Mn2Si), with the difference in thermal EMF (E800) of 8.0 mV it amounted to 2.4 and 2.8 mm. Despite these facts, the lack of penetration in the joint for ferromagnetic materials is bigger: 42 mm versus 27, and 39 mm versus 35. This means strong deflection of electron beam over the pair of ferromagnetic parts, which is confirmed by the measurements of deflection on the weld top.

The suggested method for measuring magnetic field in control points over welded parts during EBW allows obtaining data on distribution of magnetic field induction for verification of numerical models of the process. A special attention should be paid to the protection from noise in sensitive components.

Acknowledgments
The research was carried out at National Research University National Research University Moscow Power Engineering Institute with the financial support of the Russian Science Foundation (project No. 18-19-00652).

References:
[1] Wei P S and Lii T W 1990 *J. Heat Transfer* **112(3)** 714-20
[2] Ziolkowski M and Hartmut B 2009 *COMPEL: Int. J. for Computation and Math. in Electrical and Electronic Eng.* **28(1)** 140-53
[3] Dragunov V K and Chepurin M V 2001 *Welding Production* **12** 8–16
[4] Blakeley P I and Sanderson A 1984 *Welding J.* **63(1)** 42–9
[5] Nazarenko O K 1982 *Automatic welding* **1** 33–9
[6] Watanabe K, Shida T, Susuki M *et al.* 1975 *J. Jap. Welding Soc.* **44(2)** 121–7
[7] Dragunov V K, Myakishev Y V, Goncharov A L and Sliva A P 2006 *Welding Int.* **20(10)** 811–5
[8] Dragunov V K, Sliva A P, Goncharov A L and Chepurin M V 2014 *Method of dissimilar metallic materials electron-beam welding* (Russian Federation: Patent No 2534183)
[9] Laptenok V D, Druzhinina A A, Murygin A V and SereginYu N 2016 *IOP Conf. Ser.: Mater. Sci. Eng.* **122** 012021
[10] Murphy J T 2009 *Method of electron-beam welding and weldments, constructed by this method* (Russian Federation: Patent No. 2346795)

IOP Conf. Series: Materials Science and Engineering **681** (2019) 012041 doi:10.1088/1757-899X/681/1/012041

Rapid welding technology for polyethylene gas pipes at temperatures below standard

N P Starostin[1], O A Ammosova[2]

[1]Doctor of Technical Sciences, Head of Laboratory, Institute of Oil and Gas Problems SB RAS, Yakutsk, Russia
[2]Candidate of Technical Sciences, Senior Researcher, Institute of Oil and Gas Problems, SB RAS, Yakutsk, Russia

E-mail: oammosova@gmail.com

Abstract. Rapid polyethylene pipes welding technology at temperatures below standard is proposed. This technology consists of preheating, equalization to acceptable temperature, reflow in regulated mode, and cooling under heat insulation layer or in heat-insulating chamber. Additional welding parameters, including heating and temperature equalization duration, heat-insulating layer thickness or heat-insulating chamber dimensions, are calculated from condition of ensuring heat process flow according to welding pattern at acceptable temperature. For mathematical model adequacy to actual heat welding process, it is proposed to consider polyethylene phase transformation heat in temperature range. When solving problem using end-to-end counting method, effective heat capacity coefficient is introduced, which is determined using temperature dependence of polymer material crystallinity degree. Crystallinity degree in welding process is determined by differential scanning calorimeter. Results of calculating cooling rates for polyethylene pipes electrofusion welding at various temperatures are given. Significant increase of cooling rate in heat-affected zone during welding at low temperatures, at which fine-crystalline structure is formed, which causes weld material plasticity, is shown. It is shown that when welding polyethylene pipes in low ambient temperatures, preheating, temperature equalization and cooling with thermal insulation with calculated parameters lead to thermal process in heat-affected zone according to welding patterns at acceptable temperature.

1. Introduction

Welding work during installation and repair of gas pipelines made of polyethylene pipes is recommended to be carried out at an ambient air (AA) temperature from minus 15 to 45 °C. At lower air temperatures, welding is recommended to be carried out in lightweight, heated structures with maintaining temperature in allowable range. At the same time, for welding, it is necessary to withstand the welded pipe sections for a long time for temperature control, which is unacceptable in emergency situations.

When welding polyethylene pipes at low temperatures during reflow process of welding surfaces, peripheral areas of welded joint are heated slightly due to low thermal conductivity of polymer material. In addition, at low AA temperature, convective heat transfer from free surfaces of the joint increases. These two factors contribute to increase in molecular heat transfer and increase in cooling rate in heat-affected zone (HAZ) at precipitation stage. It is known that at high rates of melt cooling,

IOP Conf. Series: Materials Science and Engineering **681** (2019) 012041 doi:10.1088/1757-899X/681/1/012041

crystal growth occurs slowly, and fine-crystalline structure is formed, which causes plasticity of polyethylene and low bonding strength [1].

In order to obtain welded joint at low temperatures, strength of which is not lower than during welding at conditions of permissible air temperatures, it is necessary to ensure that heat process in HAZ proceeds according to pattern, typical for welding at conditions of temperature from regulated range. Such temperature field dynamics can be obtained by preheating, reflowing in standard mode and cooling in a heat-insulating chamber or under a layer of heat-insulating material. When preheating, it is necessary to ensure close to uniform temperature distribution in permissible range by equalizing temperature. It is more effective to determine duration of heating and equalization of temperatures by calculation, simulating a non-stationary thermal process. Based on mathematical modeling, parameters of heat-insulating chamber or layer thickness of heat-insulating material are also determined. In mathematical modeling of heating and cooling of welded joint, it is necessary to consider absorption of heat during polymer material melting and its release during crystallization.

2. Mathematical model

It is known that polyethylene material for pipes has crystallinity degree of 55-65% [2]. When simulating thermal process in products made of polymeric materials, in order to consider heat of phase transformation, Avrami-Kolmogorov crystallization kinetics equation is usually used, the parameters of which are determined from DSC thermograms [3–5]. When developing technology of welding PE pipes at low temperatures, it is necessary to ensure cooling of product according to predetermined pattern of changes in the temperature field, without describing nucleation and growth of crystals. In this case, thermograms of differential scanning calorimeter (DSC) can be used directly for determination of phase transformation heat in mathematical model. In heat equation describing a non-stationary temperature field in welded joint, using shock-capturing method, an effective heat capacity coefficient is introduced [6]. The heat equation is written as:

$$\left(c(T) + \rho^{-} L^{100\%} \frac{dX_C}{dT} \right) \frac{\partial T}{\partial t} = div\big(\lambda(T)\, grad T\big), \; x \in D, \; 0 < t \le t_m, \tag{1}$$

where Q – heat source power; $L^{100\%}$ – enthalpy of phase transition of completely crystalline polymer; X_C – degree of crystallinity; $c(T) = \rho^{+}c^{+} + X_C(T)\cdot\left(\rho^{-}c^{-} - \rho^{+}c^{+}\right)$, $\lambda(T) = \lambda^{+} + X_C(T)\cdot\left(\lambda^{-} - \lambda^{+}\right)$; $c^{-}, \rho^{-}, \lambda^{-}$ and $c^{+}, \rho^{+}, \lambda^{+}$ – specific heat, density and thermal conductivity for solid and liquid phases of pipe material, respectively.

Degree of crystallinity X_C is determined by formula:

$$X_C(T) = \begin{cases} X_C^{\infty}, & T \le T_1, \\ X_C^{\infty} - \dfrac{\int_{T_1}^{T} q(u)\,du}{L^{100\%} v_T}, & T_1 < T < T_2. \\ 0, & T \ge T_2, \end{cases} \tag{2}$$

Here $q(T)$ is the dependence of heat flux on temperature, referred to unit mass of substance, recorded by differential scanning calorimeter (DSC); X_C^{∞} is the maximum degree of crystallinity of studied polymeric material; v_T is the temperature change rate.

Differentiating integral over variable upper limit in function (2), we have:

$$\chi(T) = L^{100\%} \frac{dX_C}{dT} = \begin{cases} 0, & T \le T_1, \\ \dfrac{-q(T)}{v_T}, & T_1 < T < T_2, \\ 0, & T \ge T_2, \end{cases} \tag{3}$$

Then equation (1) is written as:

$$\left(c(T) + \rho^- L^{100\%} \chi(T)\right)\frac{\partial T}{\partial t} = div\left(\lambda(T) grad T\right), \; x \in D, \; 0 < t \le t_m, \tag{4}$$

When welding parts with built-in heater, equation (4) is supplemented by condition of concentrated heat source. On other boundaries we have traditional boundary conditions. The non-stationary temperature field is determined by solving equation (4) with given initial and boundary conditions by the finite element method [7].

When butt welding with heating tool, temperature of the heating tool is set at welded ends of pipes. The task of determining temperature field is solved by the finite difference method [8].

Convective heat exchange α at the contact of the pipe and the coupling with thermal insulation is determined from the expression [9]:

$$\frac{1}{\alpha} = \frac{1}{\alpha_0} + \frac{h_{iz}}{\lambda_{iz}}, \tag{5}$$

where h_{iz} is the thickness of the thermal insulation layer, λ_{iz} is the thermal conductivity of the insulation material, α_0 is the heat exchange with the environment.

3. Data for calculations

The design scheme of pipe wall and coupling is presented in figure 1. Triangulation of computational domain D was performed using the Gmsh grid generator [10]. Problem (4) was solved by finite element method using free-access program Dolfin/FEniCS [11]. Visualization of results was implemented using the Paraview package [12].

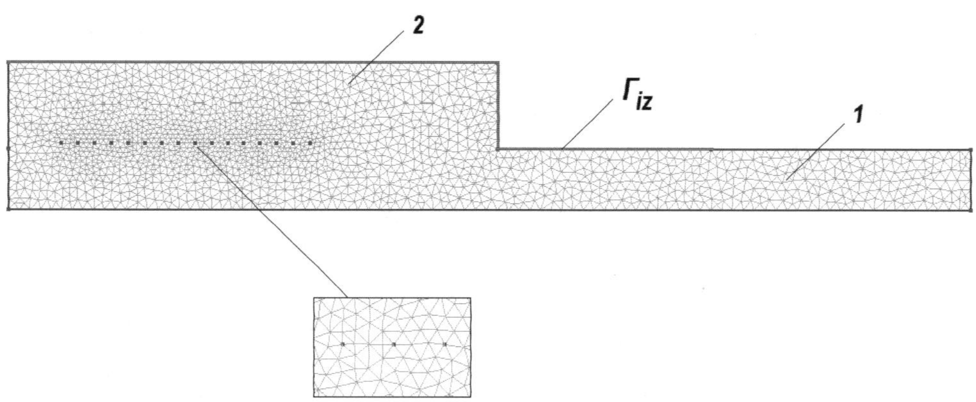

Figure 1. Design scheme: 1 - pipe wall; 2 - coupling wall.

Simulation of temperature fields during electrofusion welding of polyethylene pipes was carried out for brand PE 100 GAZ SDR 11 63×5.8 at various ambient temperatures from +20 to -50 °C. Thermophysical properties of pipe and coupling material: thermal conductivities are $\lambda^- = 0.46$; $\lambda^+ = 0.24$ W/(m·K); densities are $\rho^- = 950$; $\rho^+ = 800$ kg/m^3; heat capacities $c^- = 1900$; $c^+ = 2100$ J/(kg·K) [13]. 3. Data for calculations: $r_1 = 0.0257$ is inner pipe radius; $r_{tr} = 0.0315$; $r_{muf} = 0.0397$ is outer pipe and coupling radii, m; heating elements are located at radius r_{sp}= 0.0321 m; z_1 = 0.005; z_2=0.028; z_{muf} = 0.046; z_{tr} = 0.09 m. Spirals in coupling are made of copper: thermal conductivity λ_{Al} = 238.2 W/(m·K); density ρ_{Al} = 2675 kg/m^3; heat capacity c_{Al} = 951.3 J/(kg·K) [13]. There are 32 coils in the coupling. Welding voltage was taken to be U = 32 V; R = 1.6 Ω; β = 4.30 · 10^{-3} 1/°C. Duration of heating and cooling for standard welding at 20 °C is 70 seconds and 8 minutes, respectively.

4. Calculation results

Main idea of proposed approach lies in ensuring changes in temperature field close to dynamics, obtained at permissible air temperatures, in the area of welded sleeve joint, when welding polyethylene pipes at low ambient temperatures. According to calculations, at air temperatures below minus 15 °C, the size of melted zone decreases, which indicates smaller volume of melt produced (Figure 2). Size of heat-affected zone (HAZ), in which structural changes occur in welded materials, during heating and melting, has significant effect on formation of welded joint. Usually, HAZ is determined experimentally by examining structural indicators. Let us take commonly used assumption that structural changes occur at temperatures above softening temperature of material. Mathematically, task is reduced to finding the curve farthest from the welded joint, at each point of which maximum temperature is reached equal to material softening temperature of 80 °C.

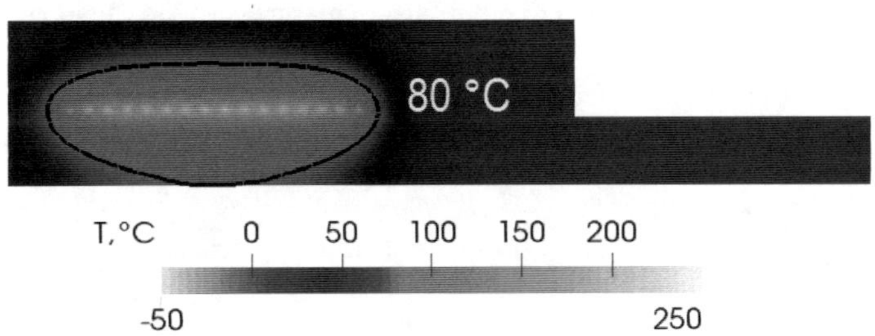

a

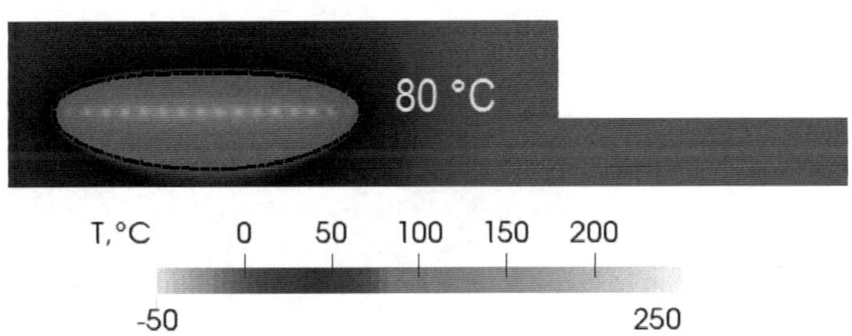

b

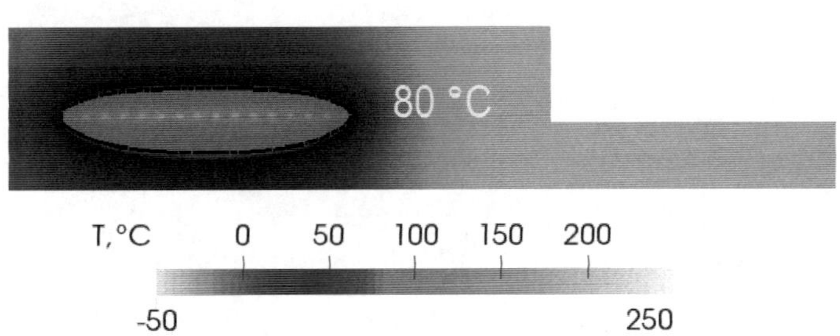

c

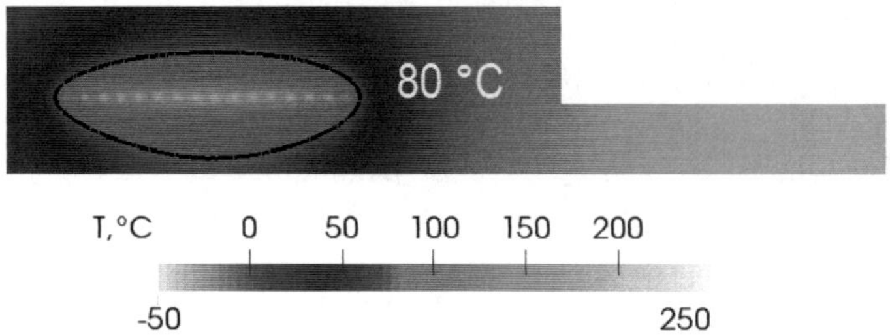

d

Figure 2. The temperature change in cross section of coupling and pipe after heating at different AA temperatures: a) 20; b) minus 15; c) minus 50; d) minus 50 °C after preheating and levelling.

In order to increase depth of penetration at air temperatures below minus 15 °C, it is proposed to carry out preheating and leveling prior heating. Preheating of welded surfaces of pipe and coupling is performed using heating spiral of coupling. Further, leveling is carried out to establish more uniform temperature field in coupling and pipe. Duration of heating and leveling, as well as power supplied to the voltage spiral, are determined by calculation. In order to maintain required cooling rate, outer surface of coupling with pipe is covered with insulating material. Thickness of covering material is determined by calculation.

The quality of welded coupling connection is significantly influenced by cooling rate of material. Figure 3 shows the time dependences of temperatures during cooling of welded joint for 8 minutes at various ambient temperatures.

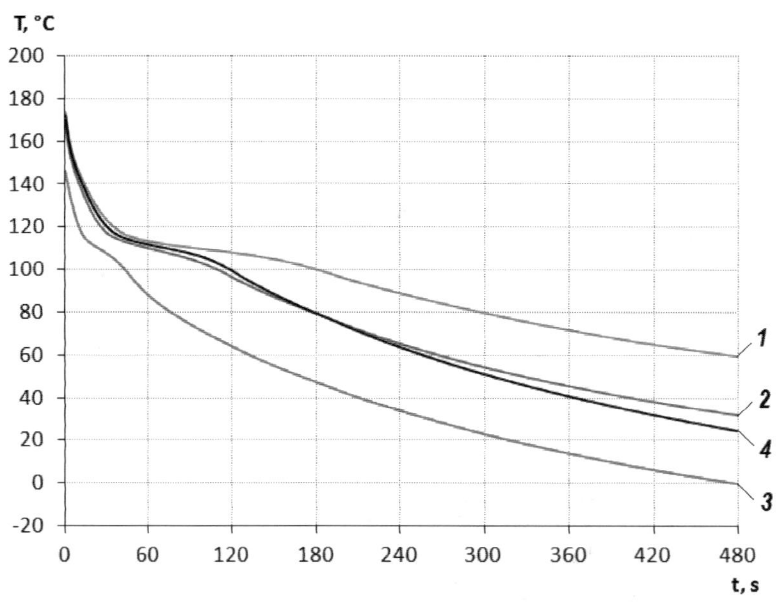

Figure 3. Time dependences of temperatures during cooling of welded joint for 8 minutes at various ambient air temperatures: -15, 20, -50 and -50 °C (heating, leveling, cooling under thermal insulation).

Cooling rate at temperature of 20 °C (curve 1), in the range of 120–115 °C is $V_{120\text{-}115}$=25 °C/min, in the range of 115–110 °C $V_{115\text{-}110}$=7 °C/min, in the range of 110–105 °C $V_{110\text{-}105}$=5 °C/min, in the range of 105–100 °C $V_{105\text{-}100}$=8 °C/min, in the range of 100–95 °C (interval in which intensive crystallization occurs)$V_{100\text{-}95}$=10 °C/min. Note that the rate of $V_{100\text{-}80}$=10 °C/min lasts until the end of the crystallization process.

At AA temperature of minus 15 °C (lower limit of AA temperatures allowed for welding without using light heated shelters) cooling rates are: $V_{120\text{-}115}$=40, $V_{115\text{-}110}$=13, $V_{110\text{-}105}$=10, $V_{105\text{-}100}$=14, $V_{100\text{-}95}$=20, $V_{95\text{-}90}$=18, $V_{90\text{-}85}$=17, $V_{85\text{-}80}$=16 °C/min (curve 2).

At AA temperature of minus 50 °C welding without heating and cooling without heat insulation leads to significant decrease in temperature ($V_{120\text{-}115}$=85, $V_{115\text{-}110}$=33, $V_{110\text{-}105}$=26, $V_{105\text{-}100}$=37, $V_{100\text{-}95}$=46, $V_{95\text{-}90}$=40, $V_{90\text{-}85}$=33, $V_{85\text{-}80}$=28 °C/min) and temperature distribution curves lie outside allowable range of temperature variations (curve 3).

At low temperature conditions, the use of preheating and heat-insulating chamber with wall thickness of 2 cm results in cooling rate at crystallization stage, that is characteristic for welding at acceptable AA temperature (curve 4).

5. Conclusion

- When welding polyethylene pipes PE 100 GAZ SDR 11 63×5.8 at permissible ambient temperatures, cooling rate of weld material in range 115-120 °C is 25, in range 85-115 from 5 to 10 °C/min. When welding at minus 50 °C, cooling rate of weld, corresponding to the same temperature ranges, is 85 and from 28 to 46 °C
- When electrofusion welding of polyethylene pipes, preheating is performed using built-in heater with subsequent temperature equalization with estimated duration, melting in regulated mode and cooling under insulation layer of calculated thickness leads to cooling of weld material at speed that guarantees high-quality weld.

Acknowledgments
The research was carried out within the state assignment of FASO of Russia (project No. AAAA-A17-117040710038-8 dated April 7, 2017).

References

[1] Kaygorodov G K and Kargin V Yu 2001 *Truboprovody i ekologiya* **2** 13–14
[2] Gorilovskiy M, Kalugina E, Ivanov A and Satdinova F 2005 *Plasticheskie massy* **4** 9–12
[3] Chebbo Z, Vincent M, Boujlal A, Gueugnaut D and Tillier Y 2015 *Polym. Eng. Sci.* **55** 123–31
[4] Spina R, Spekowius M and Hopmann C 2018 *Thermochim. Acta* **659** 44–54
[5] Kulikova T and Trufanov N 2008 *Vychislitelnaya mekhanika sploshnykh sred* **1**(2) 38–52
[6] Samarskiy A A and Vabischevich P N 2004 *Vychislitelnaya teploperedacha* (Moscow: Editorial URSS) p 784
[7] Abedian A, Parvizian J, Düster A and Rank E 2014 *Appl. Math. Mech.* **35(**10) 1239–1248
[8] Samarskii A A 2001 *The theory of difference schemes* (New York: Basel. Marcel Dekker Inc)
[9] Isachenko V P, Osipova V A, Sukomel A S 1969 *Heat Transfer* (Moscow: Energija) p 440
[10] Software GMSH. URL: http://geuz.org/gmsh/
[11] Logg A, Mardal K A and Garth N 2011*Wells Automated Solution of Differential Equations by the Finite Element Method.* The FEniCS Book
[12] Software package PARAVIEW. URL: http://www.paraview.org/
[13] A P Babichev, N A Babushkina, A M Bratkovskiy and etc. 1991 *Fizicheskie velichiny Spravochnik* ed I S Grigoreva and E Z Meyltkhova (Mosciw: Energoatomizdat) p 1232

Modeling heat input when friction stir welding

V Statsenko[1], A Sukhorada[2] and V Lelyukhin[2]

[1] Senior Production Editor, FEFU, Vladivostok, RU; [2] Production Assistant, FEFU, Vladivostok, RU

E-mail: vladsta@mail.ru, alexeyman_09@mail.ru, lelv0@mail.ru

Abstract. An important element of friction stir welding technology is the determination of the temperature of the material in the mixing zone, which can be determined by calculating the amount of heat introduced into the welding zone. To determine this value, experimental studies were carried out on a stand in which the material to be welded (aluminum alloy AMg5) is modeled as an experimental tube with a diameter of 20 mm, and the tool (from tool steel P6M5) is modeled as a working plate. On the stand, studies of the dependence of the frictional moment and heat-liberation value were performed. According to the obtained experimental data, the heat-liberation dissipation and heat power on the working tool are calculated, the dependences of these values on the radius of the working tool are obtained.
To increase the amount of heat input in the center of the contact area, a tool with a drive with two concentrically located (one inside the other) shafts is proposed, while the inner shaft rotates at a higher speed than the outer one. According to the obtained experimental data, the dependences of heat release for each shaft and the total heat release for a tool with two shafts at an external shaft rotation speed $\omega = 63\ s^{-1}$ and different rotation speeds of the internal shaft are calculated. In total, for the external and internal shafts, the total heat release value is about 2 times higher than the total heat release value using a single shaft.
The purpose of this work is to find the values of heat release in friction stir spot welding, as well as the development of friction spot welding tools with higher heat release.

1. Description of the technical solution and experimental technique

Friction stir welding (STD) refers to hardphase welding methods, so it is important to know the temperature conditions. Experimental measurement of temperature in the heating zone is associated with certain difficulties, since the tool at this moment rotates and closes the heating zone with a flat end surface. In most publications [1-3], devoted to the calculation of the temperature distribution in the process of SST, they are based on solving the heat-transfer equation and offer numerical calculation methods. But for calculations, the main necessary parameter is the amount of heat energy supplied to the heating zone (heat release).

The distribution of heat release over the radius of the contact area of the tool with the workpiece during classical friction welding has a parabolic shape [7]. This is due to the uneven distribution of the linear speed of rotation of the tool face over the radius of the heating area; it varies from zero in the center to a maximum at the outer radius. The change in frictional moment is not considered.

To study the dependence of the moment of friction and heat release along the radius of a rotating tool, a laboratory stand was designed and manufactured [4].

Its main element is a cylindrical working tube made of aluminum alloy AMg5, which outer diameter is 20 mm, wall thickness is 2 mm, length is 40 mm. This tube simulates the material of the plates being welded. It rotates on the surface of the working plate of tool high-speed steel R6M5, which simulates a tool for friction stir spot welding. Due to the change in the speed of rotation of the working tube, it is possible to set different linear speeds of rotation of the end surface of the tool relative to the surface of the steel plate, corresponding to the speeds of rotation of the real tool at different radii.

A working plate of R6M5 tool high-speed steel is fixed on the desktop of the stand through the gasket. The worktable is fixed on the movable part of the pressure ball bearing, its fixed part is fixed on the base. When the working tube is rotated and pressed with a certain force to the working plate, the frictional moment is determined by the force measured by the electronic dynamometer and the length of the rod (shoulder).

The experiments on heat release were carried out on a VM127 universal vertical milling machine. When measuring the frictional moment, the compression force of the tube relative to the plate is set, which maintains the temperature of the contact point in all modes.

During the experiments were determined the heat losses by thermal conductivity along the rod, on which the working tube is fixed, as well as from the working plate through the gasket to the working table, and losses due to convection from the surface of the rotating working tube into the environment. After the necessary measurements were made, the results of calculations of these heat losses showed that in different modes the heat conduction losses along the tool stem are 8–10%, to the work table it does not exceed 9–15%, heat abstraction by convection from the cylindrical surface of the experimental tube is 2.5– 4.5%. These losses are considered in the calculations of heat release.

2. Results of experiments and calculations

Experiments on measuring heat release were carried out at the speed of rotation of the spindle of the milling machine in the range of 400–1600 rpm (42–167 s^{-1}). At the same time, by regulating the pressing force of the working tube and the working plate, the temperature of the contact place in all modes was kept constant - 335 ± 15.

As a result of the experiments performed, the dependence of the frictional moment change with an increase in the speed of rotation of the working tube was obtained Figure 1a.

These data made it possible to calculate the volume power of heat release during friction, which is directly proportional to the rotational moment M_{fr} and the speed of rotation ω and inversely proportional to the contact area of the working tube and plate Figure 1b. The data presented on this graph were obtained considering heat losses.

Analysis of these graphs shows that with a uniform increase in the rotational speed, the intensity of the friction moment is different - more significantly reduced at low speeds (up to 100 s^{-1}) and less significantly at speeds of rotation more than 100 s^{-1}. The calculated dependence of the volume power of heat release at the same time has a complex appearance with an increase of this value at rotational speeds in the ranges of 40-83 s^{-1} and 105-167 s^{-1} and its decrease at rotational speeds of 83-105 s^{-1}. This dependence is determined by the fact that with increasing rotation speed, the friction moment decreases with different intensities, the contact area remains constant.

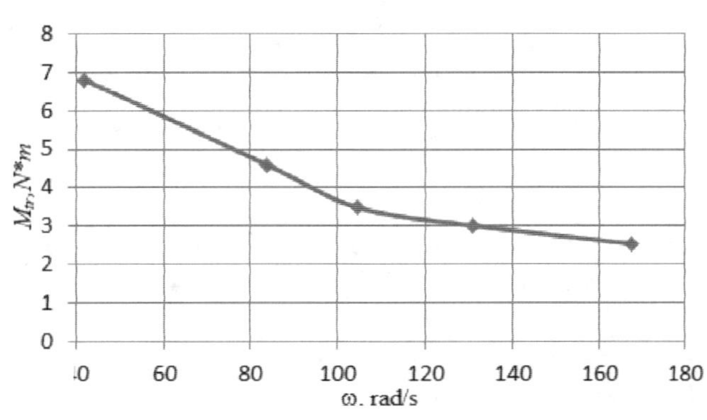

a

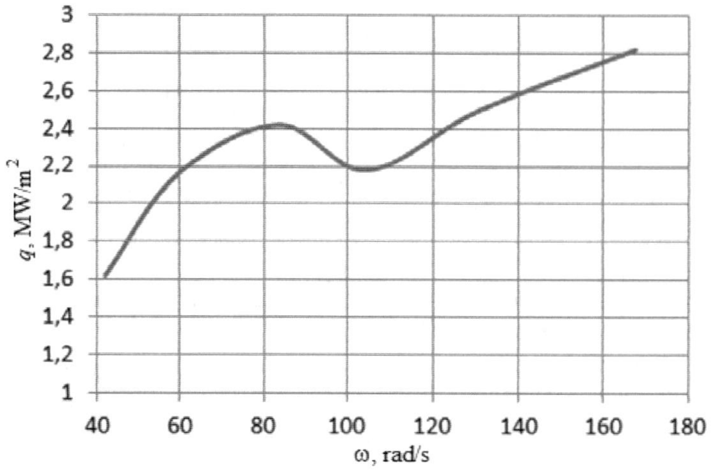

b

Figure 1. Dependence of frictional moment (a) and heat-liberation value (b) on the speed of rotation of the working tube.

According to these data, for the friction stir spot welding tool, the specific and total heat release calculations for each ring section 2 mm wide were made, and then the total heat release on the tool. The results of the calculation of the dependence of the total thermal power (heat release) on the entire contact area on the tool radius are presented in Figure 2. These data allow us to estimate the dimensions (diameter) of a tool for a given heat power and its rotational speed. So, on a tool with a radius of 8 mm, at a rotation speed of 63 s^{-1}, heat release of 0.4 kW can be obtained, and at a speed of s^{-1} - 0.6 kW.

The obtained data confirm the literature data [1-3] on the minimum heat release in the central part of the heating spot.

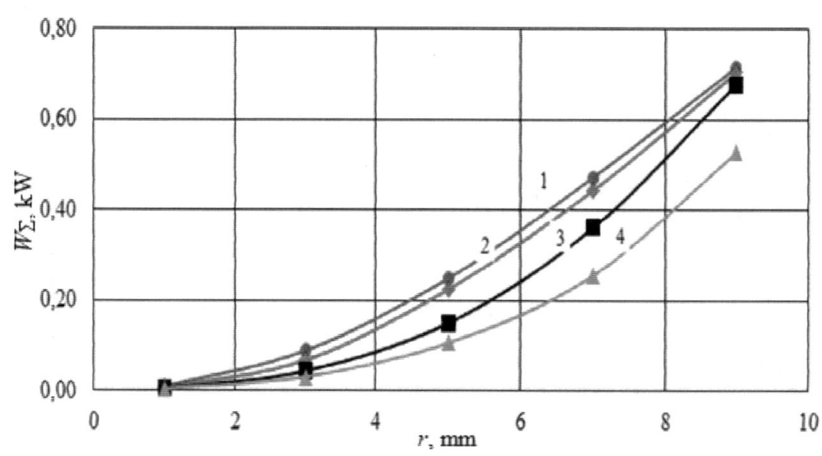

Figure 2. Dependence of total heat release on tool radius
1 - ω = 167 s^{-1}; 2 - 126 s^{-1}; 3 – 84 s^{-1}; 4 - 63 s^{-1}

To increase the amount of heat input in the center of the contact area, it is proposed to use a tool with a drive with two concentrically located (one inside the other) shafts, while the inner shaft rotates at a higher speed than the outer shaft Figure 3 [9].

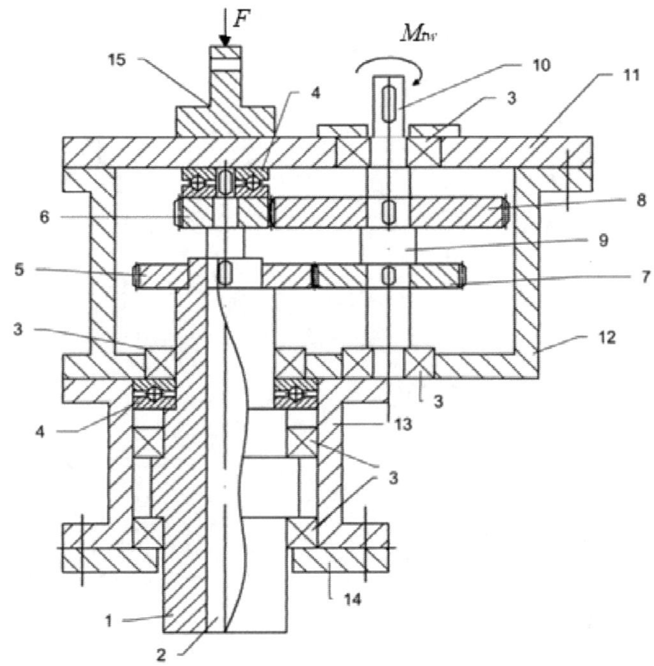

Figure 3. Tool for friction stir spot welding.

The tool for friction stir spot welding has an outer shaft 1, an inner shaft 2, radial bearings 3, thrust bearings 4, gears 5, 6, 7, 8, shaft 9, input shaft section 10, top cover 11, body 12, glass 13, the bottom cover 14, the compression device 15.

Under the action of the rotational moment of the M_{tw} shaft 9 using the gears 7 and 8, as well as the gears 5 and 6 drives the shafts 1 and 2. Since these gears have different diameters, this ensures the rotation of the inner shaft is about 2 times larger than outer. Due to this, a higher heat abstraction occurs on the internal shaft. The compression device 15 ensures the pressing of the shafts 1 and 2 to the welding point.

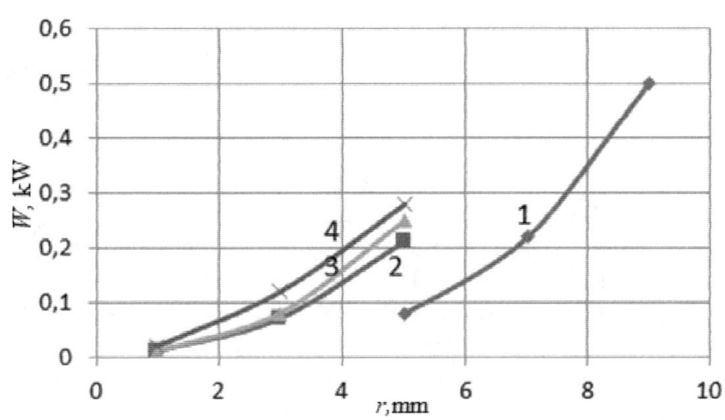

a

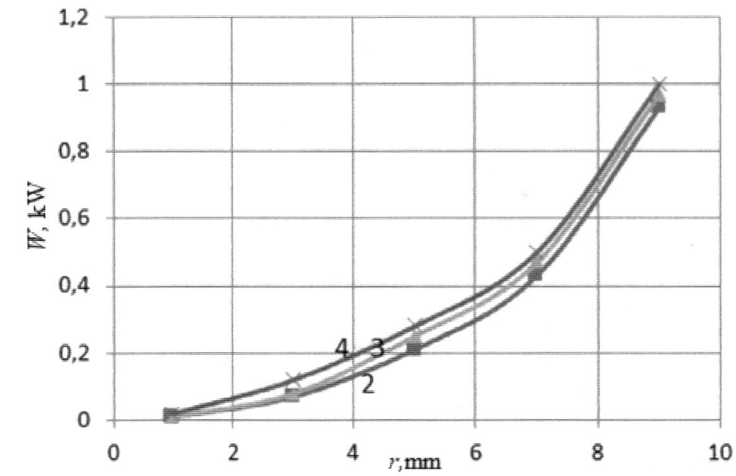

b

Figure 4. Dependence of heat release for each shaft (a) and total heat release (b) for a tool with two shafts at the speed of rotation of the external shaft 1 - $\omega = 63\ s^{-1}$; rotation speed of the inner shaft: 2 - $\omega = 105\ s^{-1}$, 3 - $\omega = 157\ s^{-1}$ and 4 - $\omega = 209\ s^{-1}$

According to the obtained experimental data, the dependences of heat release for each shaft and the total heat release for a tool with two shafts at an external shaft rotation speed $\omega = 63\ s^{-1}$ and different rotation speeds of the internal shaft are calculated, the results of calculations are presented in figure 4.

Analysis of these graphs shows that the total heat release from the tool has a more uniform distribution, and in total for the external and internal shafts the value of the total heat release is higher than the value of the total heat release using a single shaft. So, at the speed of rotation of the external shaft - 63 s^{-1}, and the internal one - 209 s^{-1} for a working tool with a radius of 9 mm, the total heat release is 2 times more than for one shaft at the same rotational speed.

3. Conclusion

Thus, in friction stir spot welding of AMg5 aluminum alloy using an R6M5 high-speed steel tool at its rotational speed of 167 s-1, in order to bring heat power of 0.7 kW, it is necessary to use a tool with a diameter of 18 mm, for heat power of 0.4 kW enough to choose a tool with a diameter of 12 mm. Similar data can be obtained for other speeds of rotation of the working tool. When using a tool with two shafts rotating at different speeds, you can get an increase in heat release by 2 times with a constant contact area.

References

[1] Kotlyshev R.R., Shuchev K.G., Kramskoy A.V. Calculation of temperatures for friction stir welding of aluminum alloys, Bulletin of the DGTU (2010) 693-699.

[2] Medvedev A.Yu., Pavlinich S.P., Atroshchenko V.V., Markelova N.I. Modeling of the temperature field for linear friction welding, Bulletin of the USATU (2010). 76-81.

[3] Statsenko V.N., Negoda E.N., Suhorada A.E., Polutsky K.A. Analysis of technology by friction stir weiding, Vestnik of the engineering school of the FEFU. (2017).

[4] Rzaev R.A., Chularis A.A., Dzhalmukhambetov A.U., Atuev Sh.M. Dynamic model of temperature distribution in the metal for friction stir welding, Fundamental research. (2016) 47-55.

[5] Grujicic M., Arakere G., Yalavarthy H.V. He T., Yen C.F. and Cheeseman B.A. Modeling of AA5083 Material Microstructure Evolution During Butt Friction-Stir Welding, Journal of Materials Engineering and Performance. (2010) 672–684.

[6] Martin J. Pushing the boundaries – friction stir goes deeper than before, TWI Connect. January/February. P. (2006).
[7] Vill V.I. Svarka metallov treniem, Mashinostroenie, Leningr, otd-nie, Russian, 1970.
[8] Thomas W. M., Nicholas E. D., Needham J. C., Murch M. G., Temple-Smith P., and Dawes C. J., International Patent Application No. PCT/GB92/02203; GB Patent Application No. 9125978.8; U.S Patent No. 5,460,317. 1991.
[9] Statsenko V.N., Sukhorada A.E., Romanova V.V., Tool for friction stir welding. Patent for utility model No. 186699, application No. 2018118628, priority of the invention 05.22.2018, registration date 01.29.2019

IOP Conf. Series: Materials Science and Engineering **681** (2019) 012043 doi:10.1088/1757-899X/681/1/012043

Hybrid method of controlling arc welding process with consumable electrode and related processes

V S Sidoruk[1], S Yu Maksimov[1], D N Krazhanovsky[1] and Y N Saraev[2]

[1]E.O.Paton Electric Welding Institute of the NAS of Ukraine,
[2]Institute of Strength Physics and Materials Science of Siberian Branch of RAS, Tomsk, Russia

Abstract. For the development of a classical method of mechanized arc welding with a self-regulation of the process of melting the electrode supplied to the welding zone at a constant speed, the following combination was proposed: on the one hand, a digital power source with automatic control and programming of volt-ampere characteristics VAC, which is a two-step one in our case: a – for increased arc length (with a low current), b – for shortened arc (with a high current); on the other hand, self-regulation of the melting process by the corresponding reaction of the power source. Moreover, the electrode melting occurs in a programmed oscillatory mode between the extreme values of the arc length: maximum and minimum. Such a combination is called a hybrid method of controlling the process of arc welding with a consumable electrode.

1. Introduction
The basis of arc fusion welding, as well as related processes (surfacing, soldering, remelting), is the melting of a metal electrode under the action of an arc. Arc melting of an electrode is a non-stationary process: formation, detachment from the end of the electrode and transfer of molten metal droplets through the interelectrode (arc) gap; instability of arc length, etc. There is a problem of controlling this process to maintain its stability. The typical phenomena which mean the loss of stability are an excessive increase in the length of the arc until its natural break and a decrease in the length of the arc until a short circuit of the electrode on the workpiece.

2. Electrode melting process
Two possibilities of maintaining stability are known: active control and self-regulation of the electrode melting process. *Active control* is known in two variants: manual and automatic.

Automatic control, as a rule, reproduces the algorithms of manual control by means of automation. In this case, in the welding complex, in addition to the welding unit and power source, there is a third component: process control unit.

Self-regulation differs from active control in the fact, that welding complex **does not contain** a control unit. The stability of the process is maintained due to the fact, that power source is designed in such a way, that it spontaneously reacts to the situation in the arc, and it is unambiguous: to every change in the length of the arc, and accordingly, to its voltage, it reacts by changing the current directed in the opposite direction. As the arc length decreases (at a constant electrode feed rate), the source increases the current, due to which the electrode melting rate grows and the arc length is increased to a preset value. In the case of an increase in the value of the arc length against a preset one, the response of the source is the opposite: the current is decreased, and accordingly, the melting rate of the electrode is decreased. Such reaction of the source occurs due to the specificity of its external (volt-ampere) characteristics, which is expressed graphically in the form of a curve in the coordinates I – U ($VAC_{s.}$).

The process of self-regulation of the arc takes place also during manual arc welding. Below the characteristic $VAC_{s.}$ are presented for manual arc welding (figure 1 a) and for *mechanized* arc welding with self-regulation (figure 1 b).

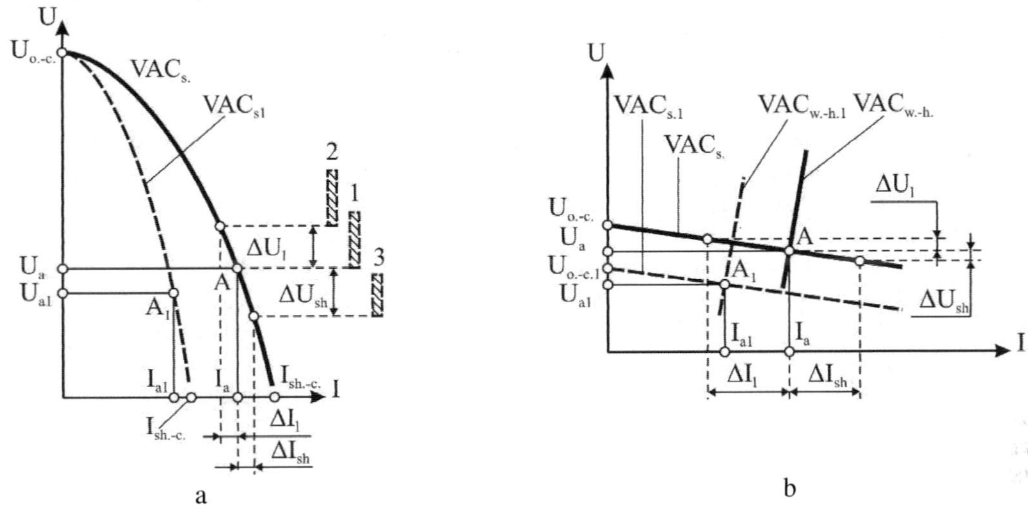

Figure 1. $VAC_{s.}$ typical for manual (a) and *mechanized* arc welding with self-regulation (b) Designations: $VAC_{s.}$ – volt-ampere characteristics of welding current source; $U_{o.-c.}$ – open-circuit voltage of current source; $I_{sh.-c.}$ – short-circuit current of current source; A – point on the $VAC_{s.}$ curve, corresponding to the accepted mode; U_a – preset arc voltage; I_a – preset arc current; ΔU_{sh} – change in arc voltage at its shortening; ΔU_l – change in arc voltage at its elongation; ΔI_{sh} – change in arc current at its shortening; ΔI_l – change in arc current at its elongation; $VAC_{w.h.}$ – volt-ampere characteristics of welding head.

In manual arc welding, welder maintains a preset length of the arc and, accordingly, its voltage constant. The function of the source is maintaining the arc current constant. It is provided by steeply falling VAC_s. At very significant fluctuations in the length (voltage) of the arc ΔU, the arc oscillations ΔI are negligible. The element of *self-regulation* is the constancy of the current, which is evident.

The preset arc current I_a is established at the source by a corresponding change in the short-circuit current I_{sh-c}. A dotted line in figure 1a shows VAC_s in case of decreasing I_a to the value I_{a1}. The arc voltage U_{a1} corresponds to it.

In figure 1b, the preset welding mode **with self-regulation** is indicated by a point A – the intersection of the curve of VAC_s and $VAC_{w.h.}$ (welding head). Here, the preset arc current is established at the welding head: the larger the cross-sectional area of the electrode and its feed rate, the larger the preset welding current I_a (and vice-versa). In this case, even with a slight decrease in the arc length and, accordingly, in its voltage $\Delta U_{sh,}$ the welding current ΔI_{sh} increases significantly, due to to which the melting rate of the electrode grows and the length of the arc is increased and vice versa. This occurs because VAC_s represents a flat falling curve. Here, the self-regulation of the electrode melting rate is an oscillatory process, which we previously called self-modulation [1].

The preset arc current I_{a1} on the welding head is reduced by a corresponding change in the electrode feed rate. A dotted line in figure 1b shows $VAC_{w.h.1}$ for reducing I_a to the value I_{a1}. To this value, a set reduction in the open-circuit voltage of the power source to $U_{o.-c.1}$ corresponds in order to provide the necessary arc voltage reduction to U_{a1}.

IOP Conf. Series: Materials Science and Engineering **681** (2019) 012043 doi:10.1088/1757-899X/681/1/012043

Historically it formed so that, at first, **automatic control** of the electrode melting process was developed. As in manual arc welding, its function consisted in maintaining a constant arc length. The automatic control system continuously measured the arc voltage (arc length indicator), which is an adjustable parameter. With an increase in the arc length, the system generated an electrical signal supplied to the motor of the electrode wire feed mechanism to increase its feed rate (regulating effect). After the arc voltage was restored at a preset value, the signal to increase the wire feed rate was stopped. When reducing the length of the arc, the system generated a signal to reduce the electrode wire feed rate. As a result, the average wire feed rate was equalized with the rate of its melting. As in manual arc welding, the arc current was increased at the power source. As far as with an increase in the arc current, the melting rate of the electrode wire was increased, the system accordingly increased its feed rate. The priority of this method of welding belongs to the company General Electric Company, P.Nobel (1920) [2, p.180]. The systems for the process control, described above were objectively in a state of interaction, competing and simultaneously complementing each other.

Automatic process control provides an accurate performance of the preset algorithm and is characterized by an excellent quality of the welded joint. Its main drawback is the complexity of the device and a high cost of the equipment, as well as additional costs for maintenance of the equipment. This method of control is challenging during the manufacture of critical structures, as well as in the case of special welding processes.

Self-regulation of the electrode melting process is free from the disadvantage mentioned above. However, it lacks its main advantages like adaptation to the welding conditions and the ability to fulfil the special technology requirements. The range of mode parameters here is essentially narrower. And here is another disadvantage of self-regulation principle: as B.E. Paton showed, self-regulation is possible only at a very significant current density on the electrode [3]. Based on the production experience, we found that the minimum possible current density on the electrode is 30 A/mm^2.

An explanation to this fact is given in figure 1b, where dotted lines show the positions of VAC_s and $VAC_{w.h.}$ for the case of an excessively low value I_{a1}. Here, the open-circuit voltage of the power source is so low that the sum $U_{a1}+\Delta U_{el}$ exceeds the value $U_{o.-c.1}$, and the arc is broken. This fact limits the technical capabilities of the method and increases its cost.

It became necessary to develop such an algorithm for self-regulation of electrode melting, at which the stability of the welding process would be high if the current density on the electrode decreased, at least as in manual arc welding (7 A/mm^2).

It was decided to use the oscillatory nature of the electrode melting process [1]. It was proposed to perform electrode melting in an oscillation melting mode which is expanded in the amplitude of control, i.e. at an increase in the arc length (voltage) to higher than a preset value, to power it with the current, at which the melting rate is substantially lower than the feed rate and to increase the current to a value, at which the melting rate of the electrode is much higher than its feed rate after the inevitable decrease in the arc length (voltage), lower than a preset pulse [4]. Such current manipulations occur due to the spontaneous reaction of the source, having a specific VAC_s (figure 2).

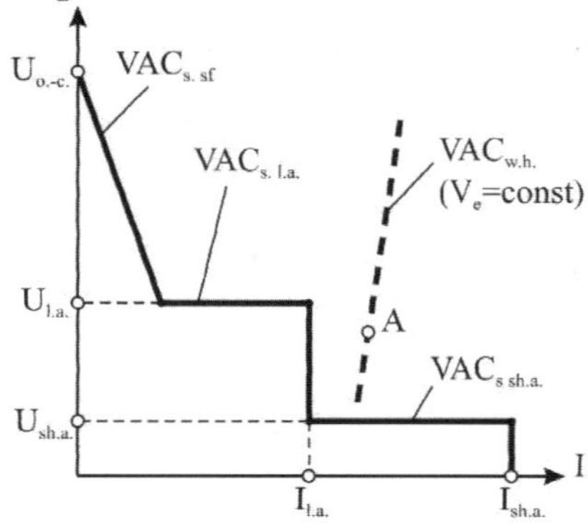

Figure 2. VAC_{source} for realizing oscillatory nature of the electrode melting process
Designations: $VAC_{s\ sf}$ is a steeply falling region of volt-ampere characteristics; $VAC_{s\ l.a.}$ is a rigid area corresponding to the parameters of a «long» arc; $VAC_{s\ sh.a.}$ is a rigid area corresponding to the parameters of a «short» arc; $VAC_{w.h.}$ is the volt-ampere characteristics of a welding head; $I_{l.a.}$ is the current of a «long» arc; $I_{sh.a.}$ is the current of a «short» arc; $U_{l.a.}$ is the «long» arc voltage; $U_{sh.a.}$ is the «short» arc voltage.

If it is necessary to reduce the mean current value, the current of a "long" arc $I_{l.a.}$ should be reduced up to several amperes due to the "supporting" action of the steeply falling $VAC_{s\ s.f.}$. At the same time, the minimum required current value of a short arc $I_{sh.a.}$ is maintained, which is sufficient to accelerate the melting of the electrode and to prevent a short circuit.

The feature of VAC of such a source consists in the fact, that the voltage $U_{sh.a.}$ can be reduced to a value lower than the sum of the cathode and anode voltage drops. In this case, the arc does not burn, and the electrode melts in a metal pool due to the contact melting phenomenon [5], and also through a drop of molten metal at the end of the electrode.

For the experiments, the source LET-500 [6] with a digital control was used, where $VAC_{s.}$ of any shape can be programmed.

Here, the **combination** of *self-regulation of the electrode melting process* due to the specific reaction of the power source to the situation in the arc and *of automatic control* of the source is clearly seen. Such a combination can be called a *hybrid method of process control.* It allows providing the stability of the electrode melting process **in almost the whole range of welding/surfacing technology**: from manual arc welding, including welding with a fire-crack electrode at a low current density, to a mechanized arc welding with both long and short arc in any environment, including the process of electrode melting with a controlled transfer of molten metal droplets during short circuits.
The disadvantage of this method is an increased spattering of electrode metal caused by the explosion of a liquid metal bridge between the end of the electrode and a molten metal droplet absorbed by the metal pool under the electrodynamic influence of the pinch effect. This disadvantage was radically reduced by the **automatic control** of all stages of the process of a droplet formation and its transfer into a metal pool. The features of such formation and transfer were well studied by that time [7], mainly at the E.O.Paton Electric Welding Institute of the NAS of Ukraine.

In the world, a significant number of methods of arc welding with automatic control is known [2, 8]. As a rule, they all develop the method developed at the PWI by A.G. Potapievsky [9], where a current source is used, having combined volt-ampere characteristics consisting of two regions: "upper" – steeply falling (at elevated voltages) and "lower" – flat falling.
Let us recall the most "advanced" methods with automatic control.

1 – "adaptive pulsed processes of welding and surfacing" [10], proposed by the Institute of Strength Physics and Materials Science of the Siberian Branch of the Russian Academy of Sciences, patented in 1984;

2 – STT [11] - "surface tension metal transfer", patented by the Lincoln Electric Company in 1988, it has been implemented since 1994;

3 – CMT [12, 13] - "cold metal transfer", known since 2002

The physics of the process is the most adequately reflected in the work (1). In the work (2), the distinctive "brand" point is the term "surface tension transfer", which is perceived as an obvious semantic "tension", since the main "driver" of a drop "transfer" here is the electrodynamic pinch effect. In the work (3), "for the sake of the brand", another liberty was admitted – the introduction of the term "cold metal", although the cooling of metal is not performed, but the electrode is pulled away from the metal pool. Another distinctive point is a rhythmic change in the electrode polarity.

In the listed works it is not mentioned about the volt-ampere characteristics, except for the work (1), where the use of serial production power sources with standard $VAC_{s.}$ and about the presence of thyristor pulsed regulators in the welding circuit, controlling the algorithms of adaptive changes of electrical parameters at all stages of electrode melting.

In the work [14] it is reported about $VAC_{s.}$ of a special shape in the form of a multilink line consisting of a set of straight lines intersecting at different angles. The automatic control system activates the VAC regions depending on the electrode melting stage.

The mentioned technical solutions for automatic control of the welding/surfacing process are related to a comparatively narrow area of technology: in shielding gases at a low current density on the electrode for the cases of formation and transfer of electrode metal droplets. For the technologies with fine-drop and jet transfer of electrode metal, they are not suitable. In small-scale and individual production, the priority is given so far to the technology with self-regulation of the electrode process and its latest version: with pulsed self-regulation.

3. Conclusion

Recently, interesting perspectives are opened up for hybrid methods of process control in which universal inverter power sources with digital control are used, on which VAC of any desired shape is programmed. This is the most promising approach for today for solving different problems arising in production. The rich possibilities of such an approach have not yet been disclosed. In particular, by reprogramming of $VAC_{s.,}$ a pure arc process with pulses on a short arc without sh. c. (see oscillogram figure 3), arc process on a long arc, having its advantages, arc process with jet electrode metal transfer, etc. can be obtained.

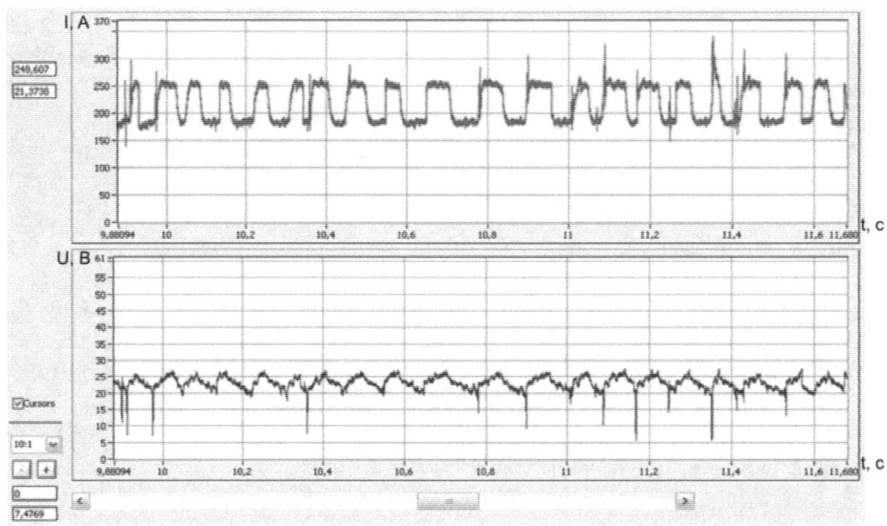

Figure 3. Oscillogram of the welding process with pulses on a short arc without short circuits.

Acknowledgments

The work was performed within the framework of an agreement on scientific and technical cooperation between the E.O. Paton Electric Welding Institute NAS of Ukraine and Institute of strength physics and materials science SB of RAS, and the program of basic scientific research of state academies of sciences for 2013-2020, direction III.23.2.1.

References

[1] Sidoruk V S, Dudko D A and Tsvigun A F 1987 Automodulation of current in submerged arc welding *Avtomaticheskaya svarka* **5** 25-28
[2] Pogrebisky D M 2016 Welding of Metals Classification Brief History, Development. *INSTY PRINTS Jerusalem*
[3] Paton B E 1952 Self – regulation of arc welding using consumable electrode *Avtomaticheskaya Svarka* **1** 38-45
[4] Paton B E Krivtsun I V Sidoruk V S Maksimov and S Yu Kolisnyk G F 2012 *PU Patent 104214*
[5] Paton B E Sidoruk V S 1991 *RU Patent 2021086*
[6] Vladimirov A V Habuzov V A Lebedev V A Maksimov S Yu and Golyshev A A 2011 Universal power sourses for arc welding and plazma cutting *Avtomaticheskaya Svarka* **1** 41-46
[7] Saraev Yu N 1988 Control of electrode metal transfer in CO_2 welding with arc short circuits

(Overview) *Avtomaticheskaya Svarka* **12** 16-22

[8] 2012 Controlled Gas Metal Arc Welding Processes an Assessment on Development *Petromin Pipeliner*

[9] Potapievskiy A G 2007 Welding in shielded gas with a consumable electrode M.: Mashinostroenie 240

[10] Saraev Yu N 1995 Development of adaptive pulse technological processes of welding and surfacing Thesis for the degree of Doctor of Technical Sciences

[11] Bruce D 2003 Assessing the Benefits of Surface Tension Transfer ® Welding to Industry *Journal of Industrial Technology* **19 (4)** 2-7

[12] Hacke H and Himmelbauer K 2005 The CMT-Process-A Revolution in Welding Technology *IIW Doc. №XII–1875–05*

[13] Kazmaiar J Stiglbauer W and Egerland S 2010 CMT Advanced – An Arc Process Revolution *Scheet Metal Welding Conference XIV, Livonia, Ml, May 11th – 14th*

[14] Zhimihng O Yong W Masao U and Manabu T 1999 New Concept for the Characteristic of an Arc Welding Power Source (Report II) *Trans. JWRI* **28 (1)** 26-38.

Modeling of laser welding of plates made of dissimilar metals with the use of a composite insert

V P Shapeev[1, 2], **A N Cherepanov**[1], **V I Isaev**[1]

[1] Khristianovich Institute of Theoretical and Applied Mechanics SB RAS, Institutskaya 4/1, Novosibirsk, Russia, 630090
[2] Novosibirsk State University, Pirogova 1, Novosibirsk, Russia, 630090

E-mail: vshapeev@gmail.com

Abstract. This paper deals with computer simulations of laser welding of plates made of dissimilar metals using a composite intermediate insert. They are performed with the previously developed 3D numerical model implemented with the collocations and least squares method. The use of the composite insert when welding dissimilar alloys prevents the formation of brittle intermetallic phases that adversely affect the strength of the joint. Based on the proposed mathematical model, the features of butt-welding of titanium and stainless steel plates and aluminum and titanium plates are investigated. The rational thermal conditions of laser welding and the geometric parameters of the welds are determined.

1. Introduction

Laser welding of metals is widely used in modern industries. It has several advantages over other types of welding: high speed, flexibility of control and environmental friendliness of the process, low deformation of the parts, etc. The demand for laser welding necessitates the study of this complex physical process and the possibilities of its application. In industrialized countries, new laboratory and industrial installations, teams for scientific research in the field of laser processing of metals are being created. Representative scientific conferences are devoted to its problems; the accumulated experience is described in numerous publications and reviews [1,2].

Recently, there has been increased interest in the development of laser welding technology for dissimilar metals [4-8]. At that, new methods of researching the process are applied and new technological methods are being developed [9-15]. Welded parts and structures made of dissimilar metals, due to the features of their operational properties, are widely used in the production of space, aviation, and other modern equipment. Parts and structures made with the combined use of various metals demonstrate lightness, strength, and high corrosion resistance. A promising way to connect such parts is laser welding. The development of such a technology is primarily hindered by the specifics of the physical process, which is characterized by high temperature and rate in a very small region of the laser beam action on the metal, the formation of a plasma torch. All this causes technical and methodological difficulties for a purely experimental study and optimization of the laser processing of metals and alloys.

Welding of dissimilar metals has its own difficulties and specifics, which are superimposed on the specifics of the laser welding process itself. Despite numerous studies in the field of dissimilar metals welding, the problem of laser welding of alloys based on aluminum, titanium, and steel still needs further study. The key challenge in the welding of such alloys is related to the formation of brittle intermetallic compounds which reduce the strength of the welded joint significantly. For example, titanium and aluminum are chemically active metals with very different physicochemical and mechanical properties. When welding with fusion of titanium and aluminum alloys, brittle $TiAl_3$-type intermetallic compounds are formed in the welded joint region. Moreover, when welding titanium with steel, due to the low solubility of iron in α-titanium, brittle intermetallic compounds of TiFe, $TiFe_2$, Ti_2Fe types and eutectics of various compositions are formed. To prevent the formation of intermetallic compounds during welding of dissimilar metals, various technological methods are used. One of the effective techniques in the production of critical parts is the use of an intermediate insert. There is a need to optimize the process of laser welding of dissimilar metals. It is necessary to choose the optimal laser radiation power and welding speed, and the optimal parameters of various composite intermediate inserts. In the matter of choosing the optimal welding parameters, an expensive physical experiment can be substantially supplemented and partially replaced by calculations using adequate mathematical models, which make it possible to obtain various quantitative characteristics of the welding process. The combination of field experiments and numerical modeling in the development of the technology under consideration can accelerate the search for its optimization. Therefore, in view of the demand for laser welding in industry, its mathematical modeling is relevant [16–25] in addition to studying it using physical experiment methods. Using a mathematical model that adequately describes the properties of the process under study, it is possible to calculate its parameters that are interesting from the perspective of technology. In this regard, numerical modeling can significantly supplement the often expensive physical experiment, in which it is difficult to measure some important process parameters.

In this paper, we propose numerical 3D models for calculating the process of laser butt welding of plates made of dissimilar metals, using composite inserts obtained by explosion welding of thin layers of dissimilar metals. An essential circumstance of the developed and simulated technology is that explosion welding gives durable compounds of various metals with good quality welds and allows one to obtain sufficiently strong composite inserts. This technology is proposed to be used for welding plates of metals, direct welding of which with each other presents certain technological difficulties.

Two cases of dissimilar metals welding were simulated using the numerical collocation and least squares (CLS) method developed by the authors [26–27]. For modeling, modifications of the thermophysical model of laser welding, previously proposed by the authors and used for the numerical simulation of welding plates of various homogeneous metals, were formulated. The previous versions of the model were previously described in sufficient detail in [14–15, 19, 21–22, 24]. In particular, they describe in detail the self-consistent model of the vapor channel, which is different from [17]. The depth and shape of the channel in this model are consistent with the amount of heat received from the laser beam by the surface of the channel and heat fluxes from it to the outside of the welding zone. In some of them, the presence of a heat transfer rate increased due to melt convection in the weld pool was modeled, as in [21, 25], simply by an artificially increased heat conductivity coefficient. In others, convection of the melt in the weld pool was described by the Navier – Stokes equations [22]. Here, a simple method was used to simulate the increased rate of heat conductivity in the weld pool. The difference of the modified versions of this model from [19,21,22] is that for each metal and melt in the pool it uses its own thermophysical parameters in all the equations and boundary conditions. For example, this is taken into account when choosing the coefficients of thermal conductivity and heat capacity, melting heat, and other parameters. An essential detail of the model used here is that in the heat transfer from the laser beam to the welding zone, in its heat exchange with other parts of the welded product and with the external environment, heat fluxes were taken into account, which, in comparison with those not taken into account, determine the most significant quantitative characteristics of the temperature distribution in the product during the welding process.

Here, in the first case, the welding of plates made of steel and titanium alloys is simulated, in the second case, the welding of plates made of aluminum and titanium alloys. In the first case, the insert consisted of layers of steel, copper, niobium, and titanium. In the second one, the insert contained aluminum and titanium layers. In both cases, the use of intermediate insert between dissimilar metals to be welded is intended to prevent their mixing in the weld pool and the formation of brittle intermetallic compounds in the weld.

Calculations based on the proposed models make it possible to select the thicknesses of the layers in the intermediate insert, the laser radiation power, and the welding speed at which "dagger" penetration of the welded joints is achieved. In addition, it is important to make sure that with the calculated welding mode in the weld pools there is no mixing of metals, leading to the formation of intermetallic compounds. Finally, it is important to make sure that with the selected mode during the welding process, a temperature distribution established in the product does not damage the explosion welds in the insert.

2. Schemes for welding dissimilar metals with intermediate inserts

To obtain high-quality welds when welding a titanium alloy with stainless steel, it seems promising to use intermediate layers of other alloys. Thus, earlier, various metal inserts (copper, bronze, etc.) were used for this purpose at the Institute of Theoretical and Applied Mechanics SB RAS. The most effective was an insert made of a copper plate, the use of which made it possible to obtain joints with tensile strengths of the order of 350 MPa. A significant increase in the strength of the welded joint of steel plates with titanium was provided by the use of composite inserts obtained by explosion welding of a layered composite made of thin layers of four metals. Plates of copper and niobium were taken as the two middle layers in the insert, and plates of steel and titanium as the outer ones. In the welded joint, the titanium layer of the insert fits tightly with the welded titanium plate, and the steel layer of the insert with the plate of steel (figure 1). During welding, the laser beam passes first along one of the joints, and then along the other. With this method of joining plates, steel is welded to steel, and titanium to titanium. With the given parameters of the welding process, it is possible to choose the thicknesses of the outer layers in the composite insert so that the molten steel does not mix with titanium and thereby do not create conditions for the formation of brittle intermetallic phases in the weld.

For welding a titanium plate of rectangular shape with aluminum, a composite insert was used, cut from a bimetallic plate obtained by explosion welding of titanium and aluminum layers 3 mm thick. The width of the rectangular insert was taken equal to the thickness of the plates being welded. In this case, the welding had two stages as well: the titanium plate was butt welded with titanium layer of the insert, and the aluminum plate with aluminum layer of the insert (figure 9).

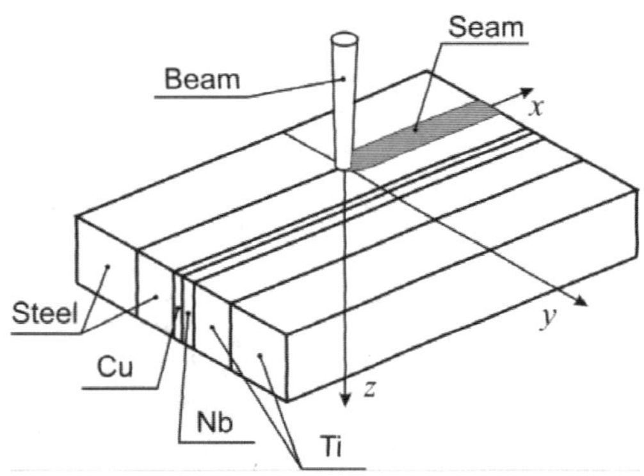

Figure 1. Scheme of welding plates of steel and titanium with an intermediate composite insert.

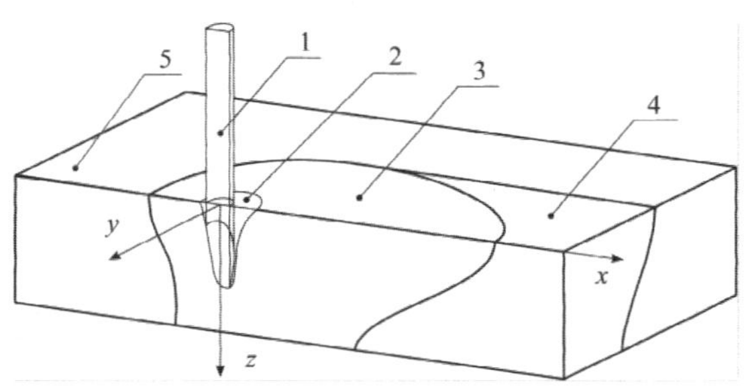

Figure 2. Scheme of the welding zone. 1 – laser beam, 2 – vapor-gas channel, 3 – melt pool, 4 – welding seam, 5 – solid phase.

3. Numerical simulation of the welding process

The described welding process is a complex set of simpler interacting physical processes. It depends on the values of many parameters, first of all, on the power of laser radiation, welding speed, the insert layers thicknesses and their thermophysical parameters. Due to the specifics of laser welding, this process is difficult both for experimental research and numerical simulation. It is necessary to use a sufficiently universal numerical method for solving boundary value problems for partial differential

IOP Conf. Series: Materials Science and Engineering **681** (2019) 012044 doi:10.1088/1757-899X/681/1/012044

equations in order to ensure a sufficiently accurate approximation not only of differential equations, but also of nonlinear boundary conditions, including those at curvilinear boundaries.

In view of the presence of discontinuity surfaces of the thermal conductivity coefficient at the boundaries between different metals and their different phases, for the convergence of the numerical solution obtained using this method, it is necessary that it possess conservatism, that is, comply with the conservation laws with some accuracy. Otherwise, as known, the numerical solution obtained by any methods does not converge on any grid – it either has an inevitable finite error, or diverges. There are discontinuities of the thermal conductivity coefficient in the computational domain cased by the different properties of welded metals in liquid and solid states. For adequate handling of this the numerical algorithm must be conservative. I.e., it must represent conservation laws on the computational grid. In this sudy we use a fairly universal method of collocation and least squared, which is capable to support all the functionality mentioned above as is was demonstrated before when solving various differential and integral equations [26-27].

This paper contains the simulations of welding of an austenitic steel plate with a titanium one (figure 1, figure 3 - figure 8), as well as welding a titanium plate with an aluminum plate (figure 9 - figure 11) using the corresponding inserts.

3.1. Welding of titanium and steel plates

Below are some results of numerical modeling of the butt welding of VT1-0 titanium alloy and 12Kh18N10T steel rectangular plates with thickness of t = 3 mm with a composite insert in between consisting of layers of copper and niobium with thickness of 0.2 mm. The laser power W = 2.5 kW, the welding speed in steel V_w = 1 m/min and in titanium V_w = 1.8 m/min. In numerical experiments, the temperature fields, the position of the internal boundaries in the welded joint, the shape and depth of the vapor channel were determined. The figures show rectangular coordinate system associated with the laser beam and one of the welded joints: the origin is at the intersection of the upper surface of the plates with the axis of the laser beam, the z axis is directed along the beam, the x axis is along the joint.

Figures 3–7 show the calculated positions of the isotherms corresponding to the melting points: 1 – titanium, 1941 K; 2 – steel, 1775 K, 3 – copper, 1356 K. Isotherm 4 corresponds to temperature of 650 K. For clarity, only a part of the computational domain containing the isotherms under consideration is shown. The straight lines show the boundaries of the copper and niobium inserts before welding, denoted by symbols Cu and Nb, respectively. Parts of the computational domain occupied by titanium and steel before the welding are denoted by Ti and Steel, respectively. The focus position of the laser beam in all calculations z_F = 4 mm.

The black region containing point (0, 0, 0) in figure 4 corresponds to the steam channel. The values of the thermophysical parameters with which the calculations were carried out were taken from well-known sources [28, 29] and others.

Figure 7 shows the calculated position of the welds boundaries in steel and titanium. For comparison, figure 8 presents the morphology of the cross section of the weld obtained in the experiment. Note that from the position of the melting isotherms of the metals being welded, it is easy to determine the cross-sectional area of the molten part of each metal and to calculate their quantitative ratio in the mixture in the weld pool and weld. The calculation provides these parameters with sufficient accuracy. A certain morphological difference between the calculated shape of the weld boundaries and the experimental one is obviously due to the lack of consideration in the model version used here of melt flows in the weld pool, which affect the conditions of local heat transfer in the melt zone and the shape of the pool walls. Quantitatively, the volume and, qualitatively, the picture of the melted region of the joint are in a satisfactory agreement with the experiment.

Here we omit the discussion of the various details that are obtained from the calculation, including the melt cooling rate, which affects the crystal structure of the weld. However, we note that in the calculated welding mode, such a temperature distribution is obtained in the plates and in the insert that the weld between copper and niobium does not melt and retains strength. Moreover, in the weld pool there is no mixing of the molten steel and titanium.

IOP Conf. Series: Materials Science and Engineering **681** (2019) 012044 doi:10.1088/1757-899X/681/1/012044

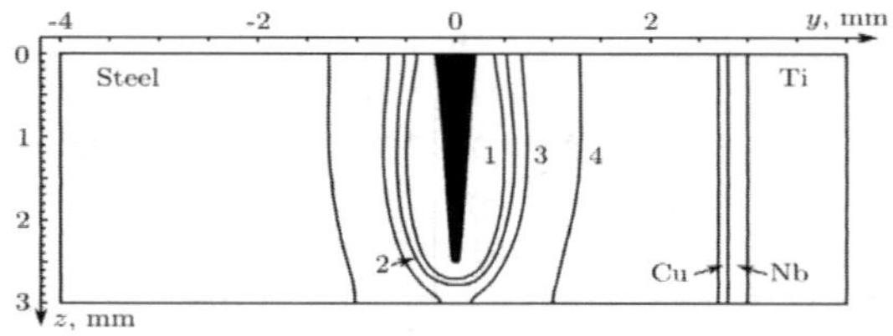

Figure 3. Calculated temperature distribution in the cross section x=0, the beam moves in steel. Isotherms: 1 – 1941 K (Tl for titanium), 2 – 1775 K (Tl for steel), 3 – 1356 K (Tl for copper), 4 – 650 K.

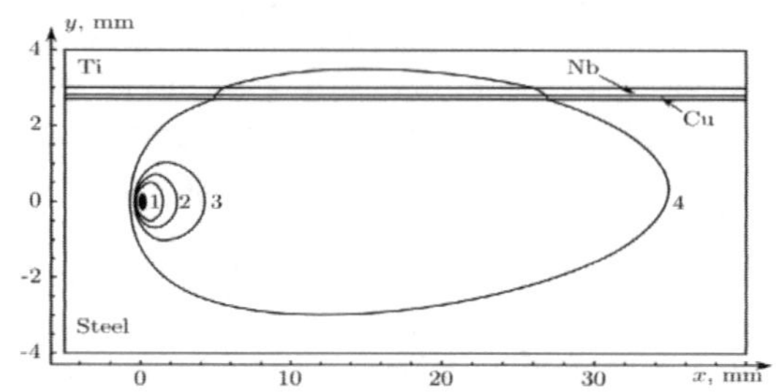

Figure 4. Calculated temperature distribution in the cross section z = 0. The beam moves in steel. Isotherms: 1 – 1941 K (Tl for titanium), 2 – 1775 K (Tl for steel), 3 – 1356 K (Tl for copper), 4 – 650 K.

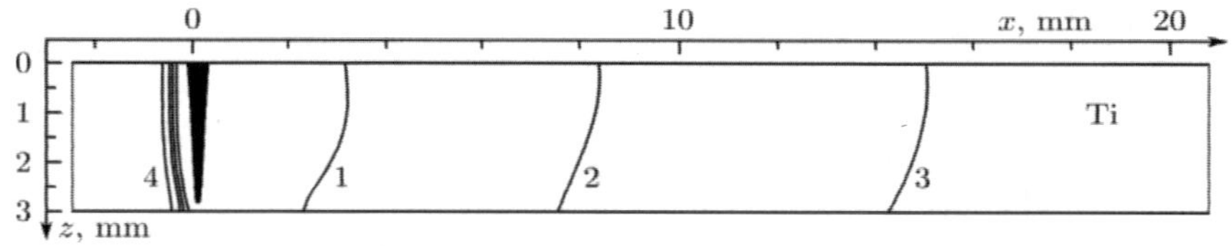

Figure 5. Calculated temperature distribution in the cross section y=0. The beam moves in titanium. Isotherms: 1 – 1941 K (Tl for titanium), 2 – 1775 K (Tl for steel), 3 – 1356 K (Tl for copper), 4 – 650 K.

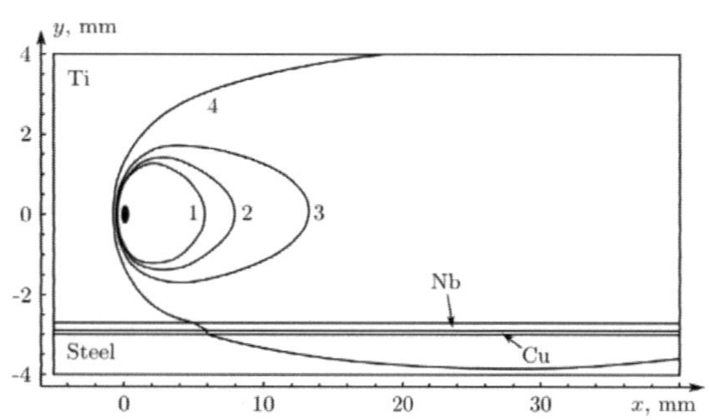

Figure 6. The temperature distribution in the cross-section z=0. The beam moves in titanium. Isotherms: 1 – 1941 K (Tl for titanium), 2 – 1775 K (Tl for steel), 3 – 1356 K (Tl for copper), 4 – 650 K.

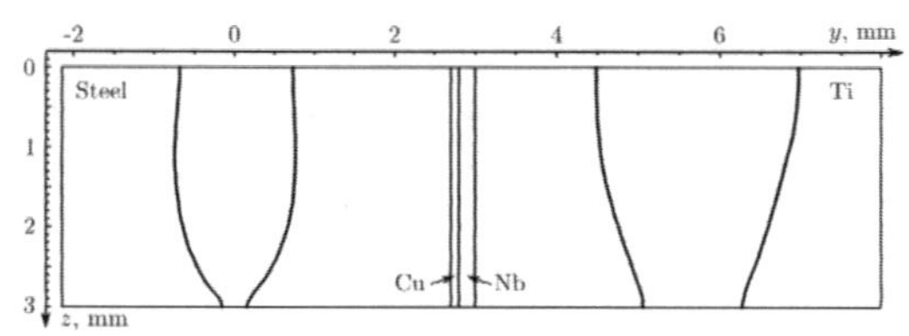

Figure 7. The calculated position of the boundaries of the welds in steel and titanium.

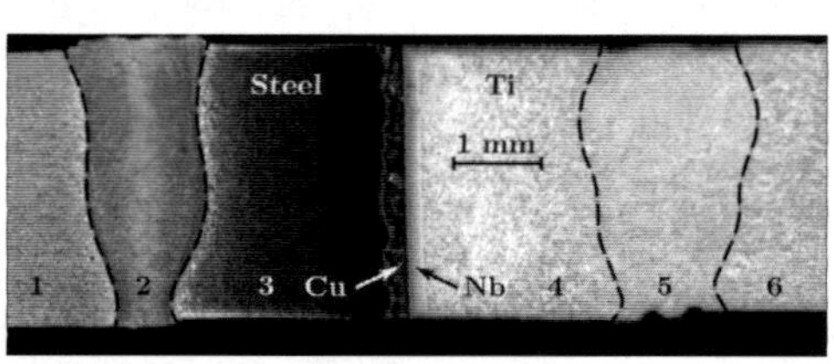

Figure 8. Photograph of a section of the welded joint after welding using a composite insert.

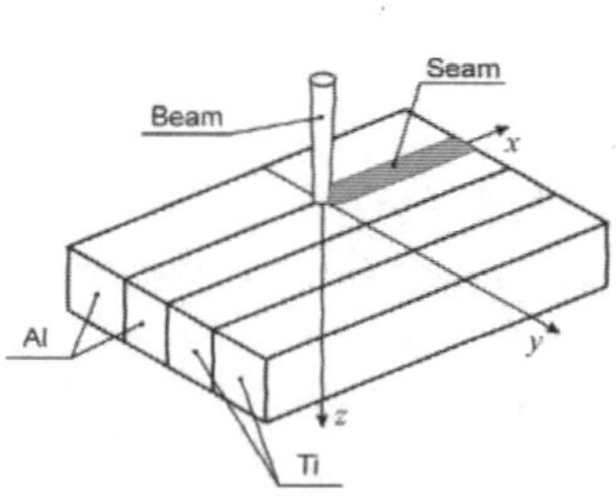

Figure 9. Laser welding scheme for aluminum and titanium plates with an intermediate insert.

3.2. Welding of plates made of titanium and aluminum

In the scheme (figure 9), the aluminum plate is located in the region y < 0. Aluminum layer of insert 3 mm wide is joined to it in the region y > 0 along the x axis, which, in turn, is explosion-welded to layer of titanium 3 mm wide, and the insert fits tightly with the titanium plate. All layers have width of 3 mm, so their upper and lower surfaces coincide. The surfaces of the plates are blown with an inert gas to protect the liquid metal from oxidative processes.

Below are some results of the numerical simulation of butt welding of VT1-0 titanium alloy and AD1 aluminum plates with thickness t = 3 mm with a composite insert in between consisting of aluminum and titanium layers with thickness of 3 mm and 2 mm, respectively. Laser power W = 3.1 kW, speed Vw = 3.5 m/min for welding of aluminum; and W = 1.9 kW, Vw = 1.8 m/min for welding of titanium. In the numerical experiments, the temperature fields, the position of the internal boundaries in the welded joint, the shape and depth of the vapor channel were determined.

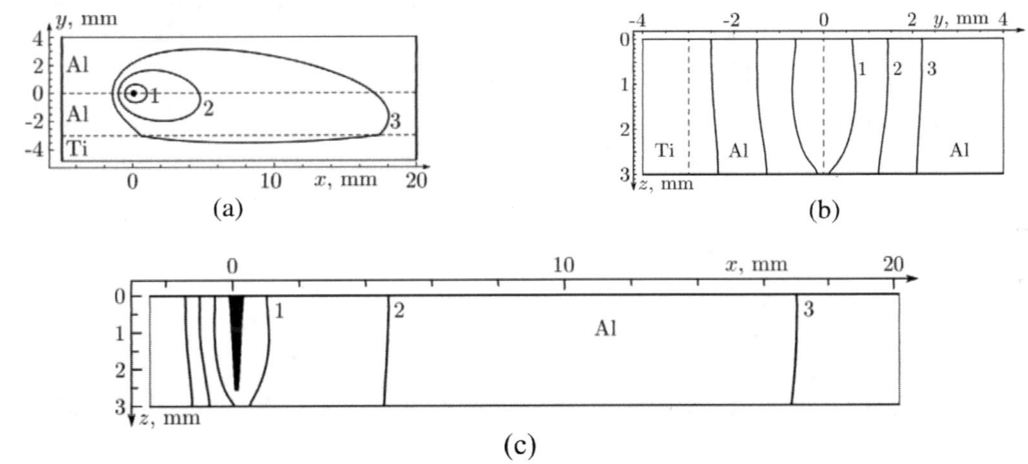

Figure 10. The calculated isotherms for butt welding of aluminum plate with layer of aluminum in the insert in cross-sections z = 0 mm (a), x = 0.1 mm (b) y = 0 mm (c); the welding speed 3.5 m/min, laser power W = 3.1 kW; 1 – aluminum melting temperature 873 K, 2 – 700 K, 3 – 500 K.

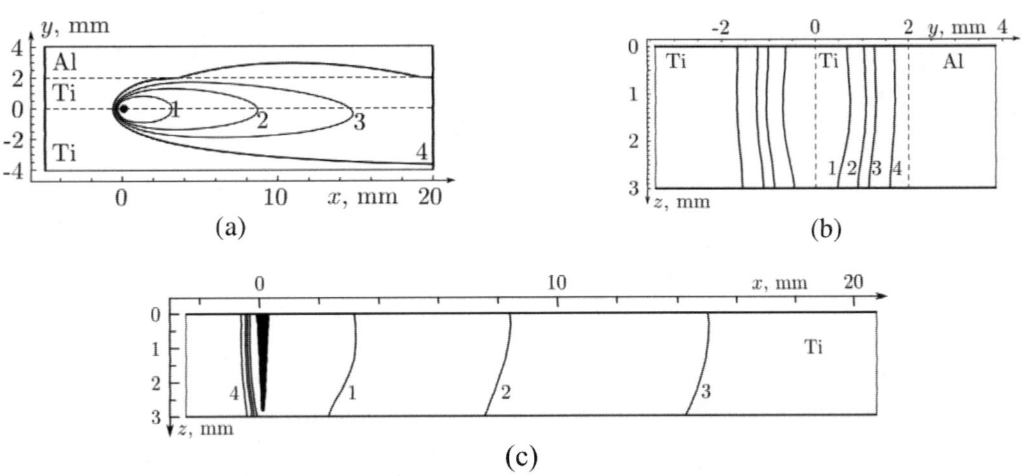

Figure 11. The calculated isotherms for butt welding of titanium plate with titanium layer in the insert in cross-sections z = 0 mm (a), x = 0 mm (b), y = 1.2 mm (c); the welding speed 1.8 m/min, laser power W = 1.9 kW.

The figures show the calculated positions of the isotherms corresponding to the melting points of: aluminum (figure 10) and titanium (figure 11). The black region containing point (0, 0, 0) corresponds to the vapor channel. The values of the thermophysical parameters with which the calculations were carried out were taken from well-known sources [28, 29] and others.

Note that according to the position of the melting isotherms of the welded metals, it is easy to determine the time of the melt cooling and some parameters of the welds. In this case, as in the previous one, the main conclusion from the temperature distribution pattern is that in the calculated welding mode, the welding seam obtained by the explosion in the insert between the layers of aluminum and titanium does not melt.

4. Conclusion

The models of laser welding of dissimilar metals plates with an intermediate composite insert in between used in the work allow us to predict various parameters of the welded joint that are important for the technology. Intermediate composite inserts obtained by explosion welding of thin layers of metals can be used for laser welding of critical parts and structures of dissimilar metals.

Acknowledgments

The research was carried out within the framework of the Program of Fundamental Scientific Research of the state academies of sciences in 2013-2020 (№ 0323-2018-0012) and partially supported by the Russian Foundation for Basic Research RFBR project codes 18-08-00571.

References

[1] Zhao H, White D and DebRoy T 1999 *International Materials Reviews* **44** (6) 238
[2] Dowden J and Schulz W 2017 *The Theory of Laser Materials Processing Heat and Mass Transfer in Modern Technology (Bristol: Springer)* p 432
[3] Rathod M and Kutsuma M 2003 *Quarterly J. Jap. Weld. Soc.* **2** 282
[4] Gnauk J, Wenke R and Frommeyer G 2005 *Materials Science & Engineering: A* **490** 413-414
[5] Cherepanov A, Afonin Y and Orishich A 2009 *Tyazheloe Mashinostroenie* **8** 24–26 [in Russian]
[6] Sidyakin V, Arbuzov V and Horstov V 2009 *Mir tehniki i tekhnologii* **10** 40–45 [in Russian]
[7] Casalino G, Mortello M and Peyre P 2016 *Procedia CIRP* **41** 992-997
[8] Tomashchuk I, Sallamand P, Méasson A, Cicala E, Duband M and Peyre P 2017 *J. of Mater. Process. Tech.* **245** 24-36
[9] Yao C, Xu B, Zhang X, Huang J, Fu J and Wu Y 2009 *Optics and Lasers in Engineering* **47** 807-814
[10] Kaleko D 2012 *The Paton welding journal* **10** 27
[11] Vazquez R, Koch H and Otto A 2014 *Physics Procedia* **56** 1334
[12] Dal M and Fabro R 2016 *Optics & Laser Technology* **78** 2
[13] Pinkerton A 2016 *Optics & Laser Technology* **78** 25
[14] Cherepanov A N, Orishich A M, Pugacheva N B and Shapeev V P 2015 *Thermophysics and Aeromechanics* **22 (2)** 135
[15] Cherepanov A N, Shapeev V P and Isaev V I 2015 *High Temperature* **53 (6)**, 841
[16] Rosenthal D 1941 *Welding J.* **20** 220
[17] Kroos J, Gratzke U and Simon G 1993 *J. of Phys. D: Appl. Phys.* **26 (3)**, 474
[18] Kar A and Mazumder J 1995 *J. of Appl. Phys.* **78 (11)** 6353
[19] Shapeev V P and Cherepanov A N 2006 *Vychislitel'nye tekhnologii* **11 (6)** 102 [in Russian]
[20] Roy G G, Elmer J W and DebRoy T 2006 *J. Appl. Phys.* **100 (3)** 34903
[21] Cherepanov A and Shapeev V 2009 *Vychislitel'nye tekhnologii* **14 (3)** 96-106 [in Russian]
[22] Isaev V, Cherepanov A and Shapeev V 2010 *Thermophysics and Aeromechanics* **17 (3)** 419 [in Russian]
[23] Anca A, Cardona A, Risso J and Fachinotti V D 2011 *Appl. Math. Model.* **35** 688
[24] Cherepanov A N and Shapeev V P 2013 *Thermophysics and Aeromechanics* **20 (2)** 237

[25] Saldi Z S, Kidess A, Kenjeres S, Zhao C and Kleijn C R 2013 *Intern. J. of Heat and Mass Trans* **66** 879
[26] Isaev V and Shapeev V 2010 *Computat. Math. and Math. Phys.* **50** 1670
[27] Shapeev V P 2016 *MMCP 2015 Proceedings (EPJ Web of Conferences)* **108** 01009
[28] Zinoviev V 1989 *Thermophysical properties of metals at high temperatures* (Moscow: Metallurgiya) p 384 [in Russian]
[29] Babichev A, Babushkina N and Bratkovskii A 1991 *Physical Quantities: Handbook* (Moscow: Energoatomizdat) p 1232 [in Russian]

Post-treatment of arc-welded aluminum alloy joints by laser shock peening

A A Tareva[1], D M Melnikov[1] and M A Melnikova[1]

[1] Bauman Moscow State University, 105005, 2nd Baumanskaya str., Moscow, Russia

E-mail: Daenoor@gmail.com

Abstract. The possibility to improve aluminum alloy welded joints with laser shock penning was demonstrated. It was shown that microhardness of peened area in heat affected zone is higher than one without post-treatment. Also in this area the residual tensile stresses turns into residual compressive stresses. Laser peening process is proposed to carry out using low-energy lasers. In this case were analysed different types of ablative overlays.

1. Introduction

In recent years laser shock peening (LSP) became an attractive way to create a local plastic deformation for various implements [1]. LSP is a cold surface modification process that uses high power pulsed laser radiation [1 – 4]. This is an important post processing method for metal parts. Now it is extensively used to enhance the fatigue lifetime of jet engine fan, compressor blades and recently in aircraft structures [5]. LSP is the most suitable alternative process among the existed shot peening technology [6]. From that point of view, LSP can be put into commercially practice in the small industries by reducing a cost of the experiment. For this purpose, methods for LSP using low-energy kHz lasers have been recently developed [6].

The laser peening process is illustrated on Figure 1. It uses high power density laser pulses (5-10 GW/cm^2) which are located at the metal surface, coated with an ablative film and transparent tamping layer (usually water). As laser beam passes through the transparent layer and hits the surface of the material, the thin layer of the ablative layer is vaporized. The vapour continues to absorb the remaining laser energy, heats up and then ionizes into plasma. A rapidly expanding plasma is limited in expansion due to water counterstand. It creates a high surface pressure, which propagates into the material as a shock wave [7, 8].

A shock wave that propagates deep into the metal is caused by a sudden pressure leap. Plastic deformation of the material occurs when shock wave pressure exceeds Hugoniot elastic limit [9, 10]. The strain rate reaches about 10^6 s^{-1}. Thereby, the number of dislocations and transfer speed increases. That leads to changes in microstructure and mechanical properties (hardness and strength) of the material. The total amount of cold work is much lower than after shot peening despite the high strain rates [11]. Shock wave causes maximum of residual compressive stresses at the surface and decreasing deep in material. The larger the magnitude of the shock pressure, the deeper the plastic deformation develops [12].

LSP can be used for small-scale processing despite the traditional methods such as shot or ultrasonic peening, for example, micro dents [13], fatigue crack mouth [14] or welded joints [15].

Welded joints have negative effect on the reliability of manufactured products. One of the main problem is high residual tensile stresses.

The weak areas of welded joints are fusion zone (FZ), transition zone (TZ) and heat affected zone (HAZ) [15]. The compressive residual stresses after LSP can significantly increases fatigue life and fatigue strength by inhibiting the initiation and propagation of cracks [16]. The positive effect of LSP has been demonstrated in increase of fatigue life, wear resistance, etc. [17] for aluminum, steels, and titanium alloys.

The aim of this work is to study the technological features of LSP applied to possibility to improve the properties of welded joints of aluminum alloys.

2. Experimental procedures

Spot diameter, number of shots and laser intensity are parameters that influence LSP process. The process of laser shock peening is schematically shown in Fig.1.

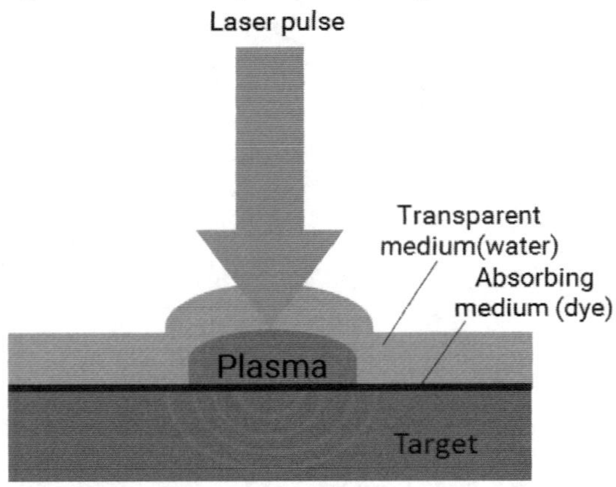

Figure 1. Scheme of laser shock peening.

It is necessary to perform the following procedures for generation of laser shock wave. Target surface is covered with an ablation film. Evaporation of this layer leads to the formation of a plasma flame. The expansion of the flame leads to short-pulse pressure waves formation. An ablative layer also prevents base metal from melting and laser ablation. Energy transparent overlay (tamping layer) is applied to increase the pressure of the shock wave by inhibiting the expansion of the plasma.

The right choice of coatings is important for this purposes. Except the absorption coefficient, the ablative layer must be resistant to multiple shocks due to the specifics of low-energy lasers that require small spot size and, therefore high repetition rates, to maintain sufficient power density. It is impossible to update the ablative film after each shot with kHz frequencies.

Black paint and 50 μm thick aluminum foil were used as an ablative coatings, water – as a tamping overlay. The effect of irradiation with overlaps was investigated by a single or multiple laser shots. Effectiveness studies of coatings were carried out on pure nickel samples (as a model material).

LSP effect on welded joints was investigated on arc welded aluminum alloy samples of Al-6Mg. There used two measurement methods. X-ray method was used to determine the residual stresses. This method is based on deformation measuring of the crystal lattice from the displacement of the diffraction lines. The microhardness studies were carried out on an EmcoTest Durascan 20.

The laser source for aluminum samples was a solid-state laser "Solar LQ 829" with a wavelength of 532 nm, pulse energy up to 250 mJ, pulse duration 20 ns (FWHM) and a repetition frequency of 10 Hz.

3. Results and discussion

The LSP results with different ablative films are shown in the Fig. 2. Measurements were carried out in the center of dimples that are formed after LSP. In cases where the coating was damaged after laser shots, microhardness measurements were carried out beside the spot center in non-melted area. The best results were obtained for aluminum tape attached with adhesive tape. This ablative layer does not disintegrate even after 4 shots and provides high microhardness results. Black paint is an effective coating, but it can withstand only one laser shot and then it is necessary to replace it. The same happens with glue and foil combination.

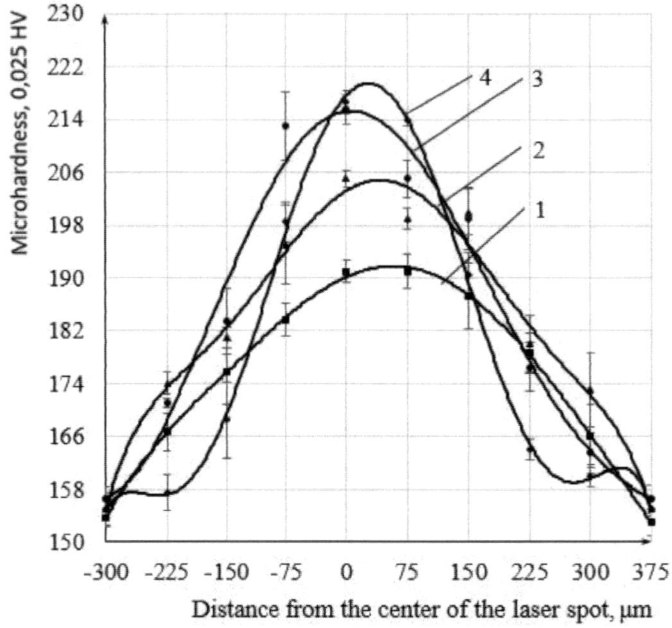

Figure 2. Distribution of external microhardness of Ni sample after LSP with different coatings (1 – Al foil and adhesive tape; 2 – Al foil and adhesive tape (2 shocks); 3 – glue with Al foil; 4 – black dye. Energy pulse 180 mJ, spot diameter – 200 μm.

The photos of coatings are shown in Figure 3., The tape carbonizes after one or two shots, does not transmit laser radiation and is able to withstand up to 4 laser pulses. This is an important advantage since the treatment of the welded joint involves overlapping and coating needs to be renewed constantly.

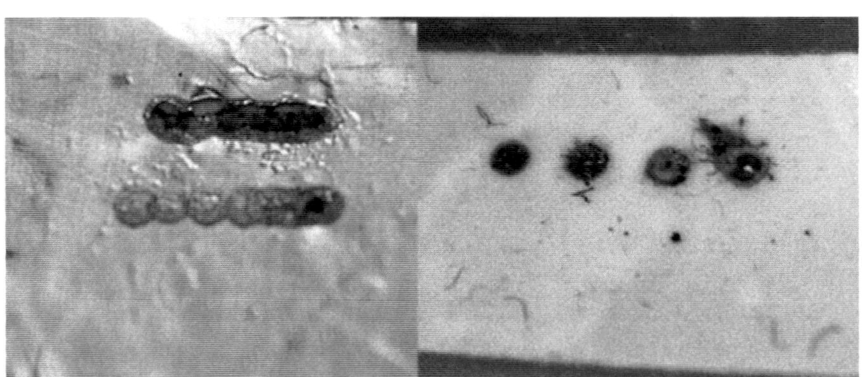

Figure 3. Photos of adhesive tape with aluminum foil (left) and adhesive tape (right) after multiple laser shot

In this work we obtained the increase of hardness for aluminum alloys. We reached hardness values 70% higher than on untreated samples with the same parameters 1,8 GW/cm^2, four repetitions of penning and 0,5 mm spot diameter. Figure 4 shows the distribution of internal hardness across the sample. The character of the curve is sufficiently stable: microhardness decreases with the raise of distance from surface. The maximum of microhardness is at the 50 μm depth and the minimum effect is at the value of 500 μm.

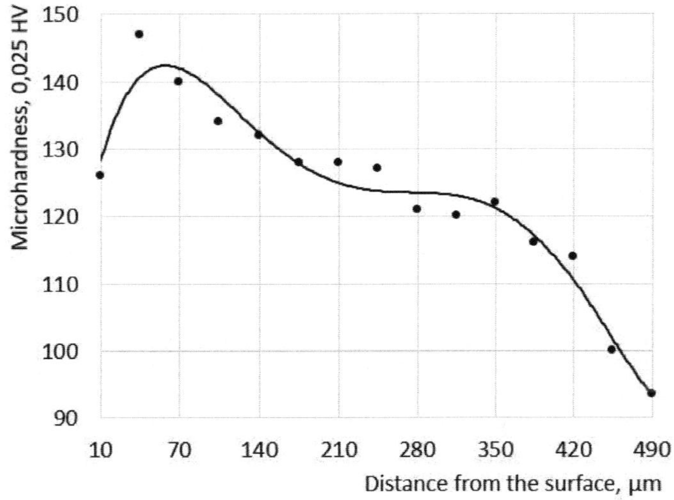

Figure 4. Microhardness profile of aluminum alloy after LSP (laser intensity of 1.8 GW/cm^2 and spot diameter of 0.5 mm).

Fig. 5 shows the microhardness profiles obtained before and after LSP of aluminum arc-welded joint. Laser intensity was 1.8 GW/cm^2 and spot diameter of 0.5 mm. There are two characteristic zones in welding joints indicated in Fig.5: heat affected zone and fusion zone. The results of microhardness before and after LSP are shown on the figure 5 by two curves. The increase of microhardness up to 70 % is peculiar for peened surface in transition and fusion zones. This could be explained by structural changes or residual stresses that occurs after the LSP.

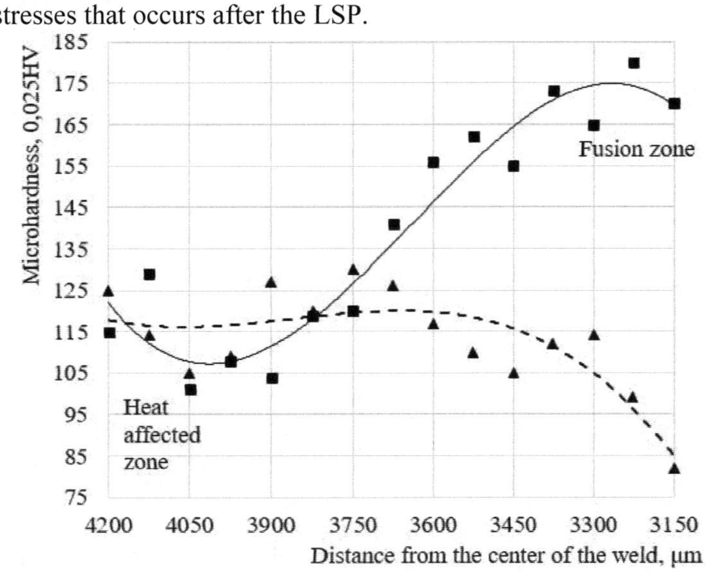

Figure 5. Microhardness profiles of aluminum arc-welded joint before and after LSP.

Residual stresses measurements were performed with the standard X-Ray diffraction method on arc-welded aluminum alloy samples in heat affected zone before and after LSP. The results of the study are shown on Fig.6. It can be observed that residual tensile stresses of 15 MPa magnitude forms after welding in transition zone. Transformation from residual tensile stress to residual compressive stress occurs after LSP treatment (the magnitude was 50 MPa). From [12] it is obvious that the change of stress sign leads to fatigue strength increase in welded structure.

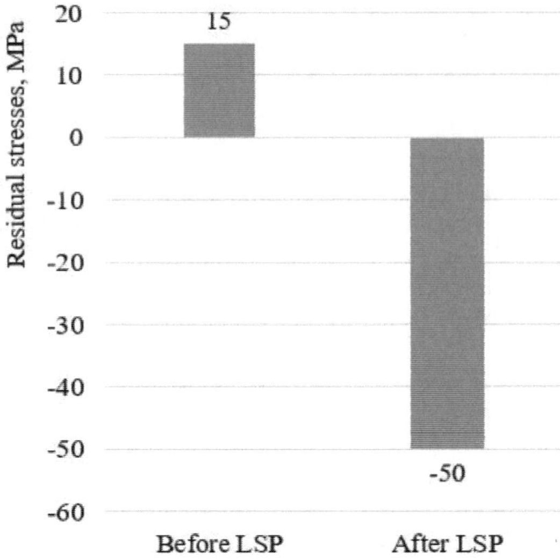

Figure 6. Average residual stress values determined at the surface in the HAZ before and after LSP of aluminum arc-welded joint.

4. Conclusion

Laser peening process of arc-welded aluminum samples has been discussed. The efficiency of ablative coatings for LSP has been studied. We have found that aluminum foil attached with adhesive tape is the best choice for complex geometries in LSP with multiple overlapping strategy. This coating is able to withstand the repeated irradiation. The results of microhardness for this coating is comparable to other coatings. It provides tight flexible contact and it is suitable for LSP of welded joints.

We also demonstrated the increase of microhardness up to 70 % after LSP for Al alloy. In addition, a change of residual stresses sign from tensile to compressive after LSP of aluminum welded joint was shown. The results allow us to conclude that laser shock peening is a promising method for post-processing of welded joints.

Acknowledgments

This work is supported by the Russian Science Foundation (№ 17-19-01706). Some experiments were carried out on the experimental basis of URF «Beam-M» (BMSTU, Moscow).

References

[1] Clauer A 2001 A historical perspective on laser shock peening *Met. Finish. News* **10** 5–6
[2] Luo K, Wang C, Li Y, Luo M, Huang S, Hua X, and Lu J 2014 Effects of laser shock peening and groove spacing on the wear behavior of non-smooth surface fabricated by laser surface texturing *Appl. Surf. Sci.* **313** 600–606
[3] Yin D, Wang D, Jing H and Huo L 2010 The effects of ultrasonic peening treatment on the ultralong life fatigue behavior of welded joints *Mater. Des.* **31** 3299–3307
[4] Tani G, Orazi L, Fortunato A, Ascari A and Campana G 2011 Warm laser shock peening – new developments and process optimization. *CIRP Ann. Manuf. Tech.* **60** 219–222
[5] Hackel L, Rankin JR, Rubenchik A, King WE, Matthews M 2018 Laser Peening: A Tool for Additive Manufacturing Post-processing *Additive Manufacturing* **24** 67-75
[6] Kalainathan S., Prabhakaran S. 2016 Recent development and future perspectives of low energy laser shock peening *Optics & Laser Technology* **81** 137–144
[7] Achintha M., Nowell D., Fufari D., Sackett E. E. and Bache M. R. 2014 Fatigue behaviour of

geometric features subjected to laser shock p eening: Experiments and modelling *Int. J. Fatigue* **62** 171–179

[8] Peyre P, Scherpereel X, Berthe L, Carboni C, Fabbro R, Béranger G and Lemaitre C 2000 Surface modifications induced in 316 L steel by laser peening and shot-peening. Influence on pitting corrosion resistance *Mater. Sci. Eng.* A **280** 294–302

[9] Peyre R, Scherpereel P, Berthe X and Fabbro L 1998 Current trends in laser shock processing *Surf. Eng.* **14** 377–380

[10] Cao Z, Che Z, Zou S and Fei Q 2011 Numerical simulation of residual stress field induced by laser shock processing with square spot *J. Shanghai Univ.* **15** 553–556

[11] Gujba A K and Medraj M 2014 Laser peening process and its impact on materials properties in comparison with shot peening and ultrasonic impact peening *Materials* **7** 7925-7974

[12] Peyre P, Fabbro R, Merrien P and Lieurade H 1996 Laser shock processing of aluminum alloys. Application to high cycle fatigue behavior *Mater. Sci. Eng.* A **210** 102–113

[13] Guo Y.B., Caslaru R. 2011 Fabrication and characterization of micro dent arrays produced by laser shock peening on titanium Ti–6Al–4V surfaces *Journal of Materials Processing* Technology **211** 729–736

[14] Keller S., Horstmann M., Kashaev N., Klusemann B. 2018 Experimentally validated multistep simulation strategy to predict the fatigue crack propagation rate in residual stress fields after laser shock peening *Int. J. Fatigue* **124** 265-276

[15] Zhang X, Yang W and Xiao R 2015 Microstructure and mechanical properties of laser beam welded Al–Li alloy 2060 with Al–Mg filler wire *Materials & Design* **88** 446-450

[16] Ruschau J, John R, Thompson S R and Nicholas T 1999 Fatigue crack nucleation and growth rate behavior of laser shock peened titanium *Int. J. Fatigue* **21** 199–209

[17] Nikolaev G A, Fridlyander I N and Arbuzov U P 1990 Welded aluminum alloys (Moscow: Metallurgiya) p 296

AUTHOR INDEX

Institute of Physics
Dirac House, Temple Back
Bristol BS1 6BE UK

ISSN: 1757-8981
ISBN 978-1-7138-0933-3

9 781713 809333

CURRAN ASSOCIATES INC.
proceedings.com